FOURIER SERIES

RNDr. ALOIS KUFNER, CSc.

RNDr. JAN KADLEC, CSc.

English translation edited by G. A. TOOMBS, B.Sc., Ph.D.
Lecturer, Department of Physics
University of Nottingham

**LONDON
ILIFFE BOOKS**

THE BUTTERWORTH GROUP
ENGLAND
Butterworth & Co (Publishers) Ltd
London: 88 Kingsway, WC2B 6AB

AUSTRALIA
Butterworth & Co (Australia) Ltd
Sydney: 20 Loftus Street
Melbourne: 343 Little Collins Street
Brisbane: 240 Queen Street

CANADA
Butterworth & Co (Canada) Ltd
Toronto: 14 Curity Avenue, 374

NEW ZEALAND
Butterworth & Co (New Zealand) Ltd
Wellington: 49/51 Balance Street
Auckland: 35 High Street

SOUTH AFRICA
Butterworth & Co (South Africa) (Pty)
Ltd Durban: 33/35 Beach Grove

English edition first published in 1971 by Iliffe Books an imprint of The Butterworth Group, in co-edition with Academia — Publishing House of the Czechoslovak Academy of Sciences, Prague

© Academia, Prague, 1971

ISBN 0 592 03944 7

Printed in Czechoslovakia

CONTENTS

Preface ... 9
Notation ... 12

CHAPTER 1 — BASIC CONCEPTS ... 1
    1.1. Functions
    1.2. Series
    1.3. Abelian summation
    1.4. Lebesgue integral
    1.5. Some formulas for trigonometric functions
    Problems

CHAPTER 2 — CONCEPT OF A FOURIER SERIES ... 19
    2.1. Solution of the equation of a vibrating string
    2.2. Fourier series of an integrable function
    2.3. Convergence of trigonometric series
    2.4. Convergence of a Fourier series
    Problems

CHAPTER 3 — HILBERT SPACE ... 51
    3.1. Introduction
    3.2. Concept of a Hilbert space
    3.3. Length of a vector and the distance between two vectors in a Hilbert space
    3.4. Complete Hilbert space
    3.5. Orthonormal sequences and the best approximation
    3.6. Orthonormalization
    3.7. Orthogonal systems
    Problems
    Appendix — The space of integrable functions
    Problems

CHAPTER 4 — SOME SPECIAL FOURIER SERIES IN SPECIFIC HILBERT SPACES   101

4.1. Some properties of the space L_2
4.2. Trigonometric Fourier series of functions of one variable
4.3. Trigonometric Fourier series with a general period
4.4. Fourier series of functions of several variables
4.5. Expansions with respect to eigenfunctions
4.6. Orthogonal polynomials

Problems

CHAPTER 5 — CALCULATION OF FOURIER SERIES   149

5.1. Expansions of various functions
5.2. Expansions of functions defined on an interval of half length
5.3. Expansions of a non-periodic function
5.4. Complex and phase forms of the Fourier series
5.5. Application of various operations with series
5.6. Application of functions of a complex variable
5.7. Conjugate series
5.8. Evaluation of various number series and integrals by means of Fourier series

Problems

CHAPTER 6 — APPROXIMATE HARMONIC ANALYSIS   196

6.1. Introduction
6.2. Krylov's method
6.3. Approximate summation of series
6.4. Numerical calculation of Fourier coefficients

Problems

CHAPTER 7 — SOME SPECIAL CRITERIA FOR CONVERGENCE   225

7.1. Gibbs phenomenon
7.2. Dirichlet's kernel
7.3. Continuous periodic functions
7.4. Principle of localization
7.5. Absolutely continuous functions
7.6. Functions of bounded variation
7.7. Fourier coefficients and the properties of the sum
7.8. Functions in $W_2^{(k)}$
7.9. Convergence of arithmetic means

Problems

CHAPTER 8 — FOURIER TRANSFORMS 260
  8.1. Fourier transforms as a limit case of Fourier series
  8.2. Inversion formula
  8.3. Fourier transforms of functions in L_2
  8.4. Properties of Fourier transforms
  8.5. Fourier transforms of distributions
  Problems

CHAPTER 9 — EXAMPLES OF THE APPLICATION OF FOURIER SERIES 292
  9.1. Classical solution of the equation for a string
  9.2. Generalized solution
  9.3. Decomposition of a tone into harmonics. Vibrations of a tuner
  9.4. Equation of heat conduction
  9.5. Laplace equation
  9.6. Some applications of Fourier series in the theory of integral equations
  9.7. Some applications of Fourier transforms
  Problems

CHAPTER 10 — SURVEY OF FOURIER SERIES AND FOURIER TRANSFORMS OF SOME COMMONLY USED FUNCTIONS 336

References 351
Bibliography 352
Index 353

Since this book was written
JAN KADLEC
died a tragic death
May this book serve as an epitaph

PREFACE

In the present book two tasks are set. First, the book is intended to serve as an introduction to the theory of Fourier series and expansions in general orthogonal systems. Secondly, it should provide a solid base for applying these series to the most common actual problems.

As far as the second task is concerned, we wish to make the following comments. Fourier series are employed in practice quite generally, but their use has very often a purely formal character. To be more specific, if a problem is solved, operations with series are performed without justifying the individual steps and the result obtained is held to be the solution of the problem without any verification. Experience shows, however, that such formal procedure may yield entirely incorrect results. Having this in mind, we would like to equip particularly that reader who is oriented primarily towards practice with a knowledge of the most fundamental propositions, which are necessary in applications and sufficient for most purposes.

Our aims have affected the choice and arrangement of the chapters. Chapters 2 to 6 contain approximately the material which is sufficient for common practical uses. Chapter 2 summarizes the most elementary and most frequently applied theorems on the convergence of trigonometric series. Chapter 3 shows that the trigonometric series are a particular case of orthogonal series in Hilbert spaces. Chapter 4 presents in essence the application of general results given in Chapter 3 to actual cases and thus enables us to expand functions in series of other than trigonometric functions. Chapter 5 has more or less the character of a cook-book and summarizes the methods of calculation and application of Fourier series which are useful in practice. Chapter 6 serves as a basis for the numerical evaluation of Fourier series and their sums.

On the other hand, Chapter 7 has a more theoretical character and is intended primarily for those readers who are interested more deeply in the mathematical aspect of Fourier series. This, however, does not mean that the propositions given in this chapter are useless in practice; it is a fact that with an increasing penetration of mathematics into all fields of human activity the deeper theoretical results become more and more significant, and are often indispensable.

Since in practice we encounter not only the Fourier series but also its "continuous analogue", i.e. the Fourier transform, we considered it convenient to augment the book by a brief chapter dealing with Fourier transforms.

Various applications of Fourier series are presented in a more or less illustrative manner in Chapter 9. In selecting the examples, however, we have focused our attention not on the trivial ones but on a certain transition to more general problems, which are of great importance in engineering sciences.

Chapters 7 to 9 thus constitute a second part of the book, let us say a particular part, where various results are imparted to the reader by presenting a number of examples. Consequently, it challenges the reader more than the first part and requires more co-operation and preliminary knowledge of the problems considered.

Finally, Chapter 10 is a survey of various formulas and tables which are useful in practical applications.

Individual chapters are augmented by problems to be solved by the reader. For some problems hints are given. All formulas, theorems, examples and problems are enumerated so that the first figure denotes the chapter and the others the order in a chapter. The chapters are subdivided into sections and subsections, if necessary.

The concept of the Fourier series may be introduced in various ways. For example, we can define it directly and study its properties, or we can use the fact that a partial sum of a Fourier series is exactly that trigonometric polynomial which gives the best approximation of a given function. This second approach, although it is not the most common one, has been adopted as a basis for our development primarily for the reason that it requires less preliminary knowledge. In spite of this we were compelled to use some propositions whose proofs reach beyond the scope of this book.

The final form of the book is the result of efforts not only by the authors but many other persons, and we wish to acknowledge their assistance. We are particularly indebted to the reviewer Assistant Professor L. MIŠÍK and scientific editor Professor M. FIEDLER who contributed with understanding and unusual care to the improvement of the manuscript, and also to our colleagues Drs. J. KOPÁČEK, J. FUKA, O. HORÁČEK and M. KUČERA, who read the text and supplied valuable advice, and to Ing. F. KUFNER, who drew the figures.

This book, having an introductory character, naturally cannot encompass everything which has been achieved in the discipline of Fourier series. How we have fulfilled our aims can be judged by the reader himself. We would greatly appreciate it if this book stimulates the reader's interest in this discipline whose importance is certainly indisputable.

A. K.
J. K.

Prague, December 1966

NOTATION

$[a, b]$	the closed interval $a \leqq x \leqq b$
(a, b)	the open interval $a < x < b$
$[a, b)$	the semi-closed interval $a \leqq x < b$
$(a, b]$	the semi-closed interval $a < x \leqq b$
$f(c + 0)$	the limit of the function $f(x)$ at the point $x = c$ from the right
$f(c - 0)$	the limit of the function $f(x)$ at the point $x = c$ from the left
$f * g$	the convolution of the functions f and g
$\hat{f}$ or $\mathscr{F}f$	the Fourier transform of the function f
$C(a, b)$	the space of functions continuous on the interval $[a, b]$
$L_1(a, b)$	the space of functions (absolutely) integrable on the interval (a, b)
$L_2(a, b)$	the space of functions square-integrable on the interval (a, b)
$L_{2,p}(a, b)$	the space of functions square-integrable on the interval (a, b) with the weight function $p(x)$

$\omega(\delta, f)$	modulus of continuity of the function f
$\frac{1}{2}a_0 + \sum_{n=1}^{\infty} (a_n \cos nx + b_n \sin nx)$	the trigonometric Fourier series
$\sum_{n=1}^{\infty} (-b_n \cos nx + a_n \sin nx)$	the conjugate series
$D_n(t) = \dfrac{1}{2\pi} \dfrac{\sin(n+\frac{1}{2})t}{\sin \frac{1}{2}t}$	the Dirichlet kernel
$F_n(t) = \dfrac{1}{2\pi(n+1)} \dfrac{\sin^2(n+\frac{1}{2})t}{\sin^2 \frac{1}{2}t}$	the Fejér kernel
inf	greatest lower bound (g.l.b.)
sup	least upper bound (l.u.b.)
$\|x\|$ or $\|x\|_H$	the norm of an element x from the Hilbert space H
(x, y)	the scalar (inner) product of two elements x, y from the Hilbert space H
C_n	the space of all n-dimensional complex vectors
R_n	the space of all n-dimensional real vectors (the n-dimensional Euclidean space)
$C = C_1$	
$R = R_1$	
δ_{ij}	the Kronecker symbol

CHAPTER 1 BASIC CONCEPTS

In this chapter, we are going to summarize and recall some concepts and statements from Mathematical Analysis which will be used in what follows. The proofs of theorems which are stated here may be found in standard works of Differential and Integral Calculus. Some further properties are mentioned in the text of the following chapters. We would like to stress the fact that the present chapter does not claim to be complete as far as the material presented is concerned.

1.1 FUNCTIONS

Let $f(x)$ be a function defined in a neighbourhood of a point c. For denoting the limit of this function from the left and from the right (provided these limits exist and are finite) we use the following symbols

$$f(c - 0) = \lim_{x \to c-} f(x) ; \quad f(c + 0) = \lim_{x \to c+} f(x) . \qquad (1.1)$$

If the finite limits $f(c + 0)$ and $f(c - 0)$ exist and are different, the function $f(x)$ is said to have a *discontinuity of the first kind* at c (the jump $|f(c + 0) - f(c - 0)|$).

Let the function $f(x)$ be defined on the interval $(-l, l)$, l being a positive number. A function $f(x)$ is called *even*, if

$$f(-x) = f(x), \qquad (1.2)$$

and *odd*, if
$$f(-x) = -f(x). \tag{1.3}$$

Example 1.1. The function $\cos \alpha x$ is an even function, the function $\sin \alpha x$ is an odd function (α being an arbitrary real number).

A function $f(x)$ defined on the interval $(-\infty, \infty)$ is called *periodic with period* T ($T > 0$), if the equality
$$f(x + kT) = f(x) \tag{1.4}$$
holds for any integer k.

Example 1.2. Both the functions $\cos \alpha x$ and $\sin \alpha x$ are periodic with period $T = 2\pi/\alpha$.

Let the function $f(x)$ be defined on a semi-closed interval of length $T > 0$, e.g. on the interval $[0, T)$. Then this function can be *periodically extended* onto the entire real interval $(-\infty, \infty)$ with period T, i.e. the function $F(x)$, periodic with period T, is defined as follows
$$F(x) = f(x - mT),$$
where m is a unique integer satisfying the inequalities $mT \leqq x < (m + 1)T$.

1.2 SERIES

Let $f_1(x), f_2(x), f_3(x), \ldots$, be a sequence of functions, defined on a set M. Denote by $s_n(x)$ the functions $\sum_{k=1}^{n} f_k(x)$; the series $\sum_{k=1}^{\infty} f_k(x)$ is said to *converge* on M to the sum $s(x)$, if for any $x \in M$, the limit $\lim_{n \to \infty} s_n(x)$ exists and is equal to $s(x)$. If the functions $s_n(x)$

converge to $s(x)$ uniformly with respect to $x \in M$, the series $\sum_{k=1}^{\infty} f_k(x)$ is said to converge *uniformly* on M.

The series is said to converge *absolutely*, if the series $\sum_{k=1}^{\infty} |f_k(x)|$ converges; if the latter series does not converge but the former does, the series $\sum_{k=1}^{\infty} f_k(x)$ is then called *conditionally convergent*. If the series $\sum_{k=1}^{\infty} |f_k(x)|$ converges uniformly, the series $\sum_{k=1}^{\infty} f_k(x)$ is called absolutely and uniformly convergent.

Let us recall the following statement: *the sum (the limit) of a uniformly convergent series (sequence) of continuous functions is a continuous function as well.*

The following statement is an important test for absolute and uniform convergence.

Theorem 1.1. *If there exists a sequence of non-negative numbers c_k such that the series $\sum_{k=1}^{\infty} c_k$ converges and $|f_k(x)| \leq c_k$ for each $x \in M$, then the series $\sum_{k=1}^{\infty} f_k(x)$ converges absolutely and uniformly on M.*

The following theorem on term by term differentation will be needed.

Theorem 1.2. *Let each function $f_k(x)$ have a derivative $f'_k(x)$ in the interval $[a, b]$ and let the series $\sum_{k=1}^{\infty} f_k(x)$ converge at least at one point x in $[a, b]$. Moreover, let the series $\sum_{k=1}^{\infty} f'_k(x)$ converge uniformly in the interval $[a, b]$. Then the series $\sum_{k=1}^{\infty} f_k(x)$ converges uniformly in $[a, b]$ to the sum $f(x)$, which is differentiable in the interval $[a, b]$ and its derivative $f'(x)$ equals $\sum_{k=1}^{\infty} f'_k(x)$.*

This statement can be written briefly as follows:

$$\left(\sum_{k=1}^{\infty} f_k(x)\right)' = \sum_{k=1}^{\infty} f_k'(x).$$

Let us have a sequence of functions $\ldots f_{-3}(x)$, $f_{-2}(x)$, $f_{-1}(x)$, $f_0(x)$, $f_1(x)$, $f_2(x)$, $f_3(x), \ldots$ and denote by $s_{mn}(x)$ the functions $\sum_{k=-m}^{n} f_k(x)$. By the sum of the series $\sum_{k=-\infty}^{\infty} f_k(x)$ we understand the double limit $\lim_{\substack{m \to \infty \\ n \to \infty}} s_{mn}(x)$.

Let us have a double sequence of functions $f_{ik}(x)$ ($i, k = 1, 2, 3, \ldots$) and denote by $\sigma_{mn}(x)$ the functions $\sum_{i=1}^{m} \sum_{k=1}^{n} f_{ik}(x)$. By the sum of the double series $\sum_{i=1}^{\infty} \sum_{k=1}^{\infty} f_{ik}(x)$ we understand the limit $\lim_{\substack{m \to \infty \\ n \to \infty}} \sigma_{mn}(x)$.

1.3 ABELIAN SUMMATION

To begin with, we define the concept of a *sequence of bounded variation*. A sequence of complex numbers $c_1, c_2, c_3, \ldots$ is of *bounded variation* if the series $\sum_{n=0}^{\infty} |c_{n+1} - c_n|$ is convergent (we put $c_0 = 0$).

Every monotonic bounded sequence is of bounded variation. Moreover, we have

Theorem 1.3. *A real-valued sequence $\{c_n\}$ is of bounded variation if and only if $c_n = a_n - b_n$, where $\{a_n\}$ and $\{b_n\}$ are bounded non-decreasing sequences of non-negative numbers.*

Let now $f_n(x)$ be a sequence of functions defined on the interval (a, b). Then

$$\sum_{n=1}^{m} c_n f_n(x) = \sum_{n=1}^{m-1} (c_n - c_{n+1}) F_n(x) + c_m F_m(x), \qquad (1.5)$$

BASIC CONCEPTS 5

where $F_k(x) = \sum_{i=1}^{k} f_i(x)$; the equality (1.5) can be verified by calculation.

Theorem 1.4. *If all the functions $|F_k(x)|$ are bounded in (a, b) by the same constant M and the sequence $\{c_n\}$ is of bounded variation and converges to zero, then the series $\sum_{n=1}^{\infty} c_n f_n(x)$ is uniformly convergent in (a, b).*

Proof. Choose $\varepsilon > 0$. We have

$$\left| \sum_{k=n+1}^{n+p} c_k f_k(x) \right| = \left| \sum_{k=n}^{n+p-1} (c_k - c_{k+1}) F_k(x) - c_n F_n(x) + c_{n+p} F_{n+p}(x) \right|$$

$$\leq \sum_{k=n}^{n+p-1} M |c_k - c_{k+1}| + M |c_n| + M |c_{n+p}|$$

$$= M \left(|c_n| + |c_{n+p}| + \sum_{k=n}^{n+p-1} |c_k - c_{k+1}| \right).$$

If n_0 is so large that $|c_n| < \varepsilon/3M$ and $\sum_{k=n}^{n+p-1} |c_k - c_{k+1}| < \varepsilon/3M$ for all $n > n_0$, then $\left| \sum_{k=n+1}^{n+p} c_k f_k(x) \right| < \varepsilon$. By the Bolzano-Cauchy criterion, the series $\sum_{k=1}^{\infty} c_k f_k(x)$ is uniformly convergent.

Theorem 1.5. *If the series $\sum_{n=1}^{\infty} f_n(x)$ is uniformly convergent and if $\{c_n\}$ is a sequence of bounded variation, then the series $\sum_{n=1}^{\infty} c_n f_n(x)$ is uniformly convergent as well.*

This theorem can be proved by using again the equality (1.5).

1.4 LEBESGUE INTEGRAL

In what follows, we shall work with integrals in the Lebesgue sense. If the reader is familiar with the Riemann integral only, he can become acquainted with both the definition and further properties of the Lebesgue integral in various books, e.g., in [9]; here we have not enough space for a detailed treatment. Yet, let us recall at least the fact that the Lebesgue integral is a generalization of the Riemann integral, that all functions which are integrable in Riemann sense (e.g. functions defined in a finite interval and having only a finite number of discontinuities of the first kind there) are also integrable in Lebesgue sense and the integrals coincide (i.e. the numbers defined by them are equal). It is necessary to emphasize that when dealing with Lebesgue integrals, the integrability is always *absolute*, i.e. a function $f(x)$ is Lebesgue integrable if and only if the function $|f(x)|$ is Lebesgue integrable.

Let us recall further that the concept of Lebesgue integration is closely related to concepts of *measurable sets* and *measurable functions*. In what follows, we will not consider any set or functions other than the measurable ones. The important concept of a *null set*, i.e. a set of measure zero, can be defined in the one-dimensional case as follows. It is a set N such that for each $\varepsilon > 0$ we can find a sequence of open intervals (a_k, b_k), $k = 1, 2, 3, \ldots$ with $\sum_{k=1}^{\infty} (b_k - a_k) < \varepsilon$ such that each point from N belongs to some of the intervals (a_k, b_k).

Example 1.3. Every closed set and every open set is measurable. Every function which is defined on a measurable set and is continuous or piecewise continuous on this set is measurable. Every finite set of points and every countably infinite set of points (i.e. a sequence of points) is a null set.

Let $f(x)$ be a (measurable) function of real variable x assuming complex values (including the value ∞), defined on the (bounded or unbounded) interval (a, b) and let the following integral exist (convergent or divergent):

$$\int_a^b |f(x)| \, dx . \tag{1.6}$$

The function $f(x)$ is said to be (absolutely) integrable on (a, b), if the integral (1.6) is a finite number. The integral (1.6) is then said to be absolutely convergent. The number (1.6) is finite if and only if the integral

$$\int_a^b f(x)\, dx = \int_a^b f_1(x)\, dx + i \int_a^b f_2(x)\, dx \tag{1.7}$$

is finite, $f_1(x)$ and $f_2(x)$ being the real and imaginary parts of $f(x)$, respectively.

Let M be a subset of the interval (a, b). The characteristic function $\chi_M(x)$ of the set M is defined as follows

$$\chi_M(x) = 1 \text{ for } x \in M, \quad \chi_M(x) = 0 \text{ for } x \notin M.$$

If the integral

$$\int_a^b \chi_M(x)\, dx = \mu(M) \tag{1.8}$$

exists, then the set M is said to have the measure $\mu(M)$. If some property holds for all $x \in (a, b)$ with the exception of a null set, then this property is said to hold "almost everywhere in (a, b)" or "almost always in (a, b)" or "for almost every $x \in (a, b)$".

If the integral (1.7) is finite, the function $f(x)$ is finite almost everywhere in (a, b), i.e. $f(x)$ is a finite number for almost every $x \in (a, b)$. If the function $f(x)$ is altered on a null set, the integral (1.7) does not change.

Let M be any measurable subset of the interval $(-\infty, \infty)$ and let the function $f(x)$ be defined in $(-\infty, \infty)$. Then we define

$$\int_M f(x)\, dx = \int_{-\infty}^{\infty} f(x)\, \chi_M(x)\, dx. \tag{1.9}$$

The dependence of the integral (1.9) on the function $f(x)$ and on the domain of integration M is described by the following theorems:

Theorem 1.6. (LEBESGUE) *Let $f_n(x)$ be functions integrable on the interval (a, b) and let the limit $\lim_{n \to \infty} f_n(x) = f(x)$ exist for almost every x in (a, b). Further, let there exist a non-negative integrable* function $g(x)$, called integrable majorant, with $|f_n(x)| \leq g(x)$ whenever n is a positive integer and $x \in (a, b)$. Then

$$\lim_{n \to \infty} \int_a^b f_n(x) \, dx = \int_a^b \lim_{n \to \infty} f_n(x) \, dx = \int_a^b f(x) \, dx. \quad (1.10)$$

Theorem 1.7. *Let* $0 \leq f_1(x) \leq f_2(x) \leq \cdots \leq f_n(x) \leq f_{n+1}(x) \leq \cdots$ *for almost every* $x \in (a, b)$ *and let the functions* $f_n(x)$ *be measurable. Then* (1.10) *again holds.*

An immediate consequence of this theorem is the following theorem on term by term integration.

Theorem 1.8. *Let the functions* $f_n(x)$ *be non-negative and measurable. Then*

$$\sum_{n=1}^\infty \int_a^b f_n(x) \, dx = \int_a^b \sum_{n=1}^\infty f_n(x) \, dx. \quad (1.11)$$

Theorem 1.9. *Let the function* $f(x)$ *be integrable on the interval* (a, b) *and let* M *be a subset of* (a, b). *Then for each* $\varepsilon > 0$ *there are* $\delta > 0$ *and* $N_0 > 0$ *with*

$$\int_M |f(x)| \, dx + \int_{(a,b) \cap (N, \infty)} |f(x)| \, dx + \int_{(a,b) \cap (-\infty, -N)} |f(x)| \, dx < \varepsilon$$

whenever $\mu(M) < \delta$ *and* $N > N_0$.

From this it follows immediately that whenever $g(x)$ is integrable on a finite interval (a, b), the function

$$f(x) = \int_a^x g(t) \, dt + C, \quad (1.12)$$

$x \in [a, b]$, is continuous on the interval $[a, b]$. Using (1.12) we get the formula

$$f(x) - f(y) = \int_y^x g(t) \, dt \quad (1.13)$$

holding for any two points x, y in $[a, b]$.

BASIC CONCEPTS 9

If, for a given function $f(x)$, there exists a function $g(x)$, which is integrable in each bounded part of the interval (a, b) and satisfies (1.13), then the function $f(x)$ is said to be *absolutely continuous* in the interval (a, b). The function $g(x)$ is then said to be a *derivative of the function $f(x)$ in the sense of an absolutely continuous function* and the notation $g(x) = f'(x)$ is used. Formula (1.13) can be then stated as follows:

$$f(x) - f(y) = \int_y^x f'(t) \, dt \, .$$

Example 1.4. The functions $\sqrt{|x|}$, $|x|$, $x \log |x| - x$ are absolutely continuous in each bounded interval $[a, b] \subset (-\infty, \infty)$ and have there derivatives in this sense $\frac{1}{2}(\operatorname{sgn} x)/\sqrt{|x|}$, $\operatorname{sgn} x$, $\log |x|$ respectively. The function $\operatorname{sgn} x = |x|/x$ for $x \neq 0$, $\operatorname{sgn} 0 = 0$, is not absolutely continuous in any interval containing the point $x = 0$ in its interior.

The above mentioned examples also show that an absolutely continuous function need not have a derivative (in the usual sense) in the entire interval (a, b). Nevertheless, the following assertion is true.

Theorem 1.10. *If $f(x)$ is an absolutely continuous function on the interval (a, b), then $f(x)$ is differentiable almost everywhere in (a, b), i.e. for almost every $x \in (a, b)$ there exists the limit*

$$\lim_{h \to 0} \frac{1}{h} [f(x + h) - f(x)] = \lim_{h \to 0} \frac{1}{h} \int_x^{x+h} f'(t) \, dt$$

and it equals $f'(x)$ almost everywhere.

Since any integrable function $f(x)$ is a derivative of its primitive function

$$\int_a^x f(t) \, dt + C$$

the limit $\lim\limits_{h \to 0} (1/h) \int_x^{x+h} f(t) \, dt$ equals $f(x)$ for almost every x.

For functions $f(x)$ and $g(x)$, absolutely continuous in a finite interval $[a, b]$, we have the following formula (*integration by parts*)

$$\int_a^b f(x) g'(x) \, dx = - \int_a^b f'(x) g(x) \, dx + f(b) g(b) - f(a) g(a)$$

$$= - \int_a^b f'(x) g(x) \, dx + [f(x) g(x)]_a^b. \qquad (1.14)$$

Remark. Let us recall the fact that for $f(x)$ continuous on a bounded interval $[a, b]$ the expression (1.7) is the usual Riemann integral. If $f(x)$ is continuously differentiable then its derivative $f'(x)$ coincides with the derivative in the sense of an absolutely continuous function.

We should turn the reader's attention to the difference between Lebesgue and improper integrals. If a function $f(x)$ is integrable in $(0, \infty)$ in Lebesgue sense (and consequently, absolutely integrable), then $\int_0^\infty f(x) \, dx = \lim_{N \to \infty} \int_0^N f(x) \, dx$. However, the limit on the right hand side can exist even if $f(x)$ is not Lebesgue integrable in $(0, \infty)$. This limit is then called the improper integral. For example, for $f(x) = (\sin x)/x$, the limit

$$\lim_{N \to \infty} \int_0^N \frac{\sin x}{x} \, dx = \frac{\pi}{2}$$

exists, but the Lebesgue integral $\int_0^\infty [(\sin x)/x] \, dx$ does not, for the number $\int_0^\infty |(\sin x)/x| \, dx$ is not finite. The integral $\int_0^\infty [(\sin x)/x] \, dx$ is then said to converge conditionally.

For the Lebesgue integral we have the usual theorems concerning the calculation of integrals, i.e. integration of a sum of functions, of a product of a function with a number, theorem on substitution etc.

We can also define double (or n-dimensional) integrals and for them we have theorems analogous to Theorems 1.6 to 1.9. The following theorem is important.

Theorem 1.11. (Fubini) *Let $f(x, y)$ be a function of two real variables and let the following integral exist and be convergent*

$$\iint_{\substack{a<x<b \\ c<y<d}} |f(x, y)| \, dx \, dy \, .$$

Then the integral $\int_c^d |f(x, y)| \, dy$ (resp. $\int_a^b |f(x, y)| \, dx$) converges for almost every $x \in (a, b)$ (resp. for almost every $y \in (c, d)$) and is integrable in (a, b) (resp. in (c, d)), and we have:

$$\iint_{\substack{a<x<b \\ c<y<d}} f(x, y) \, dx \, dy = \int_a^b \left(\int_c^d f(x, y) \, dy \right) dx = \int_c^d \left(\int_a^b f(x, y) \, dx \right) dy \, .$$

Let us recall some other properties of an integral depending on a parameter:

$$F(\alpha) = \int_M f(x, \alpha) \, dx \, ,$$

where $\alpha \in A$, A being an interval of real numbers and $f(x, \alpha)$ being measurable (as a function of x) for each $\alpha \in A$.

Theorem 1.12. I. *Let $f(x, \alpha)$ be a continuous function of α in A for almost every $x \in M$ and let a non-negative integrable function $g(x)$ (integrable majorant) exist such that for any $\alpha \in A$, $|f(x, \alpha)| \leq g(x)$ holds for almost every $x \in M$. Then the integral $F(\alpha)$ is a continuous function of α in A.*

II. *Let the integral $F(\alpha)$ exist at least for one $\alpha \in A$. Let the derivative $\partial f(x, \alpha)/\partial \alpha$ exist and be finite for every $\alpha \in A$ and for almost every $x \in M$ and let there exist a non-negative, in M integrable function $h(x)$ (integrable majorant) such that $|\partial f(x, \alpha)/\partial \alpha| \leq h(x)$ holds for every $\alpha \in A$ and almost every $x \in M$. Then the integral $F(\alpha)$ exists for every $\alpha \in A$ and we can interchange the order of integration and differentiation, i.e. the following holds for every $\alpha \in A$:*

$$F'(\alpha) = \int_M \frac{\partial f(x, \alpha)}{\partial \alpha} \, dx \, .$$

The following formulas will often be used: For $f(x)$ even

$$\int_{-l}^{l} f(x)\, dx = 2 \int_{0}^{l} f(x)\, dx \qquad (1.15)$$

and for $f(x)$ odd

$$\int_{-l}^{l} f(x)\, dx = 0. \qquad (1.16)$$

If $f(x)$ is periodic with period T, then

$$\int_{a}^{a+T} f(x)\, dx = \int_{-T/2}^{T/2} f(x)\, dx \qquad (1.17)$$

for any real number a.

Formulas (1.15) and (1.16) follow from the fact that $\int_{-l}^{l} = \int_{-l}^{0} + \int_{0}^{l}$; for an even function $f(x)$, substituting $x = -t$ and using the formula (1.2) we get $\int_{-l}^{0} f(x)\, dx = \int_{-l}^{0} f(-x)\, dx = \int_{0}^{l} f(t)\, dt$; similarly for $f(x)$ odd. The formula (1.17) can be proved by using the equality $\int_{a}^{a+T} = \int_{a}^{-T/2} + \int_{-T/2}^{T/2} + \int_{T/2}^{a+T}$: substituting now $x = t + T$ into the third integral on the right hand side, we can verify with the help of (1.4) that it is equal to the negative value of the first integral on the right hand side.

The concept of *orthogonality* is important: two (complex-valued) functions $f(x)$ and $g(x)$, defined on (a, b), are called *orthogonal*, if

$$\int_{a}^{b} f(x)\, \overline{g(x)}\, dx = 0.$$

A system of functions $f_1(x), f_2(x), f_3(x), \ldots$ is called *orthogonal*, if all its members are pairwise orthogonal, i.e. if $\int_{a}^{b} f_i(x)\, \overline{f_j(x)}\, dx = 0$ holds for all $i \neq j$.

1.5 SOME FORMULAS FOR TRIGONOMETRIC FUNCTIONS

The functions $\cos x$, $\sin x$ and $e^{ix} = \cos x + i \sin x$ are defined on $(-\infty, \infty)$ and are periodic with period 2π. There is a number of formulas involving them; some of them are stated here.

$$\cos x = 1 - \frac{x^2}{2!} + \frac{x^4}{4!} - \frac{x^6}{6!} + \ldots ;$$

$$\sin x = x - \frac{x^3}{3!} + \frac{x^5}{5!} - \frac{x^7}{7!} + \ldots ; \qquad (1.18)$$

the series on the right hand side converge uniformly on every bounded part of the real line.

$$\cos x = \frac{e^{ix} + e^{-ix}}{2} ; \quad \sin x = \frac{e^{ix} - e^{-ix}}{2i} \quad \text{(Euler's formulas)} \quad (1.19)$$

$$\cos(x + y) = \cos x \cos y - \sin x \sin y ;$$
$$\sin(x + y) = \sin x \cos y + \cos x \sin y . \qquad (1.20)$$

$$\sin mx \sin nx = \tfrac{1}{2}[\cos(m - n)x - \cos(m + n)x] ;$$
$$\cos mx \cos nx = \tfrac{1}{2}[\cos(m - n)x + \cos(m + n)x] ;$$
$$\sin mx \cos nx = \tfrac{1}{2}[\sin(m + n)x + \sin(m - n)x] . \qquad (1.21)$$

Using formulas (1.21) we can evaluate some integrals. Since

$$\int_a^{a+2\pi} \cos kx \, dx = 0 \quad \text{for every integer } k \neq 0 \text{ and}$$

$$\int_a^{a+2\pi} \sin kx \, dx = 0 \quad \text{for every integer } k$$

(a is an arbitrary real number), we have for any integers $m \neq n$

$$\int_a^{a+2\pi} \sin mx \sin nx \, dx$$
$$= \frac{1}{2} \int_a^{a+2\pi} [\cos(m-n)x - \cos(m+n)x] \, dx = 0 ;$$

$$\int_a^{a+2\pi} \cos mx \cos nx \, dx = 0 ;$$

$$\int_a^{a+2\pi} \sin mx \cos nx \, dx = 0 \quad \text{(also for } m = n\text{)}. \tag{1.22}$$

The formulas (1.22) are called formulas of orthogonality of trigonometric functions.

If $m = n$, then

$$\int_a^{a+2\pi} \cos^2 nx \, dx = \pi ; \quad \int_a^{a+2\pi} \sin^2 nx \, dx = \pi. \tag{1.23}$$

Let $A_0, A_1, \ldots, A_n, B_1, \ldots, B_n$ (resp. $C_{-n}, \ldots, C_0, \ldots, C_n$) be complex numbers. The expression

$$T_n(x) = A_0 + \sum_{k=1}^n (A_k \cos kx + B_k \sin kx)$$

$$\left(\text{resp. } \tilde{T}_n(x) = \sum_{k=-n}^n C_k e^{ikx}\right) \tag{1.24}$$

is called a *trigonometrical polynomial* (of the n-th degree).

Since the sum $\sum_{k=1}^n e^{ikx}$ is in fact a partial sum of a geometric series with the quotient $q = e^{ix}$, we have

$$\sum_{k=1}^n e^{ikx} = e^{ix} \frac{1 - e^{inx}}{1 - e^{ix}} = -\frac{e^{i(x/2)} - e^{i(n+1/2)x}}{e^{i(x/2)} - e^{-i(x/2)}}.$$

BASIC CONCEPTS

Using the equality $e^{ikx} = \cos kx + i \sin kx$ we can rewrite the latter equality as follows:

$$\sum_{k=1}^{n} (\cos kx + i \sin kx)$$
$$= \frac{\sin \left(n + \tfrac{1}{2}\right) x - \sin \frac{x}{2} + i \left(\cos \frac{x}{2} - \cos \left(n + \tfrac{1}{2}\right) x\right)}{2 \sin \frac{x}{2}}.$$

Since the real (imaginary) part of the left-hand side necessarily equals the real (imaginary) part of the right-hand side, we obtain the following important formulas

$$\sum_{k=1}^{n} \sin kx = \frac{\cos \frac{x}{2} - \cos \left(n + \tfrac{1}{2}\right) x}{2 \sin \frac{x}{2}};$$

$$\sum_{k=1}^{n} \cos kx = \frac{\sin \left(n + \tfrac{1}{2}\right) x - \sin \frac{x}{2}}{2 \sin \frac{x}{2}} \qquad (1.25)$$

holding for $x \neq 0, \pm 2\pi, \pm 4\pi, \ldots$ The last mentioned equality can be also written in the form

$$\tfrac{1}{2} + \sum_{k=1}^{n} \cos kx = \frac{\sin \left(n + \tfrac{1}{2}\right) x}{2 \sin \frac{x}{2}}. \qquad (1.26)$$

Problems

1.1. Prove Theorem 1.3.

(*Hint*: Put $a_n = \sum_{k=0}^{n-1} (c_{k+1} - c_k)^+$ and $b_n = \sum_{k=0}^{n-1} (c_{k+1} - c_k)^-$, where $a^+ = \frac{1}{2}(|a| + a)$, $a^- = \frac{1}{2}(|a| - a)$. If, conversely, $c_n = a_n - b_n$, then
$$\sum_{n=0}^{m} |c_{n+1} - c_n| \leq a_{m+1} + b_{m+1}.)$$

1.2. Prove Theorem 1.5.

(*Hint*: The sequence $\{c_n\}$ of bounded variation has a finite limit c. Use the fact that the functions $|F_k(x)|$ have the same upper bound $|M|$ and apply Theorem 1.4 to the sequence $\{c_n - c\}$.)

1.3. Prove the relation (1.16).

1.4. Using Euler's formulas (1.19) prove the equalities (1.20) and (1.21).

1.5. Let the equality $T_n(x) = \tilde{T}_n(x)$ hold in (1.24). Express the coefficients C_k and C_{-k} in terms of the coefficients A_k and B_k and vice versa.

1.6. Prove the following statement: If $T_n(x)$ is a trigonometric polynomial and c a real number, then the expression $T_n(x - c)$ is a trigonometric polynomial as well.

1.7. Show that the expressions $(\cos x)^n$ and $(\cos \frac{1}{2}x)^{2n}$ are trigonometric polynomials of the n-th degree.

(*Hint*: Use Euler's formulas (1.19).)

BASIC CONCEPTS

1.8. Find the sum $\sum_{k=0}^{n}(-1)^k e^{ikx}$; derive from it the formulas for $\sum_{k=0}^{n}(-1)^k \cos kx$ and $\sum_{k=1}^{n}(-1)^k \sin kx$.

(*Hint*: Proceed as when deriving the formulas (1.25).)

1.9. What do the sums $\sum_{k=1}^{n} \cos(2k+1)x$ and $\sum_{k=1}^{n} \sin(2k+1)x$ equal?

(*Hint*: Make use of the formulas for $\sum_{k=1}^{n} e^{ikx}$ and of the equality $\sum_{k=1}^{n} e^{i(2k+1)x} = e^{ix} \sum_{k=1}^{n} e^{2ikx}$.)

1.10. Find the arithmetical mean of the expressions (1.26) for $n = 0, 1, \ldots, N$.

(*Hint*: Find the sum $\sum_{n=0}^{N} \sin(n + \frac{1}{2})x$. Result:

$$\frac{1}{2(N+1)} \frac{\sin^2(N+1)\frac{1}{2}x}{\sin^2 \frac{1}{2}x}.)$$

1.11. Prove the inequalities: $|\alpha\beta| \leq \frac{1}{2}(|\alpha|^2 + |\beta|^2)$; $|\alpha + \beta|^2 \leq 2(|\alpha|^2 + |\beta|^2)$.

1.12. Find the sum $\frac{1}{2} + \sum_{k=1}^{n} r^k \cos kx$.

(*Hint*: Find the partial sum of the geometric series $\sum_{k=1}^{n} q^k$ with the quotient $q = re^{ix}$.)

1.13. Prove that a uniformly convergent series of continuous functions can be integrated term by term, i.e.

$$\int_a^x \left(\sum_{n=1}^{\infty} f_n(t)\right) dt = \sum_{n=1}^{\infty} \int_a^x f_n(t) \, dt.$$

(*Hint*: Use Theorem 1.12 and the equality $f_n(x) = d/dx \left[\int_a^x f_n(t) \, dt\right]$.)

1.14. Show that the system of functions e^{ikx} with k assuming all integral values is orthogonal in $(a, a + 2\pi)$, i.e.

$$\int_a^{a+2\pi} e^{ikx}\overline{e^{imx}}\,dx = \int_a^{a+2\pi} e^{ikx}e^{-imx}\,dx = 0$$

for $k \neq m$.

CHAPTER 2
CONCEPT OF A FOURIER SERIES

2.1 SOLUTION OF THE EQUATION OF A VIBRATING STRING

One of the simplest equations in mathematical physics is the equation

$$\frac{\partial^2 u}{\partial x^2} - \frac{\partial^2 u}{\partial t^2} = 0, \qquad (2.1)$$

which describes, with a certain simplification, the transverse vibrations of a string. Its solution $u(x, t)$ gives the displacement of the string at a point x and time t.

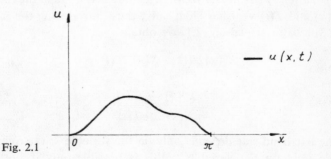

Fig. 2.1

We are going to consider a string of length π fastened at both ends (see Fig. 2.1), so that for every time instant t the function $u(x, t)$ has to satisfy the *boundary conditions*

$$u(0, t) = 0, \quad u(\pi, t) = 0. \qquad (2.2)$$

Let us note that the equation (2.1) and the conditions (2.2) do not determine the vibration of the string completely; the vibration also depends on the so-called *initial conditions*, i. e. on the displacement and velocity of the string at the initial moment $t = 0$.

As early as 1747 d'ALEMBERT had shown that the general solution of the equation (2.1) has the form

$$u(x, t) = f(x - t) + g(x + t),$$

where f and g are, up to a certain extent, arbitrary functions. (The reader can easily verify that the above function $u(x, t)$ is actually a solution of (2.1).) In 1753 D. BERNOULLI gave the solution in the form of a trigonometric series, which caused numerous disputes at the time. In what follows we are going to show how the trigonometric form of the solution can be arrived at.

Let us try to find the solution of (2.1) with boundary conditions (2.2) in a particular form, i.e. in the form of a product of two functions such that the first one depends only on the variable x and the second one only on the variable t, i.e.

$$u(x, t) = X(x) T(t). \tag{2.3}$$

This procedure is called the *Fourier method* or the method of *separation of variables*. We shall assume that none of the functions $X(x)$ and $T(t)$ vanishes identically; thus, introducing the function (2.3) into the equation (2.1) we obtain

$$X''(x) T(t) - X(x) T''(t) = 0,$$

i.e.

$$\frac{X''(x)}{X(x)} = \frac{T''(t)}{T(t)}.$$

The left hand side depends only on the variable x, the right hand side only on the variable t; this is possible only if both sides are equal to a constant, say $-\lambda$. Hence, we obtain two ordinary differential equations

$$T''(t) + \lambda T(t) = 0 \tag{2.4}$$

and

$$X''(x) + \lambda X(x) = 0. \tag{2.5}$$

From the boundary conditions (2.2) we obtain with respect to (2.3) the following conditions for the function X:

$$X(0) = 0, \quad X(\pi) = 0. \tag{2.6}$$

Let us now examine the solution of equation (2.5) with the boundary conditions (2.6). If either $\lambda < 0$ or $\lambda = 0$, the general solution of the equation (2.5) takes the form

$$X(x) = C_1 e^{\sqrt{(-\lambda)}x} + C_2 e^{-\sqrt{(-\lambda)}x}$$

and

$$X(x) = C_1 x + C_2,$$

respectively, where C_1 and C_2 are arbitrary constants. The reader can easily verify that these functions satisfy the conditions (2.6) exactly if $C_1 = C_2 = 0$ so that $X(x)$ vanishes identically; this case, however, is excluded.

Thus, to obtain a non-zero solution we have to assume that $\lambda > 0$. Then the general solution of the equation (2.5) assumes the form

$$X(x) = C_1 \cos \sqrt{(\lambda)}\, x + C_2 \sin \sqrt{(\lambda)}\, x.$$

From the boundary condition $X(0) = 0$ it follows that $C_1 = 0$, and from $X(\pi) = 0$, we have the condition

$$C_2 \sin \sqrt{(\lambda)}\, \pi = 0.$$

In order to obtain a non-zero solution we have to assume that $C_2 \neq 0$; thus, the boundary condition yields the equality $\sin \sqrt{(\lambda)}\, \pi = 0$, i.e. $\sqrt{\lambda} = k$, where k is any positive integer.

Hence, we arrive at the following conclusion: the equation (2.5) with boundary conditions (2.6) has a non-trivial (i.e. non-zero) solution, only if

$$\lambda = \lambda_k = k^2, \quad k = 1, 2, 3, \ldots;$$

then for the solution we have

$$X_k(x) = C_k \sin kx,$$

where C_k is an arbitrary non-zero constant.

The equation (2.4) now assumes the form $T''(t) + k^2 T(t) = 0$ with general solution

$$T_k(t) = A_k \cos kt + B_k \sin kt.$$

Thus, in view of (2.3), the general solution of equation (2.1) with boundary conditions (2.2) can be written as

$$u_k(x, t) = C_k \sin kx \, (A_k \cos kt + B_k \sin kt), \quad k = 1, 2, 3, \ldots, \quad (2.7)$$

where $C_k \neq 0$, and A_k and B_k are, at least at this point, arbitrary constants. Obviously, we need not include the constant C_k in the formula (2.7), because it can be incorporated in the constants A_k and B_k.

Disregarding these constants we see that there exists a sequence of solutions of the problem (2.1), (2.2). However, the reader can easily verify that not only every function of the form (2.7), but also any finite sum of functions $u_k(x, t)$ (with distinct ks) solves our problem. Hence, the function

$$u(x, t) = \sum_{k=1}^{n} (A_k \cos kt + B_k \sin kt) \sin kx, \qquad (2.8)$$

where n is any positive integer and A_k, B_k are arbitrary constants, is also a solution.

Let us now choose a point of the string, i.e. choose an x from the interval $(0, \pi)$, and examine the behaviour of the displacement at this point as a function of time. In other words, we are going to examine the function $f(t) = u(x, t)$, where x is fixed. Putting

$$a_k = A_k \sin kx, \quad b_k = B_k \sin kx,$$

we obtain from (2.8),

$$f(t) = \sum_{k=1}^{n} (a_k \cos kt + b_k \sin kt). \qquad (2.9)$$

This expression is a trigonometric polynomial; consequently, some properties of the solution $u(x, t)$ may be immediately given. For example, $f(t)$ is a periodic function with period 2π, i.e.

CONCEPT OF A FOURIER SERIES

$f(t + 2m\pi) = f(t)$ for any integer m and every t. In the expression (2.9) n is arbitrary so that the number of terms can be as large as we please. Therefore, it is quite natural to examine the series

$$\sum_{k=1}^{\infty} (a_k \cos kt + b_k \sin kt) \qquad (2.10)$$

instead of the polynomial (2.9), and consider it not only from the viewpoint of the solution of the equation of a vibrating string, but also as an independent object.

It can be shown (see Chapter 9) that under certain conditions imposed on the coefficients A_k and B_k the series obtained from (2.8) by letting $n \to \infty$ is actually a general solution of the equation (2.1) with boundary conditions (2.2).

The series (2.10) does not contain the constant term; if completed by a constant term a general trigonometric series

$$\tfrac{1}{2}a_0 + a_1 \cos t + b_1 \sin t + a_2 \cos 2t + b_2 \sin 2t + \ldots$$
$$\ldots + a_k \cos kt + b_k \sin kt + \ldots \qquad (2.11)$$

is obtained. (The first term is written as $\tfrac{1}{2}a_0$ instead of a_0 for formal reasons only; this permits us to give a single formula for all coefficients a_k.)

The problem of the vibrations of a string considered above has initiated to a large extent the development of the theory of Fourier series. However, these series are very significant in mathematical analysis alone as well as in applications.

The reader will have certainly noticed that some steps of the analysis in this section were not quite exact. For example, the relation $X''(x)/X(x) = T''(t)/T(t)$ considered above becomes meaningless at every point where either $X(x) = 0$ or $T(t) = 0$. Commenting on this we admit that the preceding considerations had a purely heuristic character and their only goal was to get the reader acquainted with the concept of a trigonometric series. A rigorous justification may be found in the literature or in the following chapters.

2.2 FOURIER SERIES OF AN INTEGRABLE FUNCTION

Let $f(x)$ be a function defined on $[-\pi, \pi]$ which is (absolutely) integrable on this interval, i.e.,

$$\int_{-\pi}^{\pi} |f(x)|\, dx < \infty .$$

Assume that the function $f(x)$ can be expanded in a trigonometric series of the form (2.11), i.e.

$$f(x) = \tfrac{1}{2}a_0 + \sum_{k=1}^{\infty} (a_k \cos kx + b_k \sin kx) ; \qquad (2.12)$$

this equality is understood in such a sense that the series on the right-hand side converges for every $x \in [-\pi, \pi]$ and has a sum equal to $f(x)$. The question is how the coefficients a_k and b_k are related to the function $f(x)$. A partial answer is given by the following proposition.

Theorem 2.1. *If the series in* (2.12) *converges uniformly to the function* $f(x)$ *on* $[-\pi, \pi]$, *then*

$$a_n = \frac{1}{\pi} \int_{-\pi}^{\pi} f(x) \cos nx \, dx ,$$

$$b_n = \frac{1}{\pi} \int_{-\pi}^{\pi} f(x) \sin nx \, dx , \qquad (2.13)$$

$(n = 0, 1, 2, \ldots)$ *where we set* $b_0 = 0$.

The proof uses essentially the orthogonality of trigonometric functions established in the preceding Chapter. Multiplying the equality (2.12) by the function $\cos nx$, we obtain

$$f(x) \cos nx = \frac{a_0}{2} \cos nx + \sum_{k=1}^{\infty} (a_k \cos kx \cos nx + b_k \sin kx \cos nx) .$$

CONCEPT OF A FOURIER SERIES

Because the series on the right hand side again converges uniformly, it can be integrated term by term. Thus we have for $n \neq 0$,

$$\int_{-\pi}^{\pi} f(x) \cos nx \, dx = \int_{-\pi}^{\pi} \frac{a_0}{2} \cos nx \, dx$$
$$+ \sum_{k=1}^{\infty} \left(\int_{-\pi}^{\pi} a_k \cos kx \cos nx \, dx + \int_{-\pi}^{\pi} b_k \sin kx \cos nx \, dx \right)$$
$$= \int_{-\pi}^{\pi} a_n \cos^2 nx \, dx = a_n \pi ,$$

since all other integrals vanish. (See (1.22) and (1.16); note that $\sin kx \cos nx$ is an odd function.) For $n = 0$ we obtain analogously

$$\int_{-\pi}^{\pi} f(x) \, dx = \int_{-\pi}^{\pi} \frac{a_0}{2} \, dx = a_0 \pi .$$

Thereby the formulas (2.13) for the coefficients a_n are proved; the formulas for the coefficients b_n can be proved in a similar way: we multiply the series (2.12) by the function $\sin nx$ and integrate again term by term. From the formulas (1.22) and (1.16) it follows that on the right hand side we obtain only one non-zero term for $k = n$, so that

$$\int_{-\pi}^{\pi} f(x) \sin nx \, dx = b_n \pi , \quad n = 1, 2, \ldots .$$

Setting $b_0 = 0$, the formula (2.13) holds for $n = 0$, too. Hence, the theorem is proved.

In the above consideration we have assumed that the function $f(x)$ can be expanded in the series (2.12). However, we can also choose the reverse approach: because the function $f(x)$ is absolutely integrable on the interval $[-\pi, \pi]$, the integrals in (2.13) exist and we can define the numbers a_n and b_n by these formulas. Consequently, we can construct formally the series

$$\frac{a_0}{2} + \sum_{n=1}^{\infty} (a_n \cos nx + b_n \sin nx) . \tag{2.14}$$

However, at this point we cannot state anything about the convergence of this series; moreover, we do not even know how its sum. if it exists, is related to the function $f(x)$.

The coefficients in the expansion of a function in a trigonometric series were established for the first time by EULER in 1777; therefore, the formulas (2.13), which are of the utmost importance, are called Euler's formulas.* The coefficients $a_0, a_1, a_2, \ldots, b_1, b_2, \ldots$ defined by these formulas will be called the *Fourier coefficients of the* (integrable) *function* $f(x)$. The series (2.14) will then be called the *Fourier series of the function* $f(x)$ and written as follows:

$$f(x) \sim \frac{a_0}{2} + \sum_{n=1}^{\infty} (a_n \cos nx + b_n \sin nx).$$

By this notation we want to stress the fact that a *formal* assignment of the series on the right hand side to the function on the left hand side is under consideration, and that an equality need not hold at all. The assigning is defined by formulas (2.13), i.e. the coefficients of the series are determined by the function $f(x)$.

Let us now establish the Fourier series of some functions.

Example 2.1. Let the function $f(x)$ be given by $f(x) = \operatorname{sgn} x$ for $x \in [-\pi, \pi]$, i.e. $f(x) = -1$ for $x \in [-\pi, 0)$, $f(0) = 0$ and $f(x) = 1$ for $x \in (0, \pi]$ (see Fig. 2.2a). Then

$$a_n = \frac{1}{\pi} \int_{-\pi}^{\pi} \operatorname{sgn} x \cos nx \, dx$$

$$= \frac{1}{\pi} \left[-\int_{-\pi}^{0} \cos nx \, dx + \int_{0}^{\pi} \cos nx \, dx \right] = 0 \text{ for } n = 0, 1, 2, \ldots;$$

---

* This terminology for the formulas (2.13), unless otherwise stated, will not be used in what follows in order to avoid confusion with the formulas (1.19) bearing the same name.

$$b_n = \frac{1}{\pi}\int_{-\pi}^{\pi} \operatorname{sgn} x \sin nx \, \mathrm{d}x = \frac{1}{\pi}\left[-\int_{-\pi}^{0}\sin nx \, \mathrm{d}x + \int_{0}^{\pi}\sin nx \, \mathrm{d}x\right]$$

$$= \begin{cases} \dfrac{4}{\pi}\dfrac{1}{n} & \text{for } n \text{ odd}, \\ 0 & \text{for } n \text{ even}. \end{cases}$$

Thus, on the interval $[-\pi, \pi]$,

$$\operatorname{sgn} x \sim \frac{4}{\pi}\left(\frac{\sin x}{1} + \frac{\sin 3x}{3} + \frac{\sin 5x}{5} + \ldots\right). \qquad (2.15)$$

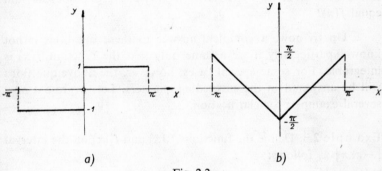

Fig. 2.2

Example 2.2. Let $f(x) = |x| - \tfrac{1}{2}\pi$ (see Fig. 2.2b). Integration by parts yields

$$a_n = \frac{1}{\pi}\int_{-\pi}^{\pi}\left(|x| - \frac{\pi}{2}\right)\cos nx \, \mathrm{d}x = \frac{1}{\pi}\left[\int_{-\pi}^{0}\left(-x - \frac{\pi}{2}\right)\cos nx \, \mathrm{d}x\right.$$

$$\left. + \int_{0}^{\pi}\left(x - \frac{\pi}{2}\right)\cos nx \, \mathrm{d}x\right] = \begin{cases} -\dfrac{4}{\pi}\dfrac{1}{n^2} & \text{for } n \text{ odd}, \\ 0 & \text{for } n \text{ even}; \end{cases}$$

$$b_n = \frac{1}{\pi}\int_{-\pi}^{\pi}\left(|x| - \frac{\pi}{2}\right)\sin nx \, \mathrm{d}x = 0 \qquad \text{for } n = 1, 2, 3, \ldots.$$

Hence, on $[-\pi, \pi]$ we have

$$|x| - \frac{\pi}{2} \sim -\frac{4}{\pi}\left(\frac{\cos x}{1^2} + \frac{\cos 3x}{3^2} + \frac{\cos 5x}{5^2} + \ldots\right). \quad (2.16)$$

The concept of the Fourier series of a function is always meaningful whenever the function $f(x)$ is integrable. It is quite natural to raise the following questions:

Is the assigned function $\sim$ Fourier series a one-to-one correspondence?

Does the series converge to a function, and if so, in what sense?

If the Fourier series of a function $f(x)$ converges, does its sum equal $f(x)$?

Up to now, a complete answer to these questions is not known, particularly if we assume only that the function $f(x)$ is integrable. For some special cases, however, the above questions will be answered in the succeeding chapters. Let us present now several examples for clarification.

Example 2.3. Define the functions $f_1(x)$ and $f_2(x)$ on the interval $[-\pi, \pi]$ as follows:

$$f_1(x) = 0 \quad \text{for all} \quad x \in [-\pi, \pi],$$
$$f_2(x) = 0 \quad \text{for all} \quad x \neq 0, \quad f_2(0) = 1.$$

It is obvious that both functions have the same Fourier series which converges uniformly on the interval $[-\pi, \pi]$, because $a_n = b_n = 0$ for every n. Its sum is equal to the function $f_1(x)$ and not to the function $f_2(x)$, since $f_2(0) \neq 0$.

This trivial example shows that the Fourier series of a function $f(x)$ need not converge to $f(x)$ even if it converges uniformly, and also the fact that two different functions may have the same Fourier series. The background for this fact is that the Fourier coefficients, defined by integrals, remain unchanged if the function is altered on a set of measure zero.

CONCEPT OF A FOURIER SERIES

Example 2.4. (FATOU) The trigonometric series

$$\sum_{n=2}^{\infty} \frac{\sin nx}{\log n} \tag{2.17}$$

converges (not uniformly) on the interval $[-\pi, \pi]$, but is not the Fourier series of its sum (see Example 5.18).

Example 2.5. (PERRON) Define the function $f(x)$ by

$$f(x) = \frac{\sin \frac{x}{2} \log \left(2 \cos \frac{x}{2}\right) - \frac{x}{2} \cos \frac{x}{2}}{2 \cos \frac{x}{2} \left[\left(\log \left(2 \cos \frac{x}{2}\right)\right)^2 + \frac{x^2}{4}\right]} \quad \text{for} \quad x \in (-\pi, \pi),$$

$$f(-\pi) = f(\pi) = 0.$$

Then we have for $x \in [-\pi, \pi]$,

$$f(x) = \sum_{n=0}^{\infty} c_n (-1)^n \sin nx,$$

where the c_is are the solutions of the system of equations

$$1 = c_i + \frac{c_{i-1}}{2} + \frac{c_{i-2}}{3} + \ldots + \frac{c_0}{i+1},$$

$$i = 0, 1, 2, \ldots;$$

however, the coefficients in the above expansion of $f(x)$ cannot be computed by formulas (2.13), because the function $f(x)$ is not integrable.

In both cases we have a trigonometric series converging on the entire interval $[-\pi, \pi]$, which, however, is not the Fourier series of its sum. (In Example 2.5 even the explicit formula for the sum is available.) These examples were chosen to stress the fact that it is necessary to distinguish between the concepts of trigonometric series and of Fourier series. The trigonometric series is any series of the form (2.14); on the other hand, the Fourier series is referred to only in connection with an integrable function, and its coefficients are given by the formulas (2.13).

Every Fourier series is, of course, a trigonometric series, but not necessarily a convergent trigonometric series. A. N. KOLMOGOROV has constructed an example of an integrable function, whose Fourier series does not converge at any point of the interval. Even the continuity of the function $f(x)$ does not guarantee the convergence of the corresponding Fourier series; there exists an example of a continuous function whose Fourier series diverges at a point.

To conclude this section, we are going to derive two other forms of the Fourier series, which are often convenient.

If the function $f(x)$ is not real-valued, the Fourier coefficients a_n and b_n are complex numbers in general. The real functions $\cos nx$ and $\sin nx$ can be expressed as complex functions by

$$\cos nx = \frac{e^{inx} + e^{-inx}}{2}, \quad \sin nx = \frac{e^{inx} - e^{-inx}}{2i}.$$

Introducing this representation into the series (2.14), we obtain

$$\frac{a_0}{2} + \tfrac{1}{2}\sum_{n=1}^{\infty} \left[a_n(e^{inx} + e^{-inx}) - ib_n(e^{inx} - e^{-inx}) \right]$$

$$= \frac{a_0}{2} + \tfrac{1}{2}\sum_{n=1}^{\infty} \left[(a_n - ib_n)e^{inx} + (a_n + ib_n)e^{-inx} \right];$$

putting

$$\frac{a_0}{2} = c_0, \quad \frac{a_n - ib_n}{2} = c_n, \quad \frac{a_n + ib_n}{2} = c_{-n}$$

$$(n = 1, 2, \ldots),$$

the series (2.14) assumes the simple form

$$\sum_{n=-\infty}^{\infty} c_n e^{inx}. \tag{2.18}$$

CONCEPT OF A FOURIER SERIES

Euler's formulas (2.13) then yield

$$c_n = \frac{1}{2\pi} \int_{-\pi}^{\pi} f(x) e^{-inx} \, dx, \quad n = \ldots, -2, -1, 0, 1, 2, \ldots. \quad (2.19)$$

Also the system of functions $\exp(inx)$, $n = \ldots, -2, -1, 0, 1, 2, \ldots$ is orthogonal; the proof of this fact is left to the reader. (See also Problem 1.14.) The series (2.18) with coefficients given by the formulas (2.19) will be called the *complex form of the Fourier series* of the function $f(x)$.

The series (2.14) can be expressed in still another form. Let $n \geq 1$ and let $h_n = \sqrt{(a_n^2 + b_n^2)}$. Furthermore, for $h_n \neq 0$ let φ_n be the solution of the trigonometric equations

$$\sin \varphi_n = \frac{a_n}{h_n}, \quad \cos \varphi_n = \frac{b_n}{h_n};$$

for $h_n = 0$ put $\varphi_n = 0$. Clearly,

$$a_n \cos nx + b_n \sin nx = h_n(\cos nx \sin \varphi_n + \sin nx \cos \varphi_n)$$
$$= h_n \sin(nx + \varphi_n).$$

Thus, choosing h_0 and φ_0 such that $h_0 \sin \varphi_0 = a_0/2$, we can write the Fourier series of the function $f(x)$ as follows

$$\sum_{n=0}^{\infty} h_n \sin(nx + \varphi_n).$$

This representation is called the *phase form of the Fourier series*.

2.3 CONVERGENCE OF TRIGONOMETRIC SERIES

In this section several simple criteria for the convergence of the trigonometric series

$$\frac{a_0}{2} + \sum_{n=1}^{\infty} (a_n \cos nx + b_n \sin nx) \quad (2.20)$$

will be given; more specifically, the uniform convergence of the series on an interval $[c, d]$ with $-\pi \leq c < d \leq \pi$ will be under consideration. The convergence at a fixed point will not be tackled, because in this case the problem reduces to the convergence of an ordinary number series, which can be tested by a number of other criteria.

Theorem 2.2. *Let $a_0, a_1, a_2, \ldots$ and $b_1, b_2, \ldots$ be two sequences of complex numbers, and let the series $\left|\dfrac{a_0}{2}\right| + \sum_{n=1}^{\infty} (|a_n| + |b_n|)$ converge. Then the series (2.20) converges absolutely and uniformly on the entire axis to a continuous function $f(x)$ and is the Fourier series of its sum.*

Proof: For the k-th term of the series (2.20) we have the estimate

$$|a_k \cos kx + b_k \sin kx| \leq |a_k||\cos kx| + |b_k||\sin kx|$$
$$\leq |a_k| + |b_k|.$$

Thus, the series $\left|\dfrac{a_0}{2}\right| + \sum_{k=1}^{\infty} (|a_k| + |b_k|)$ majorizes the series (2.20); consequently, by Theorem 1.1, (2.20) converges absolutely and uniformly and its sum is a continuous function. Therefore, for all $x \in (-\infty, \infty)$ we have,

$$f(x) = \frac{a_0}{2} + \sum_{k=1}^{\infty} (a_k \cos kx + b_k \sin kx).$$

By Theorem 2.1 the coefficients $a_0, a_1, \ldots, b_1, b_2, \ldots$ are the Fourier coefficients of the function $f(x)$, which completes the proof. In addition to this we have proved the estimate

$$|f(x)| \leq \left|\frac{a_0}{2}\right| + \sum_{k=1}^{\infty} (|a_k| + |b_k|). \tag{2.21}$$

In the remaining theorems of this section the uniform but not the absolute convergence of the series (2.20) will be considered.

CONCEPT OF A FOURIER SERIES

We shall use the concept of a sequence of bounded variation introduced in Chapter 1. Let us remind the reader of the fact that every bounded monotonic sequence of real numbers is a sequence of bounded variation.

Theorem 2.3. *Let the sequence $\{a_n\}_0^\infty$ have a bounded variation and let $\lim_{n \to \infty} a_n = 0$. Then the series*

$$\frac{a_0}{2} + \sum_{n=1}^\infty a_n \cos nx \qquad (2.22)$$

converges uniformly on every interval $[c, d]$, where either $-\pi \leq c < d < 0$ or $0 < c < d \leq \pi$.

Proof: Consider the second possibility, i.e. $c > 0$. For any positive integer k and all $x \in [c, d]$ we have by (1.26),

$$\left| \tfrac{1}{2} + \sum_{n=1}^k \cos nx \right| = \left| \frac{\sin(k + \tfrac{1}{2})x}{2 \sin \tfrac{x}{2}} \right| \leq \frac{1}{2 \sin \tfrac{c}{2}}. \qquad (2.23)$$

Now, the assertion of our theorem follows immediately from Theorem 1.4. The first possibility can be treated in an analogous manner.

Theorem 2.4. *Let the sequence $\{b_n\}_1^\infty$ have a bounded variation and let $\lim_{n \to \infty} b_n = 0$. Then the series*

$$\sum_{n=1}^\infty b_n \sin nx \qquad (2.24)$$

converges uniformly on every interval $[c, d]$, where either $-\pi \leq c < d < 0$ or $0 < c < d \leq \pi$.

The proof follows the same pattern as the proof of Theorem 2.3, but the estimate

$$\left| \sum_{n=1}^{k} \sin nx \right| = \left| \frac{\cos \frac{x}{2} - \cos (k + \tfrac{1}{2}) x}{2 \sin \frac{x}{2}} \right|$$

$$\leq \frac{\left|\cos \frac{x}{2}\right| + \left|\cos (k + \tfrac{1}{2}) x\right|}{2 \left|\sin \frac{x}{2}\right|} \leq \frac{1}{\sin \frac{c}{2}}$$

has to be applied instead of (2.23).

Theorem 2.5. *Let the sequence $\{c_n\}_{-\infty}^{\infty}$ have a bounded variation and let $\lim\limits_{n \to \infty} c_n = \lim\limits_{n \to \infty} c_{-n} = 0$. Then the series*

$$\sum_{n=-\infty}^{\infty} c_n e^{inx}$$

converges uniformly on every interval $[c, d]$, where either $-\pi \leq c < d < 0$ or $0 < c < d \leq \pi$.

The proof can be carried out in the same way as before by using the inequality

$$\left| \sum_{n=-k}^{k} e^{inx} \right| = \left| \sum_{n=-k}^{k} (\cos nx + i \sin nx) \right|$$

$$\leq 2 \left| \tfrac{1}{2} + \sum_{n=1}^{k} \cos nx \right| \leq \frac{1}{\sin \frac{c}{2}} ;$$

here we have used the fact that $\sin(-nx) = -\sin nx$ and $\cos(-nx) = \cos nx$.

Example 2.6. We are going to investigate the uniform convergence of the trigonometric series

$$\sum_{n=1}^{\infty} \frac{\cos nx}{n^\alpha}, \quad \sum_{n=1}^{\infty} \frac{\sin nx}{n^\alpha},$$

where α is a positive number. Here, the sequence $\{1/n^\alpha\}_1^\infty$ is bounded, monotonic and tends to zero. Thus, by Theorems 2.3 and 2.4 it follows immediately that the above series converge uniformly on every closed interval which is contained in $[-\pi, \pi]$ and does not contain the point $x = 0$. These series are not absolutely convergent for $\alpha \leq 1$ (see Problem 2.3); for $\alpha > 1$ the series $\sum_{n=1}^{\infty} 1/n^\alpha$ converges so that our series converge absolutely and uniformly on the entire axis by Theorem 2.2.

Theorem 2.6. *Let the sequences $\{a_n\}_1^\infty$ and $\{b_n\}_1^\infty$ converge to zero and have a bounded variation. Then the series*

$$\frac{a_0}{2} + \sum_{n=1}^{\infty} (-1)^n a_n \cos nx \qquad (2.25)$$

and

$$\sum_{n=1}^{\infty} (-1)^n b_n \sin nx \qquad (2.26)$$

converge uniformly on every interval $[c, d]$ with $-\pi < c < d < \pi$.

Proof: First, let the interval $[c, d]$ be contained in the semi-closed interval $[0, \pi)$. Using the substitution $x = y + \pi$ the series (2.25) is transformed into the series (2.22) of variable y, which converges uniformly on the interval $[c - \pi, d - \pi]$ by Theorem 2.3. Thus, the series (2.25) converges uniformly on the interval $[c, d]$.

Next, let the interval $[c, d]$ be contained in $(-\pi, 0]$. Then we use the substitution $x = y - \pi$ and again Theorem 2.3.

Finally, let the interval $[c, d]$ be contained in $(-\pi, \pi)$, and let $c < 0 < d$. As we have already proved, the series (2.25) converges uniformly in both intervals $[c, 0]$ and $[0, d]$. Hence, the series converges uniformly on the entire interval $[c, d]$.

The convergence of the series (2.26) can be proved analogously by using Theorem 2.4.

An analogous theorem can be stated for the series $\sum_{n=-\infty}^{\infty} (-1)^n c_n e^{inx}$. The formulation and proof of this theorem is left to the reader.

2.4 CONVERGENCE OF A FOURIER SERIES

In the preceding section we have considered the problem of whether and how a trigonometric series with given coefficients converges. Let now an (integrable) function $f(x)$ be given; the convergence of the corresponding Fourier series may be tested, for example, by computing the Fourier coefficients and applying some theorem on convergence of a trigonometric series. However, the evaluation of Fourier coefficients may be laborious, and moreover, sometimes the actual values of the coefficients need not be known. Then it is convenient if some conditions for the function $f(x)$ are available which guarantee the convergence of the corresponding Fourier series. Certain such conditions, useful in practice, will be given in the last theorem of this section and in Chapter 7.

The following theorem shows that functions continuous on $[-\pi, \pi]$ are uniquely determined by their Fourier coefficients. Consequently, this assertion will be true for periodic functions with period 2π which are continuous on the entire axis (i.e. which satisfy the condition $f(-\pi) = f(\pi)$); such functions will be of primary interest to us (see Theorem 2.9 and the following ones).

Theorem 2.7. *Let f and g be two functions continuous on the interval $[-\pi, \pi]$, which have the same Fourier coefficients. Then $f(x) = g(x)$ for all $x \in [-\pi, \pi]$.*

Proof: Denote $h(x) = f(x) - g(x)$. Since the Fourier coefficients of the functions f and g coincide, all the Fourier coefficients of the function h vanish, i.e.

$$\int_{-\pi}^{\pi} h(x) \cos mx \, dx = 0 \quad \text{and} \quad \int_{-\pi}^{\pi} h(x) \sin mx \, dx = 0$$

for $m = 0, 1, 2, \ldots$. Hence, we have also

$$\int_{-\pi}^{\pi} h(x) T_n(x) \, dx = 0$$

for any trigonometric polynomial T_n.

We wish to prove that $h \equiv 0$; we are going to do it by contradiction. Suppose that $h(x) \neq 0$ at least at one point, say at the point c, i.e. $h(c) = K$; without loss of generality we can assume that $K > 0$ (if $h(c) < 0$, the roles of the functions f and g can be exchanged). The point c may be chosen so that it lies inside the interval $[-\pi, \pi]$, since if $h(x) \neq 0$ for $x = -\pi$ or $x = \pi$, then, due to continuity of the function h, $h(x) \neq 0$ also for x from a right neighbourhood of the point $-\pi$ or a left neighbourhood of π.

Thus, a point $c \in (-\pi, \pi)$ exists such that $h(c) = K > 0$. Since the function h is continuous on the interval $[-\pi, \pi]$, a positive number δ exists such that $h(x) \geq K/2$ for all $x \in [c - \delta, c + \delta]$ (the number δ is chosen so small that the closed interval $I_\delta = [c - \delta, c + \delta]$ lies entirely inside the open interval $(-\pi, \pi)$; draw a sketch).

Let the function $p(x)$ be defined by

$$p(x) = 1 + \cos(x - c) - \cos \delta.$$

For $x \in I_\delta$ we have $p(x) \geq 1$, since $\cos(x - c) \geq \cos \delta$ for $|x - c| \leq \delta$; outside the interval I_δ we have $|p(x)| < 1$. Denote $T_n(x) = [p(x)]^n$; the function $T_n(x)$ is a trigonometric polynomial (see, e.g., Problem 1.6 and 1.7), so that

$$0 = \int_{-\pi}^{\pi} h(x)\, T_n(x)\, \mathrm{d}x$$

$$= \int_{c-\delta}^{c+\delta} + \int_{-\pi}^{c-\delta} + \int_{c+\delta}^{\pi} \geq \int_{c-\delta}^{c+\delta} - \left| \int_{-\pi}^{c-\delta} + \int_{c+\delta}^{\pi} \right| \quad (2.27)$$

(for the sake of brevity we have left out the integrand $h(x)\, T_n(x)$ in the above integrals). Since $|p(x)| < 1$ for x lying outside I_δ, we have for these x, $\lim_{n \to \infty} T_n(x) = \lim_{n \to \infty} [p(x)]^n = 0$, and consequently, $\lim_{n \to \infty} h(x)\, T_n(x) = 0$ for x outside I_δ. Using now Theorem 1.6 (with an integrable majorant $|h(x)|$), we obtain

$$\lim_{n \to \infty} \int_{-\pi}^{c-\delta} h(x)\, T_n(x)\, \mathrm{d}x = \int_{-\pi}^{c-\delta} \lim_{n \to \infty} h(x)\, T_n(x)\, \mathrm{d}x = 0\,,$$

and similarly for the integral with the limits $c + \delta$ and π. Hence, an n exists so large that

$$\left| \int_{-\pi}^{c-\delta} h(x)\, T_n(x)\, \mathrm{d}x + \int_{c+\delta}^{\pi} h(x)\, T_n(x)\, \mathrm{d}x \right| < \frac{K\delta}{2}\,. \quad (*)$$

Let us fix this n. For $x \in I_\delta$ we have $h(x) \geq K/2$ and $T_n(x) = [p(x)]^n \geq 1$; consequently

$$\int_{c-\delta}^{c+\delta} h(x)\, T_n(x)\, \mathrm{d}x \geq \frac{K}{2} \cdot 2\delta = K\delta\,. \quad (**)$$

Using the estimates $(*)$ and $(**)$ in (2.27), we get

$$0 \geq K\delta - \frac{K\delta}{2} = \frac{K\delta}{2} > 0\,,$$

which is the sought contradiction. Thus, necessarily $h(x) = 0$ for all $x \in [-\pi, \pi]$, i.e. $f = g$.

Using the theorem just proved, the assertion of Theorem 2.1 may be augmented as follows:

CONCEPT OF A FOURIER SERIES

Theorem 2.8. *Let $f(x)$ be a function continuous on $[-\pi, \pi]$, and let the corresponding Fourier series converge uniformly on $[-\pi, \pi]$. Then the sum of this Fourier series is equal to $f(x)$ everywhere in $[-\pi, \pi]$ and we have $f(-\pi) = f(\pi)$.*

Proof: Denote by $g(x)$ the sum of the Fourier series of the function $f(x)$. The function $g(x)$ is continuous on $[-\pi, \pi]$ and $g(-\pi) = g(\pi)$. By Theorem 2.1, the Fourier coefficients of the function $g(x)$ are equal to the Fourier coefficients of $f(x)$. Thus, by the preceding theorem, both functions are identical and the proof is completed.

Example 2.7. The function $f(x) = |x| - \pi/2$, whose Fourier series was established in Example 2.2, is continuous and satisfies the condition $f(-\pi) = f(\pi)$. Its Fourier series converges uniformly by Theorem 2.2; thus, from Theorem 2.8 it follows that the equality symbol instead of the symbol $\sim$ can be written in (2.16), i.e.

$$|x| - \frac{\pi}{2} = -\frac{4}{\pi}\left[\frac{\cos x}{1^2} + \frac{\cos 3x}{3^2} + \frac{\cos 5x}{5^2} + \ldots\right] \quad (2.28)$$

for every $x \in [-\pi, \pi]$.

The following theorem on the derivative of a Fourier series will be useful.

Theorem 2.9. *Let $-\pi = \alpha_0 < \alpha_1 < \ldots < \alpha_l = \pi$. Let $f(x)$ be a function continuous on the interval $[-\pi, \pi]$, let $f(-\pi) = f(\pi)$* and let a continuous derivative $f'(x)$ exist on every interval $[\alpha_i, \alpha_{i+1}]$, $i = 0, 1, \ldots, l - 1$; here, at the left endpoint the derivative from the right and at the right endpoint the derivative from the left is understood. If $a_0, a_1, a_2, \ldots, b_1, b_2, \ldots$ are the Fourier coefficients of $f(x)$ and $a'_0, a'_1, \ldots, b'_1, b'_2, \ldots$ the Fourier coefficients of the function $f'(x)$, then we have*

$$a'_0 = 0, \quad a'_n = nb_n, \quad b'_n = -na_n \quad (n = 1, 2, \ldots). \quad (2.29)$$

---

* This requirement is essential as can be seen from **Problem 2.9**.

Remark: At a point α_i the derivatives of the function $f(x)$ from the left and from the right may be different; the values of Fourier coefficients of the function $f'(x)$, however, are independent of how $f'(x)$ is defined at the points α_i (see Example 2.3 and the following remarks).

Proof of Theorem 2.9: By the formulas (2.13) we have for $n \geqq 1$

$$\pi n b_n = n \int_{-\pi}^{\pi} f(x) \sin nx \, dx = n \sum_{i=1}^{l} \int_{\alpha_{i-1}}^{\alpha_i} f(x) \sin nx \, dx \, .$$

An integration by parts yields

$$n \int_{\alpha_{i-1}}^{\alpha_i} f(x) \sin nx \, dx$$
$$= \int_{\alpha_{i-1}}^{\alpha_i} f'(x) \cos nx \, dx - f(\alpha_i) \cos n\alpha_i + f(\alpha_{i-1}) \cos n\alpha_{i-1} \, ;$$

thus, summing up,

$$\pi n b_n = \int_{-\pi}^{\pi} f'(x) \cos nx \, dx - [f(\alpha_l) - f(\alpha_0)] \cos n\pi$$
$$= \int_{-\pi}^{\pi} f'(x) \cos nx \, dx = \pi a'_n \, .$$

Hence, $a'_n = n b_n$. The proof of the equalities $b'_n = -n a_n$ and $a'_0 = 0$ is left to the reader.

The formulas (2.29) need not be memorized; it suffices to remember the fact that if a function $f(x)$ satisfies the assumptions of Theorem 2.9, then for obtaining the Fourier series of the function $f'(x)$ the series of $f(x)$ is to be differentiated term by term.

Example 2.8. The function $f(x) = |x| - \pi/2$ from Example 2.2 satisfies the assumptions of Theorem 2.9 with $\alpha_0 = -\pi$, $\alpha_1 = 0$ and $\alpha_2 = \pi$. With the exception of these three points, $f(x)$ has

CONCEPT OF A FOURIER SERIES

a derivative everywhere, which is equal to the function sgn x; thus, by Theorem 2.9,

$$\operatorname{sgn} x \sim \frac{4}{\pi}\left[\frac{\sin x}{1} + \frac{\sin 3x}{3} + \frac{\sin 5x}{5} + \ldots\right]$$

in agreement with the result of Example 2.1.

In the following example the theorem on differentiation of a series is applied for integrating a series. A more detailed treatment will be given in Chapter 5.

Example 2.9. Define the function $f(x)$ by

$$f(x) = \int_0^x \left(|t| - \frac{\pi}{2}\right) dt = \begin{cases} \dfrac{x^2}{2} - \dfrac{\pi}{2} x & \text{for } x \in [0, \pi] \\ -\dfrac{x^2}{2} - \dfrac{\pi}{2} x & \text{for } x \in [-\pi, 0] \end{cases}$$

i.e. $f(x) = (x^2/2) \operatorname{sgn} x - (\pi/2) x$. This function clearly satisfies the requirements of Theorem 2.9. Since $f'(x) = |x| - \pi/2$, the Fourier coefficients of the function $f(x)$ can be established from the Fourier coefficients of the function $|x| - \pi/2$; thus, we obtain by formulas (2.29)

$$\frac{x^2}{2} \operatorname{sgn} x - \frac{\pi}{2} x \sim -\frac{4}{\pi}\left(\frac{\sin x}{1^3} + \frac{\sin 3x}{3^3} + \frac{\sin 5x}{5^3} + \ldots\right).$$

Theorem 2.10. *Let* $-\pi = \alpha_0 < \alpha_1 < \ldots < \alpha_{l-1} < \alpha_l = \pi$. *Let the function* $f(x)$ *be continuous on every open interval* (α_i, α_{i+1}), $i = 0, 1, \ldots, l - 1$, *and let the finite limits* $f(\alpha_i - 0) = \lim_{x \to \alpha_i^-} f(x)$, $i = 1, 2, \ldots, l$ *and* $f(\alpha_i + 0) = \lim_{x \to \alpha_i^+} f(x)$, $i = 0, 1, \ldots, l - 1$ *exist. Furthermore, let* $f(x)$ *have a derivative* $f'(x)$ *on every open interval* (α_i, α_{i+1}), *and let the integral* $\int_{-\pi}^{\pi} |f'(x)| \, dx$ *be finite.*

Then a constant $K > 0$ exists such that for Fourier coefficients a_n and b_n of the function $f(x)$ we have the estimate

$$|a_n| + |b_n| \leq \frac{K}{n} \quad (n = 1, 2, \ldots).$$

(For the integral $\int_{-\pi}^{\pi} |f'(x)|\, dx$ the sum $\sum_{i=1}^{l} \int_{\alpha_{i-1}}^{\alpha_i} |f'(x)|\, dx$ is understood.)

Proof. First, let us estimate the value of

$$\pi a_n = \int_{-\pi}^{\pi} f(x) \cos nx\, dx, \quad n \geq 1.$$

Putting $x_k = (k/n)\pi$ for $k = -n, -n+1, \ldots, -1, 0, 1, \ldots, n-1, n$, we obtain

$$\pi a_n = \sum_{k=-n+1}^{n} \int_{x_{k-1}}^{x_k} f(x) \cos nx\, dx.$$

Denote P the set of all k such that the interval $[x_{k-1}, x_k]$ contains a point from the system $\alpha_0, \alpha_1, \ldots, \alpha_l$; the set P contains at most $2l + 2$ numbers, because each α_i lies in at most two such intervals. Denote R the set of all remaining integers k with $-n < k \leq n$. Then we have

$$\pi a_n = \sum_{k \in P} \int_{x_{k-1}}^{x_k} f(x) \cos nx\, dx + \sum_{k \in R} \int_{x_{k-1}}^{x_k} f(x) \cos nx\, dx. \quad (2.30)$$

If $k \in P$, then

$$\left| \int_{x_{k-1}}^{x_k} f(x) \cos nx\, dx \right| \leq \int_{x_{k-1}}^{x_k} |f(x)|\, dx$$

because $|\cos nx| \leq 1$. The function $f(x)$ is bounded by a constant M_i on every closed interval $[\alpha_i, \alpha_{i+1}]$, since it is continuous on the open interval (α_{i-1}, α_i) and finite limits $f(\alpha_i + 0)$ and $f(\alpha_i - 0)$

exist at both endpoints. Let $M = \max(M_1, M_2, \ldots, M_l)$; then $|f(x)| \leq M$. Thus, we have for $k \in P$

$$\left| \int_{x_{k-1}}^{x_k} f(x) \cos nx \, dx \right| \leq |x_k - x_{k-1}| \cdot M = \frac{\pi}{n} M$$

and

$$\left| \sum_{k \in P} \int_{x_{k-1}}^{x_k} f(x) \cos nx \, dx \right| \leq \frac{(2l+2)\pi M}{n}. \quad (2.31)$$

If $k \in R$, an integrable first derivative exists on the interval $[x_{k-1}, x_k]$; an integration by parts yields

$$\int_{x_{k-1}}^{x_k} f(x) \cos nx \, dx = -\frac{1}{n} \int_{x_{k-1}}^{x_k} f'(x) \sin nx \, dx$$

because the values at the endpoints are zero due to the fact that $\sin nx_k = 0$. Hence, we obtain

$$\left| \sum_{k \in R} \int_{x_{k-1}}^{x_k} f(x) \cos nx \, dx \right|$$
$$\leq \frac{1}{n} \sum_{k \in R} \int_{x_{k-1}}^{x_k} |f'(x)| \, dx \leq \frac{1}{n} \int_{-\pi}^{\pi} |f'(x)| \, dx. \quad (2.32)$$

From (2.30) we obtain using (2.31) and (2.32) the inequality

$$\pi |a_n| \leq \frac{1}{n} \left[(2l+2)\pi M + \int_{-\pi}^{\pi} |f'(x)| \, dx \right].$$

Using the same procedure, we can estimate the numbers πb_n, too; it is only necessary to choose $x_k = k\pi/n + \pi/2n$ in order to make $\cos nx_k$ vanish. Summarizing these results, we have

$$|a_n| + |b_n| \leq \frac{K}{n},$$

where

$$K = 2\left[(2l + 2)M + \frac{1}{\pi}\int_{-\pi}^{\pi} |f'(x)|\, dx\right];$$

hence, the theorem is proved.

Now, we are in a position to state the main theorem of this section.

Theorem 2.11. *Let the function $f(x)$ satisfy the requirements of Theorem 2.9 and, in addition, let a continuous second derivative $f''(x)$ exist on every open interval (α_i, α_{i+1}). Furthermore, let the integral $\int_{-\pi}^{\pi} |f''(x)|\, dx$ be finite. Then a constant K exists such that*

$$|a_n| + |b_n| \leq \frac{K}{n^2} \quad (n = 1, 2, \ldots),$$

and the Fourier series of the function $f(x)$ converges uniformly to $f(x)$ on the closed interval $[-\pi, \pi]$.

Proof: The function $f'(x)$ satisfies the assumptions of Theorem 2.10; thus, for the Fourier coefficients a'_n and b'_n of $f'(x)$ we have

$$|a'_n| + |b'_n| \leq \frac{K}{n}.$$

On the other hand, by Theorem 2.9, $a'_n = nb_n$ and $b'_n = -na_n$, so that

$$|a_n| + |b_n| = \frac{|a'_n|}{n} + \frac{|b'_n|}{n}.$$

Hence, we have proved the estimate

$$|a_n| + |b_n| \leq \frac{K}{n^2}; \tag{2.33}$$

from this it follows by Theorem 2.2 that the Fourier series of the function $f(x)$ converges uniformly on the interval $[-\pi, \pi]$. Thus, its sum is equal to $f(x)$ by Theorem 2.8 and the proof is concluded.

Example 2.10. Recalling the Example 2.2 it follows that for the Fourier coefficients of the function $|x| - \pi/2$ we have the estimate (2.33)

$$|a_n| + |b_n| \leq \frac{4}{\pi}\frac{1}{n^2}.$$

Observe that for n odd the equality is attained; this fact shows that the estimate (2.33) cannot be improved in such a sense that for any function $f(x)$ satisfying the assumptions of Theorem 2.11 we would have an estimate of type (2.33) with the right hand side $1/n^\alpha$, $\alpha > 2$. However, if stronger assumptions on $f(x)$ are made, the estimate (2.33) can be improved; for example, for the function $\tfrac{1}{2}x^2 \,\mathrm{sgn}\, x - (\pi/2)\, x$ from the Example 2.9 we have $|a_n| + |b_n| = (4/\pi)(1/n^3)$ for n odd, $|a_n| + |b_n| = 0$ for n even.

Example 2.11. The function $x^2 \log |x|$ satisfies the requirements of Theorem 2.11. Hence,

$$x^2 \log |x| = \frac{a_0}{2} + \sum_{k=1}^{\infty} a_n \cos nx,$$

where

$$a_n = \frac{1}{\pi}\int_{-\pi}^{\pi} x^2 \log |x| \cos nx \,\mathrm{d}x.$$

(Prove that $b_n = 0$.) The values of the coefficients a_n cannot be computed accurately, because the above integrals can be evaluated only by approximative methods. In spite of that we are able to state something about the behaviour of the Fourier series of $x^2 \log |x|$, because Theorem 2.11 gives us the estimate $|a_n| \leq K/n^2$.

The functions, for which Theorem 2.11 guarantees the convergence of the corresponding Fourier series, are periodic and continuous; a discontinuity is admitted only for a derivative. In the next Chapter we will deal with the convergence of Fourier series of functions which are not continuous in general; however, a different kind of convergence will be under consideration. The Fourier series of functions having a finite number of discontinuities of the first kind (considered in Theorem 2.10) will be investigated more closely in Chapters 6 and 7.

At the beginning of this chapter it was stated that the Fourier series, and trigonometric series as well, are of the utmost importance in practice. Their application in engineering sciences rests mainly on the fact that the solution y of a functional relation (i.e., differential equation, integral equation etc., see Chapter 9), is sought *formally* in the form of a trigonometric series with unknown coefficients a_k and b_k, which are to be determined by substituting the series into the considered relation. Doing this, a number of operations has to be performed, for example a differentiation of y, which again are carried out purely formally. So, we differentiate or integrate a series term by term, sum up different series etc. without actually checking up whether these operations are justified. At the end we obtain relations between series and then, comparing the coefficients, certain equations for the unknowns a_k and b_k. If we succeed in solving these equations and find the coefficients a_k and b_k, we introduce them into the expansion for y and take for granted that the problem is solved. We have found an explicit form of the solution after all!

When these steps have been completed no check of the formal solution thus obtained is usually carried out. This, unfortunately, is a quite common *mistake*, and we wish to emphasize this fact here. As a matter of fact, once a *formal* solution has been found by the procedure described above, we ought to check up whether all formal operations carried out can actually be justified. Otherwise there is a danger that the formal solution is meaningless, or that we have solved a problem which is in fact unsolvable.

In order to clarify these ideas let us present the following example.

We are going to solve the differential equation $y'' = x - \pi/2$ on $[0, \pi]$ with the boundary conditions $y'''(0) = y'''(\pi) = 0$.

CONCEPT OF A FOURIER SERIES

The solution will be looked for in the form of a trigonometric series, i.e. we shall assume that

$$y = \frac{a_0}{2} + \sum_{k=1}^{\infty} (a_k \cos kx + b_k \sin kx).$$

By differentiating twice term by term, we obtain

$$y'' = -\sum_{k=1}^{\infty} (a_k k^2 \cos kx + b_k k^2 \sin kx).$$

Since

$$x - \frac{\pi}{2} = -\frac{4}{\pi} \sum_{k=1,3,5,\ldots} \frac{\cos kx}{k^2}$$

for $x \in [0, \pi]$ by Example 2.7, formula (2.28), the equation $y'' = x - \pi/2$ takes the form

$$-\sum_{k=1}^{\infty} (a_k k^2 \cos kx + b_k k^2 \sin kx) = -\frac{4}{\pi} \sum_{k=1,3,5,\ldots} \frac{\cos kx}{k^2};$$

comparing the coefficients on the left and right hand side, we obtain the following equations:

$$a_k = 0 \quad \text{for} \quad k = 2, 4, 6, \ldots,$$
$$b_k = 0 \quad \text{for} \quad k = 1, 2, 3, 4, \ldots,$$
$$a_k k^2 = \frac{4}{\pi} \frac{1}{k^2}, \quad \text{i.e.} \quad a_k = \frac{4}{\pi} \frac{1}{k^4} \quad \text{for} \quad k = 1, 3, 5, \ldots.$$

The coefficient a_0 can be arbitrary. Hence

$$y = \frac{a_0}{2} + \frac{4}{\pi} \sum_{k=1,3,5,\ldots} \frac{\cos kx}{k^4}. \tag{2.34}$$

This function satisfies the boundary condition, too; actually, since

$$y''' = \frac{4}{\pi} \sum_{k=1,3,5,\ldots} \frac{\sin kx}{k}, \tag{2.35}$$

each term vanishes at the endpoints 0 and π so that $y'''(0) = y'''(\pi) = 0$. Hence, from the formal point of view, the function (2.34) solves our problem.

This is an example of a formal procedure which does not pay any attention to whether or not all operations performed made sense. On the other hand, the procedure just presented must involve a serious mistake, because the problem, which we have "solved", has no solution at all. Actually, by differentiating the equality $y'' = x - \pi/2$ we find out that $y'''(x) = 1$ for all $x \in [0, \pi]$, and consequently, the equalities $y'''(0) = 0$ and $y'''(\pi) = 0$ cannot be satisfied by any solution.

In order to pin-point the snag in the above formal procedure, let us focus our attention on the series (2.35). Recalling the Example 2.1 we see that the expansion for the function sgn x stands on the right hand side of (2.35), which agrees with the fact that $y'''(x) = 1$ for $x \in (0, \pi)$ (see also Example 2.8); however, the sum of the series at the points $x = 0$ and $x = \pi$ is not equal to the limit of the sum from the right and the left, respectively. The sum of the series (2.35) is truly zero at the points $x = 0$ and $x = \pi$, but the series itself is not any first derivative, and consequently, not a third derivative of any function on the entire interval $[0, \pi]$.

The example presented is undoubtedly quite transparent, but shows clearly how a formal approach can lead us to erroneous results. In more complicated problems the error may be hidden much deeper and thus be more treacherous.

We should stress that we do not want to deny the importance and usefulness of formal considerations. In mathematics there are many examples where a formal or a more or less intuitive consideration yielded an important and fundamental result; however, the correctness of such considerations has always been rigorously proved afterwards. By the above example we wanted to show how the second, so often neglected part is important not only in mathematics, but also in applications.

Problems

2.1. Find the Fourier coefficients of the following functions defined on the interval $(-\pi, \pi)$.

a) $f(x) = x$;
b) $f(x) = x^2$;
c) $f(x) = \cos \alpha x$ (α is a real non-integral number);
d) $f(x) = e^{\alpha x}$ (α real);
e) $f(x) = \begin{cases} \alpha x & \text{for } x \in [-\pi, 0] \ (\alpha \text{ real}); \\ \beta x & \text{for } x \in [0, \pi] \ (\beta \text{ real}). \end{cases}$

2.2. Prove the uniform convergence of the following trigonometric series:

$$\sum_{n=2}^{\infty} \frac{\cos nx}{n \log^2 n}; \quad \sum_{n=2}^{\infty} \frac{\sin nx}{n!}.$$

2.3. Prove that the series

$$\text{(a)} \sum_{n=1}^{\infty} \frac{\sin nx}{n^\alpha}; \quad \text{(b)} \sum_{n=1}^{\infty} \frac{\cos nx}{n^\alpha} \quad (0 < \alpha \leq 1)$$

converge for all $x \in (-\pi, \pi)$ (even for any $x \neq 0$ in the case b), but do not converge absolutely.

(*Hint*: Put $x = \pi/2$ and show that the series (a) and (b) fail to converge absolutely at this point.)

2.4. Find the values of α and x such that the series

$$\sum_{n=1}^{\infty} \frac{\sin n\alpha}{n} \sin nx$$

converges. For what x does the series converge uniformly?

2.5. Prove that the sum of a uniformly convergent trigonometric series is a continuous and periodic function on $(-\infty, \infty)$.

2.6. Prove the following assertion. If $\{b_n\}$ is a sequence of bounded variation and $\lim_{n \to \infty} b_n = 0$, then the series

$$\sum_{n=0}^{\infty} b_n \sin(2n+1)x \quad \text{and} \quad \sum_{n=1}^{\infty} b_n \sin 2nx$$

converge uniformly on every closed interval $[c, d]$ which does not contain the points $x = k\pi$ with k being an integer.
(*Hint*: Proceed as in Theorem 2.4 and use the estimate from Problem 1.9.)

2.7. Using the result of the previous problem prove the convergence of the Fourier series of the function $f(x) = \operatorname{sgn} x$ on $(-\pi, \pi)$. (See Example 2.1, formula (2.15).)

2.8. Prove that the sum of the Fourier series of the function $\operatorname{sgn} x$ (see (2.15)) is equal to $\operatorname{sgn} x$ for $x \in [-\pi, \pi]$.
(*Hint*: Use the result of the preceding problem or Example 2.2 and the theorem on the term by term differentiation of a series.)

2.9. The function $f(x) = x$ satisfies the assumptions of Theorem 2.11 with the exception of the assumption $f(-\pi) = f(\pi)$. Using the results of Problem 2.1a show that the assertion of Theorem 2.11 does not hold here.

(*Remark*: This example shows that the condition $f(-\pi) = f(\pi)$ is essential for the validity of the estimate (2.33).)

2.10. Let $f(x)$ be a continuous function whose graph is a polygon and let $f(-\pi) = f(\pi)$. Show that this function can be uniformly approximated by trigonometric polynomials on $[-\pi, \pi]$, i.e. for every $\varepsilon > 0$ a trigonometric polynomial $T_n(x)$ exists such that

$$|f(x) - T_n(x)| < \varepsilon \quad \text{for all} \quad x \in [-\pi, \pi].$$

(*Hint*: Use Theorem 2.11 and take the n-th partial sum of the Fourier series of $f(x)$ for the trigonometric polynomial.)

CHAPTER 3 HILBERT SPACE

3.1 INTRODUCTION

In the preceding chapter we became acquainted with the concept of a Fourier series of a function f integrable on the interval $(-\pi, \pi)$ and established some of its properties. However, the results obtained do not state anything about the behaviour of the Fourier series, if we assume only that the function f is integrable. In order to use the chosen methods of proof, we had to assume that the function f is of a rather special kind, for example that except for a finite number of points f has a derivative of the second order. In practice however, we encounter functions for which theorems of Chapter 2 are inapplicable. Therefore, the goal of this chapter will be to develop an apparatus which permits us to expand in a series any square-integrable function, i.e. any function f with a finite integral $\int_{-\pi}^{\pi} |f(x)|^2 \, dx$. As Problems 3.1 and 3.2 show, the class of all square-integrable functions is smaller than the class of integrable functions. On the other hand, the former class is much larger than the class of functions dealt with in Chapter 2. Let us note that, for example, every bounded measurable function is square-integrable on an interval of finite length.

In this Chapter we are in fact going to present the theory of orthogonal series in a Hilbert space; these series encompass the trigonometric series as a special case. The theorems we shall prove are applicable for series whose terms are orthogonal polynomials, Bessel functions etc. Thus, in what follows we shall disregard the particular nature of the terms of a series and thus generalize the ideas we have about a trigonometric series. In order to outline this approach, assume that a sequence of complex-valued functions $f_1, f_2, f_3, \ldots$ defined on an interval $[a, b]$ instead of the traditional sequence $1, \cos x, \sin x, \cos 2x, \sin 2x, \ldots$ is given. Let f be a function whose properties will not be specified at this point. We wish to find the coefficients c_i so that the function f may be written in the form

$$f = \sum_{i=1}^{\infty} c_i f_i, \tag{3.1}$$

where the series on the right hand side converges in a certain, not yet specified sense. For determining the coefficients c_i we are going to use the same procedure as in the proof of Theorem 2.1. We multiply the equality (3.1) by the function $\bar{f}_j$ and integrate over the interval $[a, b]$. Interchanging formally the order of integration and summation, we obtain

$$\int_a^b f(x) \overline{f_j(x)} \, dx = \sum_{i=1}^{\infty} c_i \int_a^b f_i(x) \overline{f_j(x)} \, dx. \tag{3.2}$$

The function f and the sequence of functions $f_1, f_2, \ldots$ are known; consequently, also the numbers

$$e_{ij} = \int_a^b f_i(x) \overline{f_j(x)} \, dx,$$

$$d_j = \int_a^b f(x) \overline{f_j(x)} \, dx$$

are known. Observe that the number (f, g) assigned to the pair of functions f and g by

$$(f, g) = \int_a^b f(x) \overline{g(x)} \, dx$$

plays a significant role here. With the above notation we have

$$d_j = \sum_{i=1}^{\infty} e_{ij} c_i. \tag{3.3}$$

This is a system of infinitely many equations for infinitely many unknowns c_i. Such a system need not have a solution, or can have several solutions. The solvability depends mainly on the coefficients e_{ij}. In general, a solution of (3.3) is difficult to find; on the other hand, a solution can be easily established in certain particular cases. The simplest case occurs if $e_{ij} = 0$ for $i \neq j$ and $e_{ij} = 1$ for $i = j$. Then obviously $c_i = d_i$ for $i = 1, 2, 3, \ldots$ and we can write

$$f = \sum_{i=1}^{\infty} (f, f_i) f_i. \tag{3.4}$$

Thus, for solving the system (3.3) the fact that $(f_i, f_j) = 0$ for $i \neq j$ proved to be of crucial importance.

Remark 3.1. For the sequence of trigonometric functions

$$1, \cos x, \sin x, \cos 2x, \sin 2x, \ldots$$

on the interval $(-\pi, \pi)$ we have by formulas (1.22) that $e_{ij} = 0$ for $i \neq j$, but not $e_{ij} = 1$ for $i = j$. However, modifying this sequence so that we multiply each function by an appropriate constant, we achieve $e_{ii} = 1$. The reader can easily verify that here

$$\frac{1}{\sqrt{(2\pi)}}, \frac{\cos x}{\sqrt{\pi}}, \frac{\sin x}{\sqrt{\pi}}, \frac{\cos 2x}{\sqrt{\pi}}, \frac{\sin 2x}{\sqrt{\pi}}, \ldots \tag{3.5}$$

is the sought-after sequence.

The relation (3.4) shows what an expansion of a function f in functions f_i should formally look like. The series on the right hand side can be written for a rather general type of functions, namely for any function f such that the symbol (f, f_i) is meaningful for all i. Moreover, if we introduce a certain particular kind of convergence, which is closely related to the product (f, g), then we will be able to expand in a convergent series (3.4) any function f such that (f, f) is a finite number. This series converges to the function f, if the sequence $f_1, f_2, f_3, \ldots$ is rich enough in a certain sense. In particular, any function square-integrable on the interval

$(-\pi, \pi)$ can be expanded in a Fourier series converging to the function f in a certain sense.

What has been said up to this point is by no means a rigorous mathematical consideration. The only purpose of it was to show what can be done without taking account of the concrete trigonometric functions (3.5) and of the particular form of (f, g) as an integral of a product of two functions. Disregarding the special form of (f, g) we are led to the concept of a scalar product in a Hilbert space, which will be treated in the next section. The reader who might find the ensuing considerations on an abstract Hilbert space difficult can always bear in mind the preceding, particular case, i.e. the set of all real square-integrable functions with the scalar product defined simply by an integral of a product of functions as indicated above. The sequence $f_1, f_2, f_3, \ldots$ may then be understood as a different notation for trigonometric functions. As we shall see in the next section, the sequence $f_1, f_2, f_3, \ldots$ and the scalar product may also have a quite different meaning.

3.2 CONCEPT OF A HILBERT SPACE

A set H will be called a *complex linear space*, if the following conditions are satisfied:

1. For any two elements $x \in H$, $y \in H$ there is defined an element $z \in H$ denoted by $z = x + y$ and called the *sum* of the elements x and y.

2. For any element $x \in H$ and any complex number α an element $z \in H$ is defined, which is denoted by $z = \alpha x$ and called the *product* of the element x with the number α.

3. $x + y = y + x$ (commutative law).

4. $(x + y) + z = x + (y + z)$ (associative law).

5. An element $O \in H$ exists such that $x + O = x$ for all $x \in H$. The element O is called the *zero element* of H.

6. To each $x \in H$ a $y \in H$ exists such that $x + y = O$. Such an element y is called the *negative element for* x.

Let us note at this point that the zero element O and the negative element for x are determined uniquely. (See Problem 3.3.)

7. $\alpha(\beta x) = (\alpha\beta) x$ (associative law for multiplication).
8. $\alpha(x + y) = \alpha x + \alpha y$ (distributive law).
9. $(\alpha + \beta) x = \alpha x + \beta x$ (distributive law).
10. $1 . x = x$.

In Problems 3.4 and 3.5 we shall show that $0 . x = O$ and that the element $-x = (-1) . x$ is the negative element for x.

Example 3.1. Let us take the set of all complex-valued functions continuous on a closed interval $[a, b]$ for the set H. Let the sum h of two functions f and g be defined by

$$h(x) = f(x) + g(x)$$

for all $x \in [a, b]$. Define the product h of a function f with a number α by

$$h(x) = \alpha f(x).$$

It can be easily verified that the set H with the above operations becomes a linear space, whose zero element is a function vanishing identically. This space will be denoted by $C(a, b)$.

Example 3.2. Denote by $L_2(a, b)$ the set of all measurable complex-valued functions such that the integral

$$\int_a^b |f(x)|^2 \, dx$$

is finite; let the sum and the product with a number be defined in the same way as in Example 3.1. In order to show that $L_2(a, b)$ with the operations just defined is a linear space, we have to verify that the sum of two square-integrable functions is again a square-integrable function. To this purpose we are going to use the well-known inequality $|\alpha + \beta|^2 \leq 2(|\alpha|^2 + |\beta|^2)$. (See Problem 1.11.) Putting here $\alpha = f(x)$ and $\beta = g(x)$, we obtain

$$|f(x) + g(x)|^2 \leq 2(|f(x)|^2 + |g(x)|^2) .$$

Integrating this inequality over the interval (a, b), we have

$$\int_a^b |f(x) + g(x)|^2 \, dx \leq 2 \left(\int_a^b |f(x)|^2 \, dx + \int_a^b |g(x)|^2 \, dx \right).$$

Both integrals on the right-hand side are finite, and consequently, the integral on the left-hand side is also finite. This means, therefore, that the function $f + g$ is also square-integrable. Furthermore, it is quite clear that the product of a square-integrable function f with a number α is also square-integrable, because

$$\int_a^b |\alpha f(x)|^2 \, dx = |\alpha|^2 \int_a^b |f(x)|^2 \, dx .$$

The reader can easily verify that conditions 3 to 10 imposed on the operations in a linear space are also satisfied.

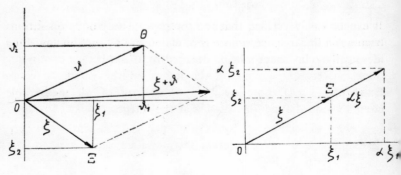

Fig. 3.1

Example 3.3. Denote by C_n the set of all n-tuples $(\xi_1, \xi_2, \ldots, \xi_n)$, i.e., finite sequences of n complex numbers ξ_i. Let $\xi = (\xi_1, \xi_2, \ldots$

… , ξ_n), $\eta = (\eta_1, \eta_2, …, \eta_n)$ and let α be a complex number. Define the sum $\xi + \eta$ as the n-tuple $(\xi_1 + \eta_1, \xi_2 + \eta_2, …, \xi_n + \eta_n)$, and the product $\alpha\xi$ as the n-tuple $(\alpha\xi_1, \alpha\xi_2, …, \alpha\xi_n)$. The set C_n with operations just defined is a linear space. The elements of this space are called the *n-dimensional (complex) vectors*, the numbers $\xi_1, \xi_2, …, \xi_n$ are called the *components* of the vector ξ.

In an analogous manner we define the set R_n consisting of all n-tuples $\xi = (\xi_1, \xi_2, …, \xi_n)$ of real numbers ξ_i, and the sum and the product. The space so obtained is called the *real n-dimensional vector space*.

For $n = 2$, the vectors from R_2 may be interpreted as the points in a plane. A vector $\xi_1 = (\xi_1, \xi_2)$ can be depicted by an orientated segment $\overrightarrow{O\Xi}$, where O is the origin of the Cartesian coordinate system and the point Ξ has coordinates (ξ_1, ξ_2). In Fig. 3.1 the addition of vectors and the multiplication of a vector by a number is shown. It is clear that these operations with vectors agree with the usual operations introduced in physics. The situation is the same for $n = 3$.

Since the elements of an abstract linear space behave in a similar way to the vectors in the spaces C_n or R_n, and as it will be shown later, the resemblance does not rest only upon properties 1 to 10, they will also be called *vectors*, and the linear space H will be called the *vector space*. The properties 1 to 10 guarantee that the vectors can be treated by the usual rules. Thus, a vector can be an n-tuple of complex (real) numbers, a continuous function, etc. depending on the nature of the vector space we consider.

A complex vector space H will be called a *complex Hilbert space**, if to each ordered pair of elements $x \in H$ and $y \in H$ there is assigned a complex number (x, y), called the *scalar product*

---

* In this section we define a Hilbert space as an incomplete space. The complete Hilbert space will be introduced in Section 3.4. However, in literature the term Hilbert space often signifies a *complete Hilbert space*; an incomplete Hilbert space is then called the *prae-Hilbert* or *unitary space*.

(or *inner product*) of x and y, such that the following conditions are satisfied:

(a) $(y, x) = \overline{(x, y)}$, where the bar denotes the complex conjugate,
(b) $(x, y + z) = (x, y) + (x, z)$,
(c) $(\alpha x, y) = \alpha(x, y)$ for any complex number α,
(d) (x, x) is real and non-negative for any $x \in H$, and $(x, x) = 0$ exactly if $x = O$.

From conditions a, b, c it follows that $(x + y, z) = (x, z) + (y, z)$, and that $(x, \alpha y) = \bar{\alpha}(x, y)$ for any complex number α.

Actually, we have $(x + y, z) = \overline{(z, x + y)} = \overline{(z, x)} + \overline{(z, y)} = (x, z) + (y, z)$ and $(x, \alpha y) = \overline{(\alpha y, x)} = \bar{\alpha}\overline{(y, x)} = \bar{\alpha}(x, y)$.

Thus, if x_i, y_j are elements in H and α_i, β_j are complex numbers, then

$$\left(\sum_{i=1}^{n} \alpha_i x_i, \sum_{j=1}^{m} \beta_j y_j\right) = \sum_{i=1}^{n} \sum_{j=1}^{m} \alpha_i \bar{\beta}_j (x_i, y_j).$$

Remark 3.2. Replacing the term *complex* by the term *real* everywhere in the definitions of the linear and Hilbert space, we obtain the definitions of the *real linear space* and *real Hilbert space*, respectively. Thus, in a real Hilbert space only multiplication by a real number is defined and the scalar product of two arbitrary vectors is a real number. For example, we have here $(x, y) = (y, x)$ and $(x, \alpha y) = \alpha(x, y)$. Considering only real functions in Examples 3.1 and 3.2, we obtain the real space $C(a, b)$ and the real space $L_2(a, b)$. Similarly, taking into account only n-tuples of real numbers in Example 3.3, we obtain the real n-dimensional vector space denoted by R_n. Observe at this point that C_1 is in fact the space of all complex numbers with the usual addition and multiplication. (For the sake of brevity, we shall further write $C_1 = C$.) Equally, R_1 (further R) is the space of all real numbers. However, when speaking about a linear or a Hilbert space in what follows, we shall always mean, unless otherwise stated, the complex space.

HILBERT SPACE

Example 3.4. Define the scalar product in C_n by

$$(\xi, \eta) = \sum_{i=1}^{n} \xi_i \bar{\eta}_i .$$

It is clear that all the requirements imposed on a scalar product are met.

In R_n we introduce the scalar product by

$$(\xi, \eta) = \sum_{i=1}^{n} \xi_i \eta_i .$$

The space R_n with the scalar product just defined is called the n-dimensional *Euclidean space*. The scalar product of vectors in R_n has the same meaning as the scalar product usually introduced in geometry. We have namely

$$(\xi, \eta) = \|\xi\| \, \|\eta\| \cos \widehat{(\xi, \eta)}$$

where $\|\xi\|$ signifies the length of the vector ξ and $\widehat{(\xi, \eta)}$ the angle between vectors ξ and η. The length of a vector ξ is given by the formula

$$\|\xi\| = \Big(\sum_{i=1}^{n} |\xi_i|^2 \Big)^{\frac{1}{2}} .$$

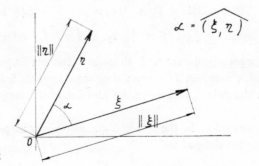

Fig. 3.2

The components $\xi_1, \xi_2, \ldots, \xi_n$ of the vector ξ are the Cartesian coordinates of ξ. For $n = 2$ the situation is sketched in Fig. 3.2.

Example 3.5. An infinite dimensional analogue of the space C_n is furnished by the space l_2. It consists of all infinite sequences of complex numbers $\xi = (\xi_1, \xi_2, \ldots)$ such that the series $\sum_{i=1}^{\infty} |\xi_i|^2$ converges, i.e. $\sum_{i=1}^{\infty} |\xi_i|^2 < \infty$. The sum of two sequences $\xi = (\xi_1, \xi_2, \xi_3, \ldots)$ and $\eta = (\eta_1, \eta_2, \eta_3, \ldots)$ is defined by

$$\xi + \eta = (\xi_1 + \eta_1, \xi_2 + \eta_2, \xi_3 + \eta_3, \ldots),$$

the product of the sequence ξ with a complex number α by

$$\alpha\xi = (\alpha\xi_1, \alpha\xi_2, \alpha\xi_3, \ldots).$$

The scalar product is defined by the formula

$$(\xi, \eta) = \sum_{i=1}^{\infty} \xi_i \bar{\eta}_i.$$

We have to verify whether the requirements imposed on a Hilbert space are satisfied. The reader can easily show that the axioms of the linear space are fulfilled. In addition to it, we have to prove that the scalar product is actually meaningful for any $\xi \in l_2$ and $\eta \in l_2$. This, however, follows immediately from the inequality

$$|\xi_i \bar{\eta}_i| \leq \tfrac{1}{2}(|\xi_i|^2 + |\bar{\eta}_i|^2)$$

(see Problem 1.11). Thus, the series $\sum_{i=1}^{\infty} \tfrac{1}{2}(|\xi_i|^2 + |\eta_i|^2)$ majorizes the series $\sum_{i=1}^{\infty} \xi_i \bar{\eta}_i$. For $\xi \in l_2$ and $\eta \in l_2$ the majorizing series converges absolutely and, therefore, the scalar product is a finite complex number. The remaining requirements follow immediately from the rules concerning the absolutely convergent series.

Example 3.6. In the space $L_2(a, b)$ (see Example 3.2) the scalar product is defined by

$$(f, g) = \int_a^b f(x) \overline{g(x)} \, \mathrm{d}x.$$

The integrability of the product $f\bar{g}$ follows immediately from the inequality

$$|f(x) \overline{g(x)}| \leq \tfrac{1}{2}(|f(x)|^2 + |g(x)|^2).$$

The conditions a, b and c are obviously satisfied. Certain complications arise with the verification of condition d. Indeed, if

$$(f,f) = \int_a^b |f(x)|^2 \, dx = 0,$$

the function $f(x)$ need not vanish identically; however, it can assume non-zero values only on a set of measure zero. Hence, we are led to introducing the following *licence*: two measurable functions f and g will be considered as *identical* on an interval $[a, b]$, if they are equal everywhere up to a set of measure zero, i.e. if they differ, for example, at finitely many points of the interval $[a, b]$. Two functions differing only on a set of zero measure are also called *equivalent with respect to the Lebesgue measure*.

Thus, let us introduce the *classes of equivalent functions* as follows: two functions belong to the same class, if their difference is a function vanishing almost everywhere (see Chapter 1.). The classes of equivalent functions can be treated in the same way as the functions themselves. The sum of two classes is defined as follows: in each class we choose a function and then sum both these functions; all such sums constitute a class, as it can be easily verified, and this class is called the sum of the former two classes. The product of a class with a complex number is defined in an analogous manner. A little thought will persuade us that the set of all classes of equivalent functions is a linear space, whose zero element is the class of all functions vanishing almost everywhere.

It is obvious that for two equivalent functions f and g the numbers $\int_a^b |f(x)|^2 \, dx$ and $\int_a^b |g(x)|^2 \, dx$ are equal. Thus we shall not define the space $L_2(a, b)$ as the set of all square-integrable functions, but as the *set of all classes of equivalent functions*. Consequently, the zero element of $L_2(a, b)$ will be the class of all functions vanishing almost everywhere.

From each class we choose a fixed function, sometimes called the representative, and shall identify all other functions from the class with the representative. In particular, the functions from the zero class can be identified with the function vanishing identically. Thus, *we shall not make any difference between a representative and the corresponding class, and shall call the classes simply functions.*

A function vanishing in $L_2(a, b)$ vanishes in fact everywhere except for a set of measure zero.

The requirement d is clearly satisfied now, because a function vanishing everywhere except for a set of measure zero was identified with the function vanishing identically. Example 2.3 shows that this identification is quite natural in investigating Fourier series.

3.3 LENGTH OF A VECTOR AND THE DISTANCE BETWEEN TWO VECTORS IN A HILBERT SPACE

As known from the preceding section, in R_n we can write

$$\|\xi\| = \left(\sum_{i=1}^{n} |\xi_i|^2\right)^{1/2} = (\xi, \xi)^{1/2}. \tag{3.6}$$

In a general Hilbert space the length of a vector is defined analogously to (3.6), i.e. by the formula

$$\|x\| = (x, x)^{1/2}. \tag{3.7}$$

The number $\|x\|$ is called the *norm* of a vector x in space H. If the circumstances necessitate that we indicate explicitly which space H is under consideration, we will also use the notation $\|x\|_H$ instead of $\|x\|$.

Example 3.7. We have

$$\|\xi\|_{R_n} = \left(\sum_{i=1}^{n} \xi_i^2\right)^{1/2},$$

$$\|\xi\|_{l_2} = \left(\sum_{i=1}^{\infty} |\xi_i|^2\right)^{1/2},$$

$$\|f\|_{L_2(a,b)} = \left(\int_a^b |f(x)|^2 \, dx\right)^{1/2}.$$

HILBERT SPACE

Theorem 3.1. *For the scalar product in a Hilbert space H we have the inequality*

$$|(x, y)| \leq \|x\| \cdot \|y\| . \tag{3.8}$$

Note that (3.8) is called the *Schwarz inequality*.

Proof: If $(x, y) = 0$, the inequality (3.8) is trivially satisfied. Thus assume that $(x, y) \neq 0$. From the property (d) it follows that

$$0 \leq (x - \alpha y, x - \alpha y)$$

for any complex number α. Hence

$$0 \leq (x, x) - \alpha(y, x) - \bar{\alpha}(x, y) + |\alpha|^2 (y, y) .$$

Putting $\alpha = (x, x)/(y, x)$, we obtain

$$0 \leq \|x\|^2 - \|x\|^2 - \|x\|^2 + \frac{\|x\|^4}{|(y, x)|^2} \|y\|^2 .$$

From this (3.8) follows immediately.

In a real Hilbert space we can define the *angle* φ between two non-zero vectors x and y by

$$\cos \varphi = \frac{(x, y)}{\|x\| \cdot \|y\|} ,$$

because the inequality (3.8) guarantees that the magnitude of the right hand side of this equation does not exceed one.

From the property (d) it follows that equality is attained in the Schwarz inequality exactly if x is a multiple of y, i.e. $x = \alpha y$. In a real Hilbert space the angle between two such vectors is $k\pi$, k integral, since its cosine is equal either to 1 or -1.

Example 3.8. In R_n the Schwarz inequality assumes the form

$$\left| \sum_{i=1}^{n} \xi_i \eta_i \right| \leq \left(\sum_{i=1}^{n} \xi_i^2 \right)^{1/2} \left(\sum_{i=1}^{n} \eta_i^2 \right)^{1/2} .$$

(This inequality was proved for the first time by Cauchy in 1821.)

In $L_2(a, b)$ the Schwarz inequality reads as follows

$$\left|\int_a^b f(x)\,\overline{g(x)}\,dx\right| \leq \left(\int_a^b |f(x)|^2\,dx\right)^{1/2} \left(\int_a^b |g(x)|^2\,dx\right)^{1/2}.$$

This inequality is also a special case of the *Hölder inequality*.

In Hilbert spaces, the concept of orthogonality of two vectors plays a role of basic importance.

Two vectors $x \in H$ and $y \in H$ will be called *orthogonal* in H, if $(x, y) = 0$.

Theorem 3.2. *The norm in a Hilbert space has the following properties*:

1. $\|x\| = 0$ *if and only if* $x = 0$. *
2. $\|\alpha x\| = |\alpha|\,\|x\|$.
3. $\|x + y\| \leq \|x\| + \|y\|$ *(triangle law)*.
4. $\|x + y\|^2 + \|x - y\|^2 = 2\|x\|^2 + 2\|y\|^2$
   (parallelogram law).
5. *If* $(x, y) = 0$, *then* $\|x + y\|^2 = \|x\|^2 + \|y\|^2$
   (Pythagoras' law).
6. *In a complex Hilbert space we have*

$$4(x, y) = \|x + y\|^2 - \|x - y\|^2 + i\|x + iy\|^2 - i\|x - iy\|^2. \quad (3.9)$$

In a real Hilbert space we have

$$4(x, y) = \|x + y\|^2 - \|x - y\|^2. \quad (3.10)$$

---

* In the definition of the linear space (p. 54), the zero element was denoted by O. In what follows, we shall make no difference between *zero* (symbol 0) and the *zero element* (symbol O) and we shall denote them both by the same symbol 0.

Proof: The assertions 1 and 2 follow immediately from the properties of a scalar product. Let us prove the inequality 3. We have

$$\|x + y\|^2 = (x + y, x + y) = \|x\|^2 + (x, y) + (y, x) + \|y\|^2$$
$$\leq \|x\|^2 + \|y\|^2 + 2|(x, y)| \leq \|x\|^2 + \|y\|^2 + 2\|x\|\|y\|$$
$$= (\|x\| + \|y\|)^2.$$

Here, we have used the Schwarz inequality. Taking the square root, we obtain the triangle law. Next, we prove 4. Observing that

$$\|x + y\|^2 = (x + y, x + y) \quad \text{and} \quad \|x - y\|^2 = (x - y, x - y),$$

we have

$$\|x + y\|^2 + \|x - y\|^2 = (x, x) + (x, y) + (y, x) + (y, y)$$
$$+ (x, x) - (x, y) - (y, x) + (y, y) = 2\|x\|^2 + 2\|y\|^2.$$

The Pythagorus' law and the relations (3.9) and (3.10) can be proved in a similar manner by a direct computation. We have only to keep in mind that $(x, y) = (y, x)$ in a real Hilbert space.

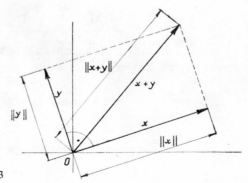

Fig. 3.3

Example 3.9. Fig. 3.3 shows that, in case of the space R_2, relation 5 reduces to the ordinary Pythagorus' theorem. Let us note that in R_n two vectors are exactly orthogonal if the cosine of the angle between these vectors is zero or if at least one vector is the zero vector. The geometrical meaning of relation 4 is apparent from Fig. 3.4.

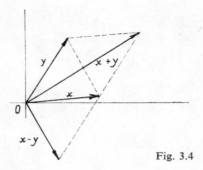

Fig. 3.4

Let us now define the concept of the *distance* (*metric*) $\varrho(x, y)$ between two vectors x and y in a Hilbert space H by the formula

$$\varrho(x, y) = \|x - y\|.$$

Example 3.10. In R_n the distance between two vectors is given by

$$\varrho(\xi, \eta) = \big(\sum_{i=1}^{n} (\xi_i - \eta_i)^2 \big)^{1/2}.$$

Thus we see that our concept of distance in a Hilbert space coincides with the commonly used concept of distance.

Example 3.11. The reader will certainly wonder what we mean by the distance between two functions in the space $L_2(0, 1)$. Here, we define

$$\varrho(f, g) = \left(\int_0^1 |f(x) - g(x)|^2 \, \mathrm{d}x \right)^{1/2}.$$

Consider the functions f_n given by

$$f_n(x) = \begin{cases} 2^n & \text{for } 0 \leq x \leq 2^{-2n}, \\ 0 & \text{for } 2^{-2n} < x < 1 \end{cases}$$

(see Fig. 3.5). It is clear that the distance of any function f_n from the zero function is equal to one. In spite of this the maximum value of f_n, which equals 2^n, tends to infinity as $n \to \infty$. This has its background in the fact that the distance between two functions in $L_2(0, 1)$ is given by the average quadratic deviation of these

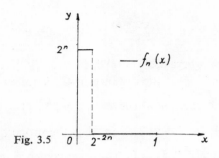

Fig. 3.5

functions. Thus, the difference of two functions having a small distance in $L_2(0, 1)$ may be arbitrarily large at some points; however, the set of points where the difference is large has to have a sufficiently small measure. (See Problem 3.27.)

Theorem 3.3. *Distance in a Hilbert space possesses the following fundamental properties*:

1. $\varrho(x, y) \geq 0$, $\varrho(x, y) = 0$ *if and only if* $x = y$.
2. $\varrho(x, y) \leq \varrho(x, z) + \varrho(y, z)$ *(triangle law)*.
3. $\varrho(x, y) = \varrho(y, x)$ *(symmetry of the distance)*.

Proof: The above properties follow immediately from the properties of a norm. Since 1 and 3 are obvious, let us prove only the triangle law. We have

$$\varrho(x, y) = \|x - y\| = \|x - z + z - y\| \leq \|x - z\| + \|z - y\|$$
$$= \varrho(x, z) + \varrho(y, z).$$

Now that the concept of distance has been introduced, we can speak about convergence in a Hilbert space.

A sequence of vectors $x_n \in H$ will be called *convergent* to a vector $x \in H$, if $\lim_{n \to \infty} \varrho(x_n, x) = 0$; this fact will be symbolized by $\lim_{n \to \infty} x_n = x$ or by $x_n \to x$ as $n \to \infty$, etc.

Example 3.12. From the inequalities

$$|\xi_i^{(m)} - \xi_i| \leq \left(\sum_{j=1}^{n} |\xi_j^{(m)} - \xi_j|^2 \right)^{1/2} = \varrho(\xi^{(m)}, \xi),$$

$i = 1, 2, \ldots, n$, where $\xi^{(m)} = (\xi_1^{(m)}, \xi_2^{(m)}, \ldots, \xi_n^{(m)})$, $\xi = (\xi_1, \xi_2, \ldots, \xi_n)$ and $\varrho(\xi^{(m)}, \xi)$ is the distance in C_n, it follows that $\lim_{m \to \infty} \xi_i^{(m)} = \xi_i$ for $i = 1, 2, \ldots, n$ whenever $\lim_{m \to \infty} \varrho(\xi^{(m)}, \xi) = 0$. Conversely, if $\lim_{m \to \infty} \xi_i^{(m)} = \xi_i$ for $i = 1, 2, \ldots, n$, then the properties of the limit imply that $\lim_{m \to \infty} \varrho(\xi^{(m)}, \xi) = 0$. Thus, convergence in the space C_n is equivalent to convergences of the respective coordinates.

Example 3.13. The sequence of functions $f_n(x) = (1 + x^2 n^2)^{-1}$ converges to zero in the space $L_2(-1, 1)$, but does not converge to zero at the point $x = 0$. From this it follows that convergence in the space $L_2(a, b)$ need not imply convergence at a point, and consequently, uniform convergence. However, uniform convergence on a bounded interval $[a, b]$ implies convergence in $L_2(a, b)$, because

$$\int_a^b |f(x) - f_n(x)|^2 \, dx \leq (b - a) \max_{x \in [a,b]} |f(x) - f_n(x)|^2 .$$

Example 3.14. Let us define a sequence of functions f_n as follows. If n is a positive integer, let k and l be integers such that $n = 2^k + l$ and $1 \leq l \leq 2^k$; then put $f_n(x) = 1$ on the interval $[(l - 1) 2^{-k}, l 2^{-k}]$ and $f_n(x) = 0$ elsewhere. The sequence f_n converges to zero in the space $L_2(0, 1)$, because $\varrho(f_n, 0) = 2^{-k/2}$, which is a number tending to zero as $n \to \infty$. On the other hand, the functions f_n do not converge to zero at any point of the interval $[0, 1]$. (We recommend the reader to draw the graph of f_n.) This example shows that convergence in $L_2(a, b)$ does not imply even convergence almost everywhere. However, in Section 3.4 we shall show that from any sequence converging in $L_2(a, b)$ we can select a subsequence converging almost everywhere.

Let us now show that, with the convergence introduced above, the scalar product is a continuous function of both arguments, i.e., we have the following proposition.

Theorem 3.4. *Let*

$$\lim_{n \to \infty} x_n = x \quad and \quad \lim_{n \to \infty} y_n = y$$

in a Hilbert space H. Then

$$\lim_{n \to \infty} (x_n, y_n) = (x, y).$$

Proof: Let $x_n \to x$ and $y_n \to y$, and let α be a complex number. Then

1. $\alpha x_n \to \alpha x$, because

$$\varrho(\alpha x_n, \alpha x) = \|\alpha x_n - \alpha x\| = |\alpha| \|x_n - x\| = |\alpha| \varrho(x_n, x);$$

2. $x_n + y_n \to x + y$, because by the triangle law we have

$$\varrho(x_n + y_n, x + y) = \|(x_n + y_n) - (x + y)\|$$
$$\leq \|x_n - x\| + \|y_n - y\| = \varrho(x_n, x) + \varrho(y_n, y);$$

3. $\|x_n\| \to \|x\|$, since the triangle law yields immediately

$$\bigl|\|x_n\| - \|x\|\bigr| \leq \|x_n - x\| = \varrho(x_n, x).$$

Hence, *in a Hilbert space the operations of addition and multiplication by a number are continuous operations and the norm is a continuous function.* The assertion of our theorem now follows immediately from the equality (3.9).

Further properties of the concepts which have been introduced may be found in the Problems.

3.4 COMPLETE HILBERT SPACE

Let us begin with an example.

Example 3.15. Let us define the scalar product in the space $C(-1, 1)$ by

$$(f, g) = \int_{-1}^{1} f(x) \overline{g(x)} \, dx .$$

The requirements imposed on a scalar product are obviously satisfied; thus, $C(-1, 1)$ with the above product is a Hilbert space. The functions f_n from Example 3.13 converge to zero function in this space, because the scalar product, and consequently also the distance, coincide with those in $L_2(-1, 1)$. Next, define functions g_n by

$$g_n(x) = \begin{cases} (1 + n^2 x^2)^{-1} & \text{for } x \geq 0 , \\ 1 & \text{for } x \leq 0 . \end{cases}$$

As $n \to \infty$, these functions converge in $L_2(-1, 1)$ to the function

$$g(x) = \begin{cases} 0 & \text{for } x > 0 , \\ 1 & \text{for } x \leq 0 . \end{cases}$$

However, the limit function $g(x)$ is not continuous, and consequently, does not belong to $C(-1, 1)$; this means that the sequence g_n does not converge to any function in $C(-1, 1)$, because, in view of Problem 3.12, a sequence can have at most one limit.

Thus, we encounter the following situation. In a Hilbert space H there exists a sequence x_n not converging in H but converging in a certain larger Hilbert space H_1 such that for $x \in H$ and $y \in H$ the scalar product in H coincides with that in H_1. In other words, a sequence convergent in a certain broader sense may not have any limit in H. Such a Hilbert space is called *incomplete*. Due to formal reasons we shall define the concept of a *complete space* in a slightly different way. Note that this definition grasps exactly the idea about completeness indicated above; this fact, however, will not be proved because this would carry us beyond the scope of the book.

Let x_n be a sequence of elements from a Hilbert space H; x_n will be called the *Cauchy sequence* (or *fundamental sequence*), if for every $\varepsilon > 0$ an integer $n_0 > 0$ exists such that for any integer $p > 0$ and $n \geq n_0$ we have

$$\varrho(x_n, x_{n+p}) \leq \varepsilon.$$

From the triangle law

$$\varrho(x_n, x_{n+p}) \leq \varrho(x_n, x) + \varrho(x_{n+p}, x) \tag{3.11}$$

it follows that every sequence converging in H is a Cauchy sequence. Actually, the right-hand side of (3.11) can be made as small as we please by choosing n sufficiently large.

A Hilbert space H will be called *complete*, if every Cauchy sequence in H is convergent in H.

The following theorem plays a significant role in the analysis and applications.

Theorem 3.5. *The space $L_2(a, b)$ is complete.*

Before turning to the proof of Theorem 3.5, we are going to prove the following auxiliary Theorem.

Theorem 3.6. *Let $f_n \in L_2(a, b)$ and let the series*

$$\sum_{n=1}^{\infty} \|f_n - f_{n+1}\|_{L_2(a,b)} \tag{3.12}$$

be convergent. Then the sequence f_n converges almost everywhere to a function f, which is the limit of f_n in $L_2(a, b)$.

Proof: Put

$$g(x) = |f_1(x)| + \sum_{n=1}^{\infty} |f_n(x) - f_{n+1}(x)|.$$

The function $g(x)$ is either finite and non-negative or is equal to $+\infty$. Put

$$g_k(x) = |f_1(x)| + \sum_{n=1}^{k} |f_n(x) - f_{n+1}(x)|.$$

By the triangle law we have

$$\left(\int_a^b |g_k(x)|^2 \, dx\right)^{1/2} \leq \|f_1\|_{L_2(a,b)} + \sum_{n=1}^{k} \|f_n - f_{n+1}\|_{L_2(a,b)}$$

$$\leq \|f_1\|_{L_2(a,b)} + \sum_{n=1}^{\infty} \|f_n - f_{n+1}\|_{L_2(a,b)} = K .$$

where K is a finite number by assumption. Hence,

$$\int_a^b |g_k(x)|^2 \, dx \leq K^2 .$$

However, the sequence of functions $|g_k|^2$ is non-decreasing and consequently, converges everywhere to the function $|g|^2$. Thus, letting k tend to infinity in the above inequality, we obtain by Theorem 1.7,

$$\int_a^b |g(x)|^2 \, dx \leq K^2 .$$

From this it follows that the function $g(x)$ is finite almost everywhere and that the series

$$f_1(x) + \sum_{n=1}^{\infty} (f_{n+1}(x) - f_n(x))$$

converges absolutely to a function $f(x)$ for almost every $x \in (a, b)$. On the other hand,

$$f_k(x) = f_1(x) + \sum_{n=1}^{k-1} (f_{n+1}(x) - f_n(x))$$

so that

$$f(x) = \lim_{k \to \infty} f_k(x)$$

almost everywhere. Moreover,

$$|f_k(x)| \leq |f_1(x)| + \sum_{n=1}^{k-1} |f_{n+1}(x) - f_n(x)| \leq g(x)$$

and

$$|f_k(x) - f(x)|^2 \leq 2|f_k(x)|^2 + 2|f(x)|^2 \leq 4|g(x)|^2$$

since $|\alpha + \beta|^2 \leq 2|\alpha|^2 + 2|\beta|^2$. Thus, the function $4|g(x)|^2$ is an integrable majorant for the sequence $|f_k(x) - f(x)|^2$; using the Lebesgue Theorem 1.6 it follows that

$$\lim_{k \to \infty} \|f_k - f\|^2_{L_2(a,b)} = \lim_{k \to \infty} \int_a^b |f_k(x) - f(x)|^2 \, dx$$
$$= \int_a^b \lim_{k \to \infty} |f_k(x) - f(x)|^2 \, dx = 0 \,.$$

This completes the proof of Theorem 3.6.

The proof of Theorem 3.5: Let f_n be a Cauchy sequence in $L_2(a, b)$. Put $\varepsilon = 2^{-k}$ and let n_k be the number n_0 from the definition of the Cauchy sequence corresponding to the chosen $\varepsilon = 2^{-k}$. If necessary, we may enlarge each n_k so as to have $n_k < n_{k+1}$ for every k. For the numbers n_k chosen in this way we obtain $\|f_{n_{k+1}} - f_{n_k}\|_{L_2(a,b)} \leq 2^{-k}$. Thus, the sequence f_{n_k} satisfies the assumptions of Theorem 3.6, so that $f = \lim_{k \to \infty} f_{n_k}$ exists in $L_2(a, b)$ and almost everywhere in the interval (a, b). On the other hand, the function f is also the limit of f_n in $L_2(a, b)$. Actually, for a chosen $\varepsilon > 0$ we can find k such that $2^{-k} + \|f - f_{n_k}\|_{L_2(a,b)} < \varepsilon$ and because

$$\|f_n - f\|_{L_2(a,b)} \leq \|f_n - f_{n_k}\|_{L_2(a,b)} + \|f - f_{n_k}\|_{L_2(a,b)}$$
$$\leq 2^{-k} + \|f - f_{n_k}\|_{L_2(a,b)}$$

for any $n > n_k$, we have also $\|f_n - f\|_{L_2(a,b)} < \varepsilon$, i.e. $f = \lim_{n \to \infty} f_n$ in $L_2(a, b)$.

Observe that the proof just carried out yields the following, additional result. *A sequence converging in $L_2(a, b)$ contains a sub-sequence which converges almost everywhere in (a, b).*

A subset H_1 of a linear space H is called the *linear subspace* of the space H, if it is a linear set, i.e., if a linear combination of any two elements from H_1 also belongs to H_1.

Let H be a Hilbert space; a linear subspace H_1 of the space H will be called *closed* in H, if every sequence of elements from H_1 converging in H is also convergent in H_1.

Example 3.16. The set of all $\xi = (\xi_1, \xi_2, ..., \xi_n)$ such that $\sum_{i=1}^{n} a_i \xi_i = 0$, where a_i are fixed complex numbers, is a linear subspace of the space C_n. From Example 3.12 and from the properties of a limit it follows that this subspace is closed in C_n.

Example 3.16 appears as a particular case of the following theorem:

Theorem 3.7. *Let S be a subset of a Hilbert space H; denote $S^\perp$ the set of all vectors $x \in H$ orthogonal to every vector $y \in S$. Then $S^\perp$ is a closed linear subspace of the space H.*

Proof: $S^\perp$ is clearly a linear subspace, because if x_1 and x_2 belong to $S^\perp$, i.e., if $(x_1, y) = 0$ and $(x_2, y) = 0$ for every $y \in S$, then we have also

$$(\alpha_1 x_1 + \alpha_2 x_2, y) = \alpha_1(x_1, y) + \alpha_2(x_2, y) = 0$$

for every $y \in S$, so that $\alpha_1 x_1 + \alpha_2 x_2 \in S^\perp$. Next, consider a sequence of vectors $x_n \in S^\perp$ such that the limit $x = \lim_{n \to \infty} x_n$ exists in H. Since $x_n \in S^\perp$, we have $(x_n, y) = 0$ for all $y \in S$; hence, by Theorem 3.4, $(x, y) = \lim_{n \to \infty} (x_n, y) = 0$ for any $y \in S$, i.e., $x \in S^\perp$. This finishes the proof.

Example 3.17. The set of all functions $f \in L_2(a, b)$ such that

$$\int_a^b f(x) \, dx = 0$$

is a closed linear subspace of $L_2(a, b)$ by Theorem 3.7. Actually,

HILBERT SPACE

the set considered is the set $S^\perp$, where S is the one-point set containing the function equalling one identically.

The space $C(-1, 1)$ is a linear subspace of the space $L_2(-1, 1)$, because each function continuous on $[-1, 1]$ is bounded there, and consequently, square-integrable. However, *it is not a closed subspace* as Example 3.15 shows.

Remark 3.3. *A closed subspace H_1 of a complete Hilbert space is again a complete space.* Actually, if x_n is a Cauchy sequence in the space H_1, it is also a Cauchy sequence in H, and consequently, a limit x exists in H. The space H_1, however, is closed so that $x \in H_1$. Hence, the vector x is the limit of the sequence x_n in H_1.

For a complete space the following proposition is true.

Theorem 3.8. *Let H_1 be a closed subspace of a complete Hilbert space H, and let $x \in H$. Then a $y \in H_1$ exists such that $x = y + z$ and $(z, s) = 0$ for any $s \in H_1$ (i.e. $z \in H_1^\perp$). The vectors y and z are determined uniquely by x.*

Proof: The theorem is trivially true if H_1 consists only of the zero element, since then $y = 0$ and $z = x$. Thus assume that H_1 contains non-zero elements. If $x \in H_1$, we set $y = x$ and $z = 0$. If $x \notin H_1$, denote $d = \inf_{y \in H_1} \varrho(x, y)$. Then a sequence $y_n \in H_1$ exists such that

$$\lim_{n \to \infty} \|y_n - x\| = d \,.$$

Denoting $z_n = x - y_n$, we have

$$z_n + z_{n+p} = 2\left(x - \frac{y_n + y_{n+p}}{2}\right);$$

since $\tfrac{1}{2}(y_n + y_{n+p}) \in H_1$, it follows that

$$\|z_n + z_{n+p}\|^2 = 4\varrho^2(x, \tfrac{1}{2}(y_n + y_{n+p})) \geqq 4d^2 \,.$$

The parallelogram law (Theorem 3.2) yields

$$\|z_n - z_{n+p}\|^2 = 2\|z_n\|^2 + 2\|z_{n+p}\|^2 - \|z_n + z_{n+p}\|^2$$
$$\leq 2\|z_n\|^2 + 2\|z_{n+p}\|^2 - 4d^2 \, .$$

Since $\|z_n\| \to d$ as $n \to \infty$, z_n is a Cauchy sequence and consequently, the limit $z = \lim_{n\to\infty} z_n$ exists due to the completeness of H. Hence, the limit $y = \lim_{n\to\infty} y_n$ exists, too. Now, we are going to show that y thus obtained satisfies the requirements stated in the theorem. We have $\|x - y\| = d$, $y \in H_1$, because H_1 is closed in H. Let $w \in H_1$, $w \neq 0$. Then $y + \alpha w \in H_1$ for any complex α so that

$$\|x - y - \alpha w\| \geq \|x - y\| = d \, .$$

Put $z = x - y$ and take the square in the previous inequality; we obtain

$$\|z\|^2 - \alpha(w, z) - \bar{\alpha}(z, w) + |\alpha|^2 \|w\|^2 \geq \|z\|^2 \, .$$

Putting $\alpha = (z, w) \|w\|^{-2}$, we obtain $|(w, z)|^2 \leq 0$. i.e. $(w, z) = 0$ for every $w \in H_1$, so that $z \in H_1^\perp$.

The uniqueness of the elements y and z can be proved by contradiction. Suppose that

$$x = y + z \quad \text{and} \quad x = y^* + z^* \, ,$$

where $y \in H_1$, $y^* \in H_1$, $z \in H_1^\perp$ and $z^* \in H_1^\perp$. From this it follows that $y - y^* \in H_1$ and $z - z^* \in H_1^\perp$, i.e. $(y - y^*, z - z^*) = 0$. Since also $y - y^* = z^* - z$, we have

$$\|y - y^*\|^2 = (y - y^*, z^* - z) = 0 \, ,$$

i.e., $y = y^*$ and $z = z^*$. Hence, the theorem is proved.

Theorem 3.8 states that from an arbitrary point in a complete space H we can "drop a perpendicular" to a closed subspace (see Fig. 3.6).

The proof of Theorem 3.8 furnishes also the existence of the so-called *minimizing element*. More precisely, we have proved that for a given $x \in H$ there exists in H_1 an element y closest to x, i.e. $\|x - y\| = \inf_{w \in H_1} \|x - w\|$.

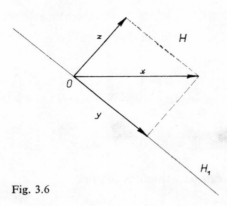

Fig. 3.6

3.5 ORTHONORMAL SEQUENCES AND THE BEST APPROXIMATION

In Section 3.3 we have defined the concept of the orthogonality of two vectors in a Hilbert space. Two functions f and g are orthogonal in the space $L_2(a, b)$, if

$$\int_a^b f(x) \overline{g(x)} \, dx = 0 \, .$$

As the formulas (1.22) indicate, the system of functions 1, $\cos x$, $\sin x$, $\cos 2x$, $\sin 2x$, ... is such that any two distinct functions from this system are orthogonal in the space $L_2(-\pi, \pi)$. If we considered these functions as elements of the space $L_2(0, \pi)$ we could easily verify that they no longer constitute an orthogonal system.

The sequence $e_1, e_2, e_3, \ldots$ is called *orthonormal in H*, if $(e_i, e_j) = 0$ for $i \neq j$ and $\|e_i\| = 1$ for all i. Thus let such an orthonormal sequence $\{e_i\}$ be given in H; denote by H_n the set of all vectors x which are linear combinations of the first n vectors $e_1, e_2, \ldots, e_n$, i.e. H_n is the set of all x having the form $\sum_{i=1}^{n} \lambda_i e_i$ with $\lambda_i \in C$.

Let a vector $y \in H$ be given and let us try to find the numbers $\alpha_1, \alpha_2, \ldots, \alpha_n$ which minimize the norm $\left\| y - \sum_{i=1}^{n} \alpha_i e_i \right\|$. If $x \in H_n$ and $x = \sum_{i=1}^{n} \lambda_i e_i$, we can write

$$0 \leq \|y - x\|^2 = (y - \sum_{i=1}^{n} \lambda_i e_i, y - \sum_{i=1}^{n} \lambda_i e_i)$$

$$= \|y\|^2 - \sum_{i=1}^{n} \lambda_i (e_i, y) - \sum_{i=1}^{n} \bar{\lambda}_i (y, e_i) + \sum_{i=1}^{n} |\lambda_i|^2 .$$

This inequality can be written as follows

$$0 \leq \|y - x\|^2 = \|y\|^2 + \sum_{i=1}^{n} [\lambda_i - (y, e_i)][\bar{\lambda}_i - (e_i, y)]$$

$$- \sum_{i=1}^{n} |(y, e_i)|^2 = \|y\|^2 + \sum_{i=1}^{n} |\lambda_i - (y, e_i)|^2 - \sum_{i=1}^{n} |(y, e_i)|^2 .$$

It is clear that the last expression attains its least value if

$$\lambda_i = (y, e_i), \quad i = 1, 2, \ldots, n .$$

For the λ_i's chosen in this way we obtain *Bessel's inequality*

$$\sum_{i=1}^{n} |(y, e_i)|^2 \leq \|y\|^2 . \tag{3.13}$$

Hence, in H_n there exists an element y_n closest to y, and

$$y_n = \sum_{i=1}^{n} (y, e_i) e_i ; \tag{3.14}$$

moreover

$$\|y - y_n\|^2 = \|y\|^2 - \sum_{i=1}^{n} |(y, e_i)|^2 . \tag{3.15}$$

HILBERT SPACE

Theorem 3.9. *The sequence of elements y_n defined by the formula* (3.14) *converges to y in H, if and only if*

$$\|y\|^2 = \sum_{i=1}^{\infty} |(y, e_i)|^2 . \tag{3.16}$$

The equality (3.16) is called *Parseval's equality*.

Proof: If $y_n \to y$ in H, then $\lim_{n \to \infty} \|y_n\|^2 = \|y\|^2$. We can easily verify that

$$\|y_n\|^2 = \sum_{i=1}^{n} |(y, e_i)|^2 ;$$

thus, letting $n \to \infty$ we obtain the equality (3.16).

Conversely, let Parseval's equality hold. Then

$$\lim_{n \to \infty} \sum_{i=1}^{n} |(y, e_i)|^2 = \sum_{i=1}^{\infty} |(y, e_i)|^2 = \|y\|^2$$

and from (3.15) it follows that $\lim_{n \to \infty} \|y_n - y\|^2 = 0$. This concludes the proof.

However, it is often difficult to verify whether or not a particular system $\{e_i\}$ satisfies Parseval's equality. Therefore, let us present another condition for the convergence $y_n \to y$ in H. We have the following proposition.

Theorem 3.10. *The elements y_n defined by the formula* (3.14) *converge to y in H exactly if to every $\varepsilon > 0$ there exists a finite linear combination x of vectors $e_1, e_2, e_3, \ldots$, i.e. $x = \sum_{i=1}^{m} \lambda_i e_i$, such that $\|x - y\| \leq \varepsilon$.*

Proof: Let $\varepsilon > 0$ and $y_n \to y$. Then $\|y_n - y\| \leq \varepsilon$ for n sufficiently large, so that we can put $x = y_n$.

Conversely, let an $x = \sum_{i=1}^{m} \lambda_i e_i$ exist such that $\|x - y\| \leq \varepsilon$.

Then, of course, $x \in H_n$ for any $n \geq m$, and consequently,

$$\|y_n - y\| \leq \|x - y\| \leq \varepsilon,$$

because y_n is the best approximation for y by the elements in H_n. Hence, $y_n \to y$ and the theorem is proved.

If the condition of Theorem 3.10 is satisfied, we say that in H the vector y can be approximated with any accuracy by a finite linear combination of vectors $e_1, e_2, e_3, \ldots$. If an arbitrary element $y \in H$ can be approximated with any accuracy by a finite linear combination of vectors $e_1, e_2, e_3, \ldots$, the system $e_1, e_2, e_3, \ldots$ is called *complete* in H.

Using Theorems 3.9 and 3.10 for every $y \in H$, we obtain the following assertion.

Theorem 3.11. *An orthonormal system $e_1, e_2, e_3, \ldots$ in a Hilbert space H is complete exactly if one of the following conditions is satisfied*:

1. $\|y\|^2 = \sum_{i=1}^{\infty} |(y, e_i)|^2$ *for every* $y \in H$.

2. $y = \sum_{i=1}^{\infty} (y, e_i) e_i$ *for every* $y \in H$, *where the series converges in H.*

3. *For every* $y \in H$ *and* $\varepsilon > 0$ *a finite linear combination* $x = \sum_{i=1}^{m} \lambda_i e_i$ *exists such that* $\|y - x\| \leq \varepsilon$.

If the Hilbert space is complete, we can give another condition for the completeness of an orthonormal system. Actually, we have the following proposition.

Theorem 3.12. *An orthonormal system of vectors $e_1, e_2, e_3, \ldots$ in a complete Hilbert space H is complete exactly if the following condition is satisfied.*

4. *If $x \in H$ is a vector such that* $(x, e_i) = 0$ *for all* $i = 1, 2, 3, \ldots$, *then* $x = 0$.

Before turning to the proof of Theorem 3.12, we shall prove the important RIESZ-FISCHER theorem.

Theorem 3.13. *Let H be a complete Hilbert space and let $\{c_i\}$ be a sequence of complex numbers such that the series $\sum_{i=1}^{\infty} |c_i|^2$ converges. Let $e_1, e_2, e_3, \ldots$ be an orthonormal system of vectors in H. Then the series $\sum_{i=1}^{\infty} c_i e_i$ converges in H and for its sum x we have $\|x\| = (\sum_{i=1}^{\infty} |c_i|^2)^{1/2}$.*

Proof: Denote $x_n = \sum_{i=1}^{n} c_i e_i$. Then for $p > 0$ we have

$$\|x_n - x_{n+p}\|^2 = \Big(\sum_{i=n+1}^{n+p} c_i e_i, \sum_{i=n+1}^{n+p} c_i e_i\Big) = \sum_{i=n+1}^{n+p} |c_i|^2 \,;$$

since $\sum_{i=1}^{\infty} |c_i|^2$ converges, the last sum in the previous equality can be made as small as we please by choosing n sufficiently large. Thus, x_n is a Cauchy sequence, and consequently, an $x \in H$ exists such that $x_n \to x$. Furthermore, we clearly have

$$\|x\|^2 = \lim_{n \to \infty} \|x_n\|^2 = \lim_{n \to \infty} \sum_{i=1}^{n} |c_i|^2 = \sum_{i=1}^{\infty} |c_i|^2\,.$$

Proof of Theorem 3.12: If the system $e_1, e_2, e_3, \ldots$ is complete, we have by condition 2 in Theorem 3.11, $x = \sum_{i=1}^{\infty} (x, e_i) e_i = 0$, because $(x, e_i) = 0$ for all i.

Conversely, if $e_1, e_2, e_3, \ldots$ is not complete, a vector y exists such that $\sum_{i=1}^{\infty} |(e_i, y)|^2 < \|y\|^2$ due to condition 1 of Theorem 3.11. By Theorem 3.13, $\sum_{i=1}^{\infty} (y, e_i) e_i$ exists. Denote $x = y - \sum_{i=1}^{\infty} (y, e_i) e_i$. Then $x \neq 0$ and

$$(x, e_i) = (y, e_i) - \sum_{j=1}^{\infty} (y, e_j)(e_j, e_i) = (y, e_i) - (y, e_i) = 0$$

for every i, which is a contradiction.

Example 3.18. As known, the system of functions $1/\sqrt{(2\pi)}$, $(\cos x)/\sqrt{\pi}$, $(\sin x)/\sqrt{\pi}$, $(\cos 2x)/\sqrt{\pi}$, ... is an orthonormal system in the space $L_2(-\pi, \pi)$. The numbers (y, e_i) from (3.13) are the Fourier coefficients from Chapter 2 multiplied by the number $\sqrt{\pi}$, only in the case of $e_1 \equiv 1/\sqrt{(2\pi)}$, a_0 has to be multiplied by $\sqrt{(2\pi)}/2$. Thus the expression (3.14) giving the best approximation assumes the form

$$\frac{\sqrt{(2\pi)}}{2} a_0 \frac{1}{\sqrt{(2\pi)}} + \sum_{k=1}^{n} \left(a_k \sqrt{\pi} \frac{\cos kx}{\sqrt{\pi}} + b_k \sqrt{\pi} \frac{\sin kx}{\sqrt{\pi}} \right) \quad (3.17)$$

which evidently is a partial sum of the Fourier series introduced in Chapter 2. Parseval's equality reads as follows,

$$\frac{1}{\pi} \int_{-\pi}^{\pi} |f(x)|^2 \, dx = \tfrac{1}{2}|a_0|^2 + \sum_{k=1}^{\infty} (|a_k|^2 + |b_k|^2). \quad (3.18)$$

To obtain the complex form of the Fourier series we have to start from the orthonormal system of functions

$$\frac{1}{\sqrt{(2\pi)}} e^{inx}, \quad n = 0, \pm 1, \pm 2, \ldots.$$

In this case, each number (y, e_i) is equal to the corresponding Fourier coefficient c_i multiplied by $\sqrt{(2\pi)}$. Equation (3.14) yields again the well-known form of a partial sum,

$$\sum_{k=-n}^{n} c_k e^{ikx}, \quad (3.19)$$

and Parseval's equality reads

$$\frac{1}{2\pi} \int_{-\pi}^{\pi} |f(x)|^2 \, dx = \sum_{k=-\infty}^{\infty} |c_k|^2. \quad (3.20)$$

We still do not know whether the partial sums of the Fourier series of a function $f \in L_2(-\pi, \pi)$ actually converge to f in the space $L_2(-\pi, \pi)$, because we do not know whether the system of trigonometric functions is complete in $L_2(-\pi, \pi)$. Shortly we shall see that it is really so; however, first we are going to prove an important proposition on the approximation of a function in $L_2(a, b)$ by continuous functions:

Theorem 3.14. *Let $f \in L_2(a, b)$. Then for every $\varepsilon > 0$ a continuous function f_h exists such that $\|f - f_h\|_{L_2(a,b)} < \varepsilon$.*

Proof: Put $f(x) = 0$ for $x \notin (a, b)$ and define the function $f_N(x)$, $N > 0$, by

$$f_N(x) = \begin{cases} f(x) & \text{for } |x| < N, \text{ provided } |f(x)| \leq N, \\ N \operatorname{sgn} f(x) & \text{for } |x| < N, \text{ provided } |f(x)| > N, \\ 0 & \text{for } |x| \geq N. \end{cases}$$

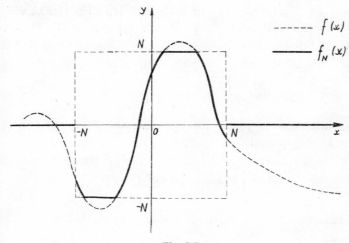

Fig. 3.7

(See Fig. 3.7.) Clearly, $|f_N(x)| \leq N$ and $|f_N(x)| \leq |f(x)|$. Thus, by the triangle law

$$|f_N(x) - f(x)|^2 \leq 4|f(x)|^2.$$

Furthermore, $\lim_{N \to \infty} f_N(x) = f(x)$ for almost every $x \in (-\infty, \infty)$.

Hence, by the Lebesgue theorem 1.6, $(4|f(x)|^2$ is an integrable majorant)

$$\lim_{N\to\infty} \int_a^b |f_N(x) - f(x)|^2 \, dx = \int_a^b \lim_{N\to\infty} |f_N(x) - f(x)|^2 \, dx = 0 ,$$

i.e. $f_N \to f$ in $L_2(a, b)$ as $N \to \infty$.

Choose a fixed N such that $\|f_N - f\|_{L_2(a,b)} < \varepsilon/2$; note that $f_N(x) = 0$ outside the interval $(-N, N)$. For any $x \in (-\infty, \infty)$ denote

$$f_h(x) = \frac{1}{h} \int_x^{x+h} f_N(t) \, dt .$$

Since $f_N(t)$ is an integrable function, we have $\lim_{h\to 0} f_h(x) = f_N(x)$ for almost all $x \in (-\infty, \infty)$. In addition, $|f_h(x)| \leq (1/h) \, hN = N$ for $|x| < N + |h|$, and $f_h(x) = 0$ for $|x| \geq N + |h|$. Thus, the function defined by

$$g(x) = \begin{cases} 4N^2 & \text{for } |x| \leq N + 1, \\ 0 & \text{for } |x| > N + 1 \end{cases}$$

is an integrable majorant for the functions $|f_h(x) - f_N(x)|^2$ for $|h| < 1$; consequently,

$$\lim_{h\to 0} \int_a^b |f_h(x) - f_N(x)|^2 \, dx = 0 .$$

The functions f_h are continuous and an h exists such that

$$\|f_h - f_N\|_{L_2(a,b)} < \frac{\varepsilon}{2} .$$

Hence

$$\|f_h - f\|_{L_2(a,b)} \leq \|f - f_N\|_{L_2(a,b)} + \|f_N - f_h\|_{L_2(a,b)} < \varepsilon$$

and Theorem 3.14 is proved.

Now, we can readily prove the following important statement.

Theorem 3.15. *The system of trigonometric functions is complete in $L_2(-\pi, \pi)$.*

Proof: Let $f \in L_2(-\pi, \pi)$; redefine f on the entire interval $(-2\pi, 2\pi)$ so as to be periodic with period 2π. (See Chapter 1.) Then evidently $f \in L_2(-2\pi, 2\pi)$. Let us now employ Theorem 3.14. For a given $\varepsilon > 0$ a continuous function f_h exists such that

$$\|f_h - f\|_{L_2(-2\pi, 2\pi)} < \frac{\varepsilon}{2}.$$

Therefore,

$$\|f_h - f\|_{L_2(-\pi, \pi)} < \frac{\varepsilon}{2}.$$

Since f is periodic with period 2π on $(-2\pi, 2\pi)$, the function f_N for N sufficiently large is also periodic there, and the same is true for $f_h(x)$. (The reader is recommended to draw a diagram.) Hence, f_h is continuous on $[-\pi, \pi]$ and we have $f_h(-\pi) = f_h(\pi)$. Recalling Problem 3.19, for f_h we can find a trigonometric polynomial g such that $|g(x) - f_h(x)| < \varepsilon/2\sqrt{(2\pi)}$ for $x \in [-\pi, \pi]$. Thus, we have

$$\left(\int_{-\pi}^{\pi} |g(x) - f_h(x)|^2 \, dx \right)^{1/2} < \frac{\varepsilon}{2},$$

and finally,

$$\|g - f\|_{L_2(-\pi, \pi)} \leq \|g - f_h\|_{L_2(-\pi, \pi)} + \|f_h - f\|_{L_2(-\pi, \pi)} < \varepsilon.$$

Summarizing the above results we see that for the function $f \in L_2(-\pi, \pi)$ we have found a trigonometric polynomial g approximating f in $L_2(-\pi, \pi)$ with the prescribed accuracy ε; this, in view of condition 3 of Theorem 3.11, shows that the system of trigonometric functions is complete in $L_2(-\pi, \pi)$.

Theorem 3.11 yields immediately the following assertion.

Theorem 3.16. *Any function $f \in L_2(-\pi, \pi)$ can be expanded in the Fourier series which converges to f in the space $L_2(-\pi, \pi)$. If a_n, b_n are the Fourier coefficients of the function f and c_n the*

Fourier coefficients of f for the complex form, then the equalities (3.18) and (3.20) hold, i.e. the series on the right hand side converge and their sums equal the values on the left hand side. If all the Fourier coefficients of $f(x)$ vanish, then $f(x) = 0$ for almost all $x \in (-\pi, \pi)$.

3.6 ORTHONORMALIZATION

Consider a sequence of non-zero vectors $x_1, x_2, x_3, \ldots$ in a Hilbert space H. Let again H_n be the set of all linear combinations $\sum_{i=1}^{n} \lambda_i x_i$ of the first n elements $x_1, x_2, \ldots, x_n$, and let us try to find the coefficients $\lambda_1, \lambda_2, \ldots, \lambda_n$ so that the corresponding combination becomes the best approximation in H_n of a given vector $y \in H$. Because $x_1, x_2, x_3, \ldots$ in general is not an orthonormal system, the results obtained in the previous section cannot be applied. Thus, we shall replace the system $x_1, x_2, x_3, \ldots$ by an orthonormal system $e_1, e_2, e_3, \ldots$ such that each H_n will be the set of all linear combinations of the first k fixed vectors $e_1, e_2, \ldots, e_k$. Then, of course, the results of Section 3.5 are applicable to the system $e_1, e_2, e_3, \ldots$. We are now going to present a procedure for constructing the sequence $e_1, e_2, e_3, \ldots$ for a given sequence $x_1, x_2, x_3, \ldots$.

Put $y_1 = x_1$. Suppose that the vectors $y_1, y_2, \ldots, y_{k-1}$ are already determined so that $(y_i, y_j) = 0$ for $i \neq j$ and define y_k by the formula

$$y_k = x_k + \sum_{i=1}^{k-1} a_{ik} y_i, \qquad (3.21)$$

where the constants a_{ik} are to be found from the conditions $(y_k, y_j) = 0$ for $j < k$. If $y_j = 0$ for some j, we choose a_{jk} arbitrarily. Forming the scalar products of the vector y_j with both sides of the equality (3.21), we obtain

$$(y_k, y_j) = (x_k, y_j) + \sum_{i=1}^{k-1} a_{ik}(y_i, y_j) = (x_k, y_j) + a_{jk} \|y_j\|^2.$$

The left hand side is to equal zero so that $a_{jk} = -(x_k, y_j)\|y_j\|^{-2}$ for $y_j \neq 0$; hence, a_{jk} is determined uniquely. Thus, the sequence $y_1, y_2, y_3, \ldots$ is determined uniquely by the recurrence formula

$$y_k = x_k - \sum_{j=1}^{k-1}{}'(x_k, y_j)\frac{y_j}{\|y_j\|^2}, \qquad (3.22)$$

where the prime on the summation sign signifies that the summation is extended over all indices j with $y_j \neq 0$.

From the construction it follows that any vector x, which is a linear combination of vectors $x_1, x_2, \ldots, x_m$, is also a linear combination of vectors $y_1, y_2, \ldots, y_m$ and vice versa, i.e. any linear combination of vectors $y_1, y_2, \ldots, y_m$ is a linear combination of vectors $x_1, x_2, \ldots, x_m$.

Deleting in the sequence $y_1, y_2, y_3, \ldots$ all the y_js equal to zero, we obtain a sub-sequence $y_1^*, y_2^*, y_3^*, \ldots$ of $y_1, y_2, y_3, \ldots$. Putting $e_j = \|y_j^*\|^{-1} y_j^*$ we obtain an orthonormal sequence in H which has all properties required at the beginning of this section. The process just described, i.e. the process permitting us to construct an orthonormal sequence $e_1, e_2, e_3, \ldots$ for a given sequence $x_1, x_2, x_3, \ldots$ is called *Gram-Schmidt's orthonormalization* of the sequence $x_1, x_2, x_3, \ldots$.

The system $e_1, e_2, e_3, \ldots$ will be complete if and only if for every $y \in H$ and every $\varepsilon > 0$ there exists a finite linear combination $x = \sum_{i=1}^{m} \lambda_i x_i$ of vectors $x_1, x_2, x_3, \ldots$ such that $\|x - y\| < \varepsilon$. Actually, any finite linear combination of vectors $x_1, x_2, x_3, \ldots$ is also a finite linear combination of vectors $e_1, e_2, e_3, \ldots$ and vice versa.

Example 3.19. Let $f_1(t) = 1$, $f_2(t) = t$, $f_3(t) = t^2, \ldots$ and let us orthonormalize this sequence in the space $L_2(-1, 1)$. As a result we obtain the sequence of *Legendre polynomials*

$$e_1(t) = \sqrt{\tfrac{1}{2}}, \quad e_2(t) = \sqrt{(\tfrac{3}{2})}\, t, \quad e_3(t) = \sqrt{(\tfrac{5}{8})}\,(3t^2 - 1), \ldots.$$

We are going to present two ways of establishing the function $e_3(t)$, for example.

(a) Assume that $e_3(t) = at^2 + bt + c$. This function is to be orthogonal to the functions f_1 and f_2, which obviously are mutually orthogonal, i.e. we should have

$$\int_{-1}^{1} (at^2 + bt + c)\,dt = 0, \quad \int_{-1}^{1} (at^3 + bt^2 + ct)\,dt = 0.$$

This yields the equations

$$\tfrac{2}{3}a + 2c = 0, \quad \tfrac{2}{3}b = 0.$$

Thus, the function $g(t) = 3t^2 - 1$ is orthogonal to both functions f_1 and f_2, and consequently, orthogonal to e_1 and e_2. Finally, it suffices to set

$$e_3(t) = \frac{g(t)}{\|g\|_{L_2(-1,1)}} = \frac{3t^2 - 1}{\left(\int_{-1}^{1} (3\tau^2 - 1)^2\,d\tau\right)^{1/2}}.$$

(b) Putting $y_1(t) = t$ and $y_2(t) = t^2$ in the formula (3.22) we obtain

$$y_3(t) = t^2 - \left(\int_{-1}^{1} \tau^2\,d\tau\right) \cdot \tfrac{1}{2} - \left(\int_{-1}^{1} \tau^2 \cdot \tau\,d\tau\right) \frac{t^2}{\tfrac{2}{3}} = t^2 - \tfrac{1}{3};$$

a normalization yields the above result.

The system of Legendre polynomials is complete in $L_2(-1, 1)$; actually, in view of Theorem 3.15 any function from $L_2(-1, 1)$ can be approximated in $L_2(-1, 1)$ by a continuous function and the latter in turn can be approximated uniformly by a polynomial on the interval $[-1, 1]$ as is known from Problem 3.21. A detailed proof is left to the reader as an exercise, because it is entirely analogous to the proof of Theorem 3.14.

Thus, any function $f \in L_2(-1, 1)$ can be written in the form

$$f(t) = \sum_{j=1}^{\infty} \left(\int_{-1}^{1} f(\tau)\, e_j(\tau)\,d\tau\right) e_j(t),$$

where the series of Legendre polynomials converges in $L_2(-1, 1)$. Moreover, every partial sum of the series on the right hand side gives in $L_2(-1, 1)$ the best approximation of the function f by a polynomial.

Finally, let us note that the orthonormalization of a system of real functions yields again a system of real functions; this fact is immediately apparent from the formula (3.22).

3.7 ORTHOGONAL SYSTEMS

Hitherto we have considered orthonormal sequences in a Hilbert space. However, our original system of functions

$$1, \cos t, \sin t, \cos 2t, \sin 2t, \ldots \qquad (3.23)$$

does not constitute an orthonormal sequence in $L_2(-\pi, \pi)$, because the norms of these functions in $L_2(-\pi, \pi)$ do not equal one.

A sequence $f_1, f_2, f_3, \ldots$ of elements from H will be called *orthogonal*, if all its elements are non-zero and $(f_i, f_j) = 0$ for $i \neq j$.

As will be seen later, it is often more convenient to establish the expansion in an orthogonal system which is not an orthonormal one.

Having an orthogonal sequence we can easily obtain an orthonormal one by putting

$$e_i = \|f_i\|^{-1} f_i . \qquad (3.24)$$

The best approximation of an element y in H by a linear combination of functions $f_1, f_2, \ldots, f_n$ will then be the element

$$y_n = \sum_{i=1}^{n} (y, e_i) e_i = \sum_{i=1}^{n} \|f_i\|^{-2} (y, f_i) f_i ; \qquad (3.25)$$

Parseval's equality assumes the form

$$\|y\|^2 = \sum_{i=1}^{\infty} \|f_i\|^{-2} |(y, f_i)|^2 . \tag{3.26}$$

An orthogonal system will be called *complete* in H, if the corresponding system (3.24) is complete in H.

Theorem 3.17. *An orthogonal system $f_1, f_2, f_3, \ldots$ is complete in H, if and only if one of the following conditions is satisfied.*

   1. *Parseval's equality* (3.26) *holds for all $y \in H$.*
   2. *For any $y \in H$,*

$$y = \sum_{i=1}^{\infty} \|f_i\|^{-2} (y, f_i) f_i ,$$

where the series on the right-hand side converges in H.

   3. *For every $y \in H$ and every $\varepsilon > 0$ a finite linear combination $x = \sum_{i=1}^{m} \lambda_i f_i$ exists such that $\|y - x\| < \varepsilon$.*

If the space H is complete, we can state an additional, equivalent condition for the completeness of a system:

   4. *If $x \in H$ is such that $(x, f_i) = 0$ for $i = 1, 2, 3, \ldots$, then $x = 0$.*

The proof can be easily carried out by using Theorems 3.11 and 3.12.

From (3.24) it follows that the series

$$\sum_{i=1}^{\infty} \|f_i\|^{-2} (y, f_i) f_i \quad \text{and} \quad \sum_{i=1}^{\infty} (y, e_i) e_i$$

are identical. They are called the Fourier series of the element y in the orthogonal system $f_1, f_2, f_3, \ldots$, and in the orthonormal system $e_1, e_2, e_3, \ldots$, respectively.

Let us recall Example 3.18, where we showed the equality (3.25) for a specific case. Although the Fourier series in the system of functions f_i given by (3.23) coincides with the Fourier series in the system e_i formed by functions

$$\frac{1}{\sqrt{(2\pi)}}, \frac{1}{\sqrt{\pi}}\cos t, \frac{1}{\sqrt{\pi}}\sin t, \frac{1}{\sqrt{\pi}}\cos 2t, \ldots,$$

the corresponding coefficients of both series are different. To avoid misunderstanding let us emphasize that the numbers (y, e_i) cannot be termed the Fourier coefficients of the function y, because in Chapter 2 the coefficients $\|f_i\|^{-2}(y, f_i)$, except for the first one, have been so named.*

Returning to the considerations made in the first section of this Chapter, we can arrive at the coefficients of a Fourier series in a different way. Actually, if $y \in H$ and the orthogonal system $f_1, f_2, f_3, \ldots$ is complete in H, then, as we know, constants c_i exist such that

$$y = \sum_{i=1}^{\infty} c_i f_i. \tag{3.27}$$

Since $(f_i, f_j) = 0$ for $i \neq j$, we have by Theorem 3.7,

$$\left(\sum_{\substack{i=1 \\ i \neq j}}^{\infty} c_i f_i, f_j\right) = 0.$$

Thus, forming the scalar products of the function f_j with both sides of (3.27), we obtain

$$(y, f_j) = (c_j f_j, f_j) + \left(\sum_{\substack{i=1 \\ i \neq j}}^{\infty} c_i f_i, f_j\right) = c_j (f_j, f_j).$$

Hence, we necessarily have

$$c_j = \|f_j\|^{-2}(y, f_j).$$

---

* *Remark:* Sometimes, however, the numbers (y, e_i) but not $\|f_i\|^{-2}(y, f_i)$ are called the Fourier coefficients; this is done mainly in cases where the authors disregard the concept of a trigonometric Fourier series.

Problems

3.1. Show that any function, square-integrable on the interval $[-\pi, \pi]$, is also absolutely integrable on this interval. (*Hint*: Use the Schwarz inequality from Example 3.8 and put $g(x) = \operatorname{sgn} f(x)$ at every point x, where $f(x)$ is finite.)

3.2. Show that the function $f(x) = |x|^{-1/2}$ is absolutely integrable but not square-integrable on the interval $[-\pi, \pi]$.

3.3. Show that in a linear space the zero element and the negative element for x are determined uniquely.

3.4. Prove that $0 \cdot x = O$, where the symbol 0 on the left hand side signifies the number zero and the symbol O on the right hand side signifies the zero element.

3.5. Show that the negative element for x is $-x$.

3.6. Determine the angle between the functions $f(t) = t$ and $g(t) = t^3$ in the space $L_2(0, 1)$.

3.7. Using the parallelogram law prove that for $\|x\| = 1$, $\|y\| = 1$, $x \ne y$ we have $\|(x + y)/2\| < 1$.

3.8. Show that the scalar product defined by (3.9) satisfies all the axioms of a Hilbert space provided the parallelogram law holds and $\|x\| = 0$ exactly if $x = 0$.

3.9. What is the distance in $L_2(-\pi, \pi)$ between any two different functions from the system (3.5)?
(*Hint*: The functions from this system are mutually orthogonal; consequently, the Pythagorus' law can be used.)

3.10. Show that in $L_2(-\pi, \pi)$ there exists a sequence of functions g_n with $\|g_n\| = 1$ such that the distance between any two different functions from this sequence is larger than one.
(*Hint*: Use the results of Problem 3.9.)

3.11. Prove that properties 2 and 3 stated in Theorem 3.2 alone imply that $\|x\| \geqq 0$.

3.12. Prove that a sequence in a Hilbert space has at most one limit.

3.13. Show that the space $P(0, 1)$ of all polynomials on the interval $(0, 1)$ is incomplete provided the scalar product is defined by the same formula as in $L_2(0, 1)$.

(*Hint*: The sequence $g_n(t) = \sum_{k=0}^{n} t^k/k!$ is a Cauchy sequence but its limit is not a polynomial.)

3.14. Let $p(t)$ be a non-negative function which is different from zero almost everywhere in the interval (a, b). Denote by $L_{2,p}(a, b)$ the space of all functions such that

$$\int_a^b p(t) |f(t)|^2 \, dt < \infty .$$

Prove that $L_{2,p}(a, b)$ becomes a Hilbert space if we define the scalar product by

$$(f, g) = \int_a^b p(t) f(t) \overline{g(t)} \, dt .$$

3.15. Prove that the Hilbert space $L_{2,p}(a, b)$ introduced in the previous problem is complete.
(*Hint*: The fact that f_n is a Cauchy sequence in $L_{2,p}(a, b)$ means that $\sqrt{(p)} f_n$ is a Cauchy sequence in the complete space $L_2(a, b)$.)

3.16. Let H be a complete Hilbert space and let the sequence of elements $x_n \in H$ be such that $(y, x_n) = 0$ for all n implies that $y = 0$. Then the sequence $e_1, e_2, \ldots, e_n, \ldots$ constructed in Section 3.6 is complete in H.
(*Hint*: Using the construction of the sequence $\{e_n\}$ show that $(y, e_n) = 0$ for all n implies that $y = 0$.)

3.17. Let $f(t)$ be a function continuous on $[-\pi, \pi]$ and let $f(-\pi) = f(\pi)$. Denote

$$\omega(\delta, f) = \max_{\substack{|x-y|<\delta \\ x,y \in [-\pi,\pi]}} |f(x) - f(y)|$$

(the so-called *modulus of continuity* of the function f). Furthermore, let n be a positive integer and let the continuous function $g_n(t)$ be defined as follows: $g_n(k\pi/n) = f(k\pi/n)$ for $k = -n, -n+1, \ldots, -1, 0, 1, \ldots, n$ and $g_n(t)$ is linear on each interval $[k\pi/n, (k+1)\pi/n]$. Show that

$$\omega\left(\frac{\pi}{n}, g_n\right) \leq 2\omega\left(\frac{\pi}{n}, f\right).$$

(*Hint*: Draw a diagram and use the triangle law.)

3.18. Let $f(t)$ and $g_n(t)$ have the same meaning as in Problem 3.17. Prove that for every $\varepsilon > 0$ an n can be found such that $|f(t) - g_n(t)| < \varepsilon$ for every $t \in [-\pi, \pi]$.

(*Hint*: Let $k\pi/n$ be the coordinate of the vertex of the polygon g_n which is closest to the point t. Then

$$|f(t) - g_n(t)| = \left|f(t) - f\left(\frac{k\pi}{n}\right) - g_n(t) + g_n\left(\frac{k\pi}{n}\right)\right|$$

$$\leq \omega\left(\frac{\pi}{n}, f\right) + \omega\left(\frac{\pi}{n}, g_n\right) \leq 3\omega\left(\frac{\pi}{n}, f\right).$$

Next use the fact that for a function f continuous on a closed interval we have $\omega(\pi/n, f) \to 0$ as $n \to \infty$.)

3.19. Show that the function f from Problem 3.17 can be uniformly approximated by a trigonometric polynomial on the interval $[-\pi, \pi]$.

(*Hint*: Use the results of Problems 3.18 and 2.10.)

3.20. Show that a trigonometric polynomial can be approximated by a polynomial on any finite interval.

(*Hint*: Use the Taylor expansions for the functions $\cos nt$ and $\sin nt$.)

3.21. Prove the **Weierstrass** theorem: For any function $f(t)$ continuous on the interval $[0, 1]$ and any $\varepsilon > 0$ a polynomial $P(t)$ exists such that

$$|f(t) - P(t)| < \varepsilon.$$

for every $t \in [0, 1]$.

(*Hint*: Extend the function $f(t)$ continuously onto the entire interval $[-\pi, \pi]$ so that $f(-\pi) = f(\pi)$; then use the results of Problems 3.19 and 3.20.)

APPENDIX –
THE SPACE OF INTEGRABLE FUNCTIONS

In the preceding parts of this chapter we have discussed only the Hilbert space. However, for a deeper investigation of Fourier series and also for a number of other problems it is convenient to be familiar with properties of a certain, very important space which is not a Hilbert space, namely of the space $L_1(a, b)$, i.e. *the space of functions which are integrable on the interval* (a, b).

We say that a function f belongs to $L_1(a, b)$, if it is absolutely integrable on the interval (a, b), i.e., if

$$\int_a^b |f(x)|\, dx < \infty\,.$$

The space $L_1(a, b)$ is clearly linear. In $L_1(a, b)$, similarly as in the case of the space $L_2(a, b)$, we shall identify functions differing on a set of measure zero. Denote

$$\|f\|_{L_1(a,b)} = \int_a^b |f(x)|\, dx\,.$$

This number is called the *norm* of the function f in the space $L_1(a, b)$. The reader can easily verify that it actually has the properties 1, 2 and 3 of a norm given in Theorem 3.2; on the other hand, it does not have properties 4, 5 and 6, because these ones are consequences of the fact that in a Hilbert space the norm is derived from a scalar product. The following example shows that the number $\|f\|_{L_1(a,b)}$ cannot be defined by means of any scalar product.

Example 3.20. Consider the following two functions in $L_1(-1, 1)$:

$$f_1(t) = \begin{cases} 0 & \text{for } 0 < t < 1; \\ 1 & \text{for } -1 < t \leq 0; \end{cases} \quad f_2(t) = \begin{cases} 1 & \text{for } 0 < t < 1, \\ 0 & \text{for } -1 < t \leq 0. \end{cases}$$

Clearly, $\|f_1\|_{L_1(-1,1)} = \|f_2\|_{L_1(-1,1)} = 1$ and $\|\tfrac{1}{2}(f_1 + f_2)\|_{L_1(-1,1)} = 1$. This, however cannot occur in a Hilbert space as Problem 3.7 shows.

Thus, the space $L_1(a, b)$ cannot be a Hilbert space. In spite of this $L_1(a, b)$ has many properties which are analogous to properties of a Hilbert space; namely, it possesses all properties implied by the conditions 1, 2 and 3 stated in Theorem 3.2, which, however, do not use the special properties of a scalar product. For example, the concept of *convergence* may be introduced in $L_1(a, b)$ by the following definition: $f_n \to f$ in $L_1(a, b)$ as $n \to \infty$ means that $\|f_n - f\|_{L_1(a,b)} \to 0$ as $n \to \infty$. Moreover, we can define a *metric* in the space $L_1(a, b)$ by $\varrho(f, g) = \|f - g\|_{L_1(a,b)}$, and the concept of a Cauchy sequence as follows: a sequence $\{f_n\}$ is called the *Cauchy sequence* in $L_1(a, b)$ if for every $\varepsilon > 0$ an n_0 exists such that $\|f_n - f_m\|_{L_1(a,b)} < \varepsilon$ for any $n > n_0$ and $m > n_0$. In an analogous way as in Theorems 3.5 and 3.6 we can prove (the reader is recommended to do it as a useful exercise) that $L_1(a, b)$ is a *complete space*, i.e. any Cauchy sequence of functions in $L_1(a, b)$ has a limit in $L_1(a, b)$. Also, carrying out in $L_1(a, b)$ the same construction as in the proof of Theorem 3.14 we can easily prove that a function $f \in L_1(a, b)$ can be approximated in $L_1(a, b)$ with any accuracy by a continuous function, i.e. for every $\varepsilon > 0$ a continuous function g exists such that $\|f - g\|_{L_1(a,b)} < \varepsilon$. From this construction the following facts are apparent. The function g can be chosen so that it is nonzero only on an interval of finite length. Moreover, if f is periodic the function g can be chosen so that it is periodic with the same period, because we can carry out a similar consideration as in the proof of Theorem 3.15. Observe also that if $f \in L_1(a, b)$ and simultaneously $f \in L_2(a, b)$, then the function f can be approximated with any accuracy in both spaces; this fact becomes particularly significant if (a, b) is an interval of infinite length.

Certain additional assertions concerning the approximation of functions in $L_1(a, b)$ and $L_2(a, b)$ will be needed in Chapters 7 and 8; these assertions will be given in the problems below.

Concluding this section, let us prove the following important theorem.

Theorem 3.18. *Let $f \in L_1(a, b)$, and for any real number λ let*

$$I(\lambda; f; a, b) = \int_a^b f(x)\, e^{i\lambda x}\, dx. \qquad (3.28)$$

Then $I(\lambda; f; a, b) \to 0$ as $|\lambda| \to \infty$.

Proof: Let $\varepsilon > 0$. In view of Theorem 1.9 there exist finite numbers a_1 and b_1 such that $a \leq a_1 \leq b_1 \leq b$ and

$$\int_a^{a_1} |f(x)|\, dx + \int_{b_1}^b |f(x)|\, dx < \frac{\varepsilon}{3}.$$

The interval $[a_1, b_1]$ is now finite; similarly as in the proof of Theorem 3.15 we can prove that a trigonometric polynomial $T_m(x)$ exists such that

$$\int_{a_1}^{b_1} |f(x) - T_m(x)|\, dx < \frac{\varepsilon}{3}.$$

Thus, it is clear that

$$|I(\lambda; f; a, b) - I(\lambda; T_m; a_1, b_1)| < \frac{2\varepsilon}{3}.$$

If the inequality $|I(\lambda; T_m; a_1, b_1)| < \varepsilon/3$ is true for $|\lambda| > \lambda_0 > 0$, we also have $|I(\lambda; f; a, b)| < \varepsilon$ for these λ, because

$$|I(\lambda; f; a, b)| \leq |I(\lambda; f; a, b) - I(\lambda; T_m; a_1, b_1)| \\ + |I(\lambda; T_m; a_1, b_1)|.$$

Thus, it suffices to prove the assertion of the theorem for the trigonometric polynomial

$$T_m(x) = \sum_{k=-m}^{m} c_k e^{ikcx} \quad \text{with} \quad c = \frac{2\pi}{(b_1 - a_1)}$$

and for the finite interval $[a_1, b_1]$. Performing the integration we obtain

$$I(\lambda; T_m; a_1, b_1)$$
$$= \sum_{k=-m}^{m} c_k \frac{1}{\lambda + ck} \left[\sin(\lambda + ck)x - i\cos(\lambda + ck)x \right]_{a_1}^{b_1}.$$

Hence

$$|I(\lambda; T_m; a_1, b_1)| \leq \sum_{k=-m}^{m} 4|c_k| (|\lambda| - c|k|)^{-1} ;$$

this shows that $\lim_{|\lambda| \to \infty} I(\lambda; T_m; a_1, b_1) = 0$. The theorem is proved.

The theorem just proved has an important consequence regarding the Fourier coefficients of an integrable function. If $f \in L_1(a, b)$, then the complex form of the Fourier series of f has coefficients

$$c_n = \frac{1}{2\pi} I(-n; f; -\pi, \pi), \quad n = 0, \pm 1, \pm 2, \ldots$$

(see (2.19)), and

$$a_n = c_n + c_{-n}; \quad b_n = i(c_n - c_{-n}) \quad (n = 1, 2, \ldots)$$

holds for the coefficients of the real form of the Fourier series. Thus, for $f \in L_1(a, b)$,

$$\lim_{|n| \to \infty} c_n = 0; \quad \lim_{n \to \infty} a_n = \lim_{n \to \infty} b_n = 0. \tag{3.29}$$

Problems

3.22. Carry out in detail the proofs of all assertions given in the Appendix.

3.23. Let either $f \in L_1(a, b)$ or $f \in L_2(a, b)$, and let the function f be zero on an interval (c, d) which is a part of (a, b). Furthermore, let $c < c' < d' < d$. Show that the function f can be approximated in the respective spaces with any accuracy by a continuous function g vanishing on the interval (c', d'). (*Hint*: For an $h > 0$, redefine the function f outside (a, b) so that the extended function will belong either to $L_1(a-h, b+h)$ or to $L_2(a-h, b+h)$, and put $g = f_h$, where f_h has the meaning given in the proof of Theorem 3.14.)

3.24. Let f be again the function from Problem 3.23, and put $a = -\pi$, $b = \pi$ here. If $\eta > 0$, show that the function f can be approximated in the respective spaces with any accuracy by a trigonometric polynomial T_n such that $|T_n(x)| < \eta$ for any x in the interval (c', d').
(*Hint*: Extend the function f periodically with period 2π onto the interval $[-2\pi, 2\pi]$, and then proceed as in the proof of Theorem 3.15. The properties follow from the construction.)

3.25. Show that a function f belonging either to $L_1(-\infty, \infty)$ or to $L_2(-\infty, \infty)$ can be approximated in the respective spaces with any accuracy by an infinitely differentiable function h which is non-zero only on an interval of finite length.
(*Hint*: Recalling Problem 3.21 prove first that the Weierstrass theorem holds for any interval $[a, b]$ of finite length; this fact may be proved by using a linear substitution, for example. Then, analogously as in the proof of Theorem 3.14, approximate the function f by a continuous function g vanishing outside the interval $[-K, K]$, $K > 0$. Now, it suffices to approximate the function g. Define the function φ by

$$\varphi(t) = \begin{cases} 1 & \text{for } |t| \leq K; \\ \exp\left\{\dfrac{(t^2 - K^2)}{|t|(|t| - 2K)}\right\} & \text{for } K < |t| < 2K; \\ 0 & \text{for } |t| \geq 2K; \end{cases}$$

clearly, φ is infinitely differentiable. By the Weierstrass theorem, the function g can be uniformly approximated by a polynomial p on $[-2K, 2K]$; hence, p approximates g

also in the space $L_1(-2K, 2K)$ or $L_2(-2K, 2K)$. Then the function $h = \varphi p$ will approximate $g = \varphi g$ and consequently, also the function f in the space $L_1(-\infty, \infty)$ or $L_2(-\infty, \infty)$.)

3.26. Show that if $f \in L_1(a, b)$ and $f \in L_2(a, b)$, then the function f can be approximated in both spaces simultaneously.
(*Hint*: The construction of the approximation does not depend on the choice of the space concerned.)

3.27. Prove that if $\int_a^b |f(x)|\, dx < \varepsilon$, then the measure of the set M consisting of those $x \in [a, b]$ with $|f(x)| \geq \sqrt{\varepsilon}$ does not exceed $\sqrt{\varepsilon}$.
(*Hint*: Prove the proposition by a contradiction.)

3.28. Let (a, b) be a finite interval. Prove that any function f either in $L_2(a, b)$ or in $L_1(a, b)$ can be approximated in the respective space with any accuracy by an infinitely differentiable function, which is non-zero only on a closed interval lying entirely in the open interval (a, b).
(*Hint*: The fact that an analogue of Problem 3.25 for a finite interval is under consideration can be used. However, first show that, for $\delta > 0$ sufficiently small, the function f can be approximated by a function f_δ defined by $f_\delta(x) = f(x)$ for $x \in (a + \delta, b - \delta)$, $f_\delta(x) = 0$ elsewhere.)

CHAPTER 4 SOME SPECIAL FOURIER SERIES IN SPECIFIC HILBERT SPACES

In the preceding chapter, the general theory of Fourier series in a Hilbert space H has been built up. The methods used there enabled us to prove some general statements concerning, first of all, the convergence of certain series formed from elements of an orthogonal system, and conditions under which the sum of the series of an element (in the sense of convergence in the norm of the space H) is actually equal to this element.

In this chapter we shall consider the general Fourier (orthogonal) series in some specific Hilbert spaces, which are formed with the aid of a specified orthonormal or orthogonal system. Various properties of these series, such as the convergence, the best approximation etc., follow from the general theory presented in Chapter 3, and thus will not be stressed in individual cases.

However, every particular case has, in addition to general features enabling us to consider it as a part of a general principle, a number of entirely specific properties. We shall encounter this situation in the present chapter; consequently, we shall sometimes use methods and results which are not related in any way to the general theory. This situation becomes particularly apparent if we investigate the *completeness* of various orthogonal systems in a chosen Hilbert space, because the completeness is a property, the proof of which uses essentially various specific properties of both the Hilbert space and the chosen orthogonal system. Let us recall, for example, that the proof of completeness of the system of trigonometric functions in the case of the Hilbert space $L_2(-\pi, \pi)$ was virtually based on the properties of the space L_2 and on some special properties of trigonometric functions.

The above mentioned special properties of various functions will be mostly used without proof, because often problems exceeding the scope of this book will be under consideration. The reader interested more deeply in these special problems will find corresponding references in each section.

4.1 SOME PROPERTIES OF THE SPACE L_2

Consider the Hilbert space $L_2(a, b)$, introduced in Chapter 3, with the scalar product

$$(f, g) = \int_a^b f(x) \overline{g(x)} \, dx \qquad (4.1)$$

and the norm

$$\|f\|_{L_2(a,b)} = \left[\int_a^b |f(x)|^2 \, dx \right]^{1/2}. \qquad (4.2)$$

Thus, convergence in the norm in this space means *convergence in the mean*. For the time being, we shall assume that the interval (a, b) is finite.

Let

$$e_1(x), e_2(x), \ldots, e_n(x), \ldots \qquad (4.3)$$

be an orthonormal system in $L_2(a, b)$. Theorems 3.11 and 3.12 give certain conditions which are necessary and sufficient for the system (4.3) to be complete in $L_2(a, b)$. We shall state now another two conditions without proof.

The first condition is due to VITALI (cf. e.g. [14]), and reads as follows.

Theorem 4.1. *The orthonormal system* (4.3) *is complete in* $L_2(a, b)$ *if and only if the equality*

$$\sum_{i=1}^{\infty} \left| \int_a^s e_i(x) \, dx \right|^2 = s - a \qquad (4.4)$$

is satisfied for all $s \in [a, b]$.

Using this condition we can verify the completeness of an orthonormal system more easily than, for example, by using Parseval's equality. Attempting to prove completeness with the aid of Parseval's equality, we have to prove that this equality holds for *every* function in $L_2(a, b)$ (cf. Theorem 3.11, condition 1); on the other hand, according to Theorem 4.1 it suffices to prove only that Parseval's equality holds for some special functions, namely for the functions $h_s(x)$ defined as follows.

$$h_s(x) = \begin{cases} 1 & \text{for } a \leq x \leq s, \\ 0 & \text{for } s < x \leq b. \end{cases}$$

The reader can easily verify that the equality (4.4) is nothing else than Parseval's equality for the functions $h_s(x)$.

Vitali's condition can be further simplified. **The following criterion shows that for verifying the completeness of a system of functions having the form (4.3) it suffices only to prove that a certain series with constant terms has a given sum. This criterion** was proved by D.P. DALZELL (cf. [14]).

Theorem 4.2. *The orthonormal system* (4.3) *is complete in* $L_2(a, b)$, *if and only if*

$$\sum_{i=1}^{\infty} \int_a^b \left| \int_a^s e_i(x) \, dx \right|^2 ds = \frac{(b-a)^2}{2}. \tag{4.5}$$

However, the practical applicability of both last mentioned criteria is, similarly as for Parseval's equality and other criteria, rather questionable. Anyway, using the mentioned relations, we can prove various interesting equalities, as shown in the following example.

Example 4.1. In Chapter 3, the following system was shown to be complete and orthonormal in $L_2(-\pi, \pi)$

$$\frac{1}{\sqrt{(2\pi)}}, \frac{\cos x}{\sqrt{\pi}}, \frac{\sin x}{\sqrt{\pi}}, \frac{\cos 2x}{\sqrt{\pi}}, \frac{\sin 2x}{\sqrt{\pi}}, \ldots . \tag{4.6}$$

Thus, equality (4.5) holds for this system and has then the form

$$\frac{1}{2\pi}\int_{-\pi}^{\pi}(s+\pi)^2\,ds$$
$$+\frac{1}{\pi}\sum_{k=1}^{\infty}\int_{-\pi}^{\pi}\left[\left(\frac{\sin ks}{k}\right)^2+\left(\frac{(-1)^k-\cos ks}{k}\right)^2\right]ds=\frac{(2\pi)^2}{2}.$$

Carrying out some elementary rearrangements we obtain the relation

$$\sum_{k=1}^{\infty}\frac{1}{k^2}=\frac{\pi^2}{6} \tag{4.7}$$

which is very useful and will be dealt with many times in what follows.

Conversely, assuming the validity of the relation (4.7) (which can be proved by other methods not related to Fourier series), we could deduce, by using Theorem 4.2, the completeness of the system (4.6).

In various considerations we shall make use of the following simple assertion:

Theorem 4.3. *Let the linear substitution*

$$t=Ax+B,\quad A>0,$$

transform the interval $[a,b]$ *onto a finite interval* $[c,d]$. *Let the functions* $e_i(x)$ *of the system* (4.3) *constitute a complete orthonormal system in the Hilbert space* $L_2(a,b)$ *and let the functions* $f_i(t)$ *be defined as follows*

$$f_i(t)=e_i\left(\frac{t}{A}-\frac{B}{A}\right).$$

Then the functions $f_i(t)$ *constitute a complete orthogonal system in the space* $L_2(c,d)$.

Proof: Using the well-known substitution rules we can immediately prove the equalities

$$\int_c^d f_i(t)\,\overline{f_j(t)}\,dt = A\int_a^b e_i(x)\,\overline{e_j(x)}\,dx \quad (i,j = 1, 2, \ldots).$$

It follows that the functions $f_i(t)$ belong to the space $L_2(c, d)$ and constitute an orthogonal system. The completeness of this system may be proved by using condition 4 of Theorem 3.17, because the space $L_2(c, d)$ is complete as shown in Chapter 3.

Let $h(t)$ be a function from $L_2(c, d)$, orthogonal to all the functions $f_i(t)$:

$$\int_c^d f_i(t)\,\overline{h(t)}\,dt = 0 \quad \text{for} \quad i = 1, 2, \ldots.$$

Putting $t = Ax + B$, we obtain

$$A\int_a^b e_i(x)\,\overline{h(Ax + B)}\,dx = 0 \quad \text{for} \quad i = 1, 2, \ldots,$$

which expresses the fact that the function $H(x) = h(Ax + B)$, belonging to $L_2(a, b)$, is orthogonal to all the functions $e_i(x)$. Since the system of all these functions is complete, we have $H(x) = 0$ for $x \in [a, b]$; thus, $h(t) = 0$ for $t \in [c, d]$, which, according to Theorem 3.17, suffices for the completeness of the system $\{f_i(t)\}$.

Remark: Moreover, from the proof it follows that $\|f_i\|_{L_2(c,d)} = \sqrt{A}$ and that the system

$$\frac{f_1(t)}{\sqrt{A}}, \frac{f_2(t)}{\sqrt{A}}, \ldots, \frac{f_n(t)}{\sqrt{A}}, \ldots$$

is a complete orthonormal system of functions in $L_2(c, d)$.

4.2 TRIGONOMETRIC FOURIER SERIES OF FUNCTIONS OF ONE VARIABLE

Let us recall that if $f \in H$ and the elements f_i constitute a complete orthogonal system in H, then f can be expressed in the form of a Fourier series

$$f = \sum_{i=1}^{\infty} c_i f_i$$

(i.e. the partial sums of the series converge to the element f in the norm of the space H), where

$$c_i = \frac{1}{\|f_i\|^2} (f, f_i).$$

Furthermore, Parseval's equality holds,

$$\|f\|^2 = \sum_{i=1}^{\infty} |c_i|^2 \|f_i\|^2.$$

4.2.1 Consider the Hilbert space $L_2(-\pi, \pi)$. As known, the system

$$1, \cos x, \sin x, \cos 2x, \sin 2x, \ldots \qquad (4.8)$$

is complete and orthogonal in $L_2(-\pi, \pi)$. Let us write the system (4.8) as follows

$$f_0, f_1, f_2, \ldots, f_n, f_{n+1}, \ldots,$$

where $f_0(x) = 1$, $f_{2n-1}(x) = \cos nx$, $f_{2n}(x) = \sin nx$ $(n = 1, 2, \ldots)$. Then we have for $f \in L_2(-\pi, \pi)$,

$$\frac{1}{\|f_0\|^2} (f, f_0) = \frac{1}{2\pi} \int_{-\pi}^{\pi} f(x) \, dx = \frac{a_0}{2};$$

$$\frac{1}{\|f_{2n-1}\|^2} (f, f_{2n-1}) = \frac{1}{\pi} \int_{-\pi}^{\pi} f(x) \cos nx \, dx = a_n;$$

$$\frac{1}{\|f_{2n}\|^2} (f, f_{2n}) = \frac{1}{\pi} \int_{-\pi}^{\pi} f(x) \sin nx \, dx = b_n \quad (n = 1, 2, \ldots)$$

and the Fourier series of the function $f(x)$ has the familiar form

$$f(x) = \frac{a_0}{2} + \sum_{n=1}^{\infty} (a_n \cos nx + b_n \sin nx). \qquad (4.9)$$

Parseval's equality then reads as follows

$$\|f\|_{L_2(-\pi,\pi)}^2 = \pi \left[\frac{|a_0|^2}{2} + \sum_{n=1}^{\infty} (|a_n|^2 + |b_n|^2) \right]. \qquad (4.10)$$

If a function $f(x)$ is defined on the entire real line and is periodic with period 2π, then the functions $f(x) \cos nx$, $f(x) \sin nx$ and $|f(x)|^2$ are also periodic with period 2π. However, because an integral of a periodic function over an interval of length equal to the period is independent of the location of the interval, we have

$$\int_a^{a+2\pi} f(x) \begin{Bmatrix} \sin nx \\ \cos nx \end{Bmatrix} dx = \int_{-\pi}^{\pi} f(x) \begin{Bmatrix} \sin nx \\ \cos nx \end{Bmatrix} dx \qquad (4.11)$$

and

$$\int_a^{a+2\pi} |f(x)|^2 \, dx = \int_{-\pi}^{\pi} |f(x)|^2 \, dx, \qquad (4.12)$$

whenever a is a real number (cf. (1.17)). From the last formula it follows that if the function $f(x)$ belongs to $L_2(a, a + 2\pi)$ then it also belongs to $L_2(-\pi, \pi)$, and vice versa.

Consider now the Hilbert space $L_2(a, a + 2\pi)$, and in this space the system (4.8). Extend a function $f(x) \in L_2(a, a + 2\pi)$ periodically with period 2π onto the entire real line (cf. Chapter 1, Section 1.1) and denote this extended function again by $f(x)$. From (4.11) it follows that the Fourier coefficients of the function f with respect to the system (4.8) in $L_2(a, a + 2\pi)$ coincide with the Fourier coefficients of f (after the periodic extension) with respect to the system (4.8) in $L_2(-\pi, \pi)$. Since the norms of f in both $L_2(a, a + 2\pi)$ and $L_2(-\pi, \pi)$ are equal by (4.12), Parseval's equality holds in $L_2(a, a + 2\pi)$, too, for it is formally the same as Parseval's equality for the system (4.8) in $L_2(-\pi, \pi)$. Thus we have proved that *the system* (4.8) *is complete and orthogonal in* $L_2(a, a + 2\pi)$ *for any real number* a, *and that the Fourier series and Parseval's equality for a function* $f(x) \in L_2(a, a + 2\pi)$ *with respect to the system* (4.8) *have the form* (4.9) *and* (4.10), respectively, i.e. the same form as in the case of the space $L_2(-\pi, \pi)$.

In what follows, this result will primarily be used for the interval $(0, 2\pi)$. From this it follows that for a periodic function with period 2π it does not matter, whether its Fourier series with respect to the system (4.8) is considered in $L_2(-\pi, \pi)$ or in $L_2(0, 2\pi)$ or, in general, in the space $L_2(a, a + 2\pi)$ — always the same series is under consideration. We should emphasize again that this is true only for *periodic* or *periodically extended* functions.

4.2.2 Consider now an *even* function $f(x) \in L_2(-\pi, \pi)$. Then the function $f(x) \cos nx$ is also even and thus (cf. Chapter 1, (1.15) and (1.16))

$$a_n = \|f_{2n-1}\|^{-2} (f, f_{2n-1}) = \frac{1}{\pi} \int_{-\pi}^{\pi} f(x) \cos nx \, dx$$
$$= \frac{2}{\pi} \int_0^{\pi} f(x) \cos nx \, dx \qquad (4.13)$$

for $n = 0, 1, 2, \ldots$ (the function f_0 is denoted here by f_{-1}). The function $f(x) \sin nx$, on the other hand, is odd and therefore $b_n = \|f_{2n}\|^{-2} (f, f_{2n}) = 0$ for $n = 1, 2, \ldots$. It follows that the Fourier series (4.9) of an even function contains only the constant term and the cosine terms; Parseval's equality has then the following form

$$\|f\|^2_{L_2(-\pi,\pi)} = \pi \left(\frac{|a_0|^2}{2} + \sum_{n=1}^{\infty} |a_n|^2 \right). \qquad (4.14)$$

If $f(x) \in L_2(-\pi, \pi)$ is *odd*, we can analogously prove that $a_0 = a_1 = a_2 = \ldots = 0$ and

$$b_n = \|f_{2n}\|^{-2} (f, f_{2n}) = \frac{1}{\pi} \int_{-\pi}^{\pi} f(x) \sin nx \, dx$$
$$= \frac{2}{\pi} \int_0^{\pi} f(x) \sin nx \, dx \qquad (4.15)$$

for $n = 1, 2, \ldots$. The Fourier series of an odd function contains the sine terms only and Parseval's equality has the following form

$$\|f\|^2_{L_2(-\pi,\pi)} = \pi \sum_{n=1}^{\infty} |b_n|^2 . \tag{4.16}$$

4.2.3 Consider now an interval whose length is a half of the period, say, the interval $(0, \pi)$. By the Hilbert space H the space $L_2(0, \pi)$ is meant throughout this section. Consider the following system of functions in this space,

$$f_0(x) = 1 ; \quad f_1(x) = \cos x ;$$
$$f_2(x) = \cos 2x; \ldots ; \quad f_n(x) = \cos nx; \ldots . \tag{4.17}$$

The reader should prove as an exercise the orthogonality of this system (in the sense of the scalar product $(f, g) = \int_0^\pi f(x) \overline{g(x)} \, dx$) and the following equalities

$$\|f_0\|^2_{L_2(0,\pi)} = \pi , \quad \|f_n\|^2_{L_2(0,\pi)} = \frac{\pi}{2} \quad (n = 1, 2, \ldots) .$$

Let $F(x)$ be a function in $L_2(0, \pi)$. The Fourier coefficients of this function with respect to the system (4.17) have the form

$$\frac{A_0}{2} = \|f_0\|^{-2} (F, f_0) = \frac{1}{\pi} \int_0^\pi F(x) \, dx ;$$

$$A_n = \|f_n\|^{-2} (F, f_n) = \frac{2}{\pi} \int_0^\pi F(x) \cos nx \, dx \tag{4.18}$$

$(n = 1, 2, \ldots)$ and the Fourier series is

$$F(x) = \frac{A_0}{2} + \sum_{n=1}^{\infty} A_n \cos nx .$$

Define now the function $f(x)$ by

$$f(x) = \begin{cases} F(x) & \text{for } x \in [0, \pi] , \\ F(-x) & \text{for } x \in [-\pi, 0] . \end{cases}$$

This function is even and belongs to the space $L_2(-\pi, \pi)$, because

$$\|f\|^2_{L_2(-\pi,\pi)} = \int_{-\pi}^{\pi} |f(x)|^2 \, dx = \int_{-\pi}^{0} |F(-x)|^2 \, dx + \int_{0}^{\pi} |F(x)|^2 \, dx$$

$$= 2 \int_{0}^{\pi} |F(x)|^2 \, dx = 2\|F\|^2_{L_2(0,\pi)}. \qquad (4.19)$$

For an even function in $L_2(-\pi, \pi)$ the formulas (4.13) are true; comparing them with the expressions (4.18) (for $f(x) = F(x)$ in $(0, \pi)$) we have

$$a_n = A_n \quad (n = 0, 1, 2, \ldots).$$

Substituting these equalities and the equality (4.9) into Parseval's equality (4.14) for the even function $f(x)$, we obtain

$$2\|F\|^2_{L_2(0,\pi)} = \pi \left(\frac{|A_0|^2}{2} + \sum_{n=1}^{\infty} |A_n|^2 \right)$$

i.e.

$$\|F\|^2_{L_2(0,\pi)} = \frac{\pi}{2} \left(\frac{|A_0|^2}{2} + \sum_{n=1}^{\infty} |A_n|^2 \right),$$

however, this is nothing else than Parseval's equality for the function $F(x) \in L_2(0, \pi)$ with respect to the system (4.17). Since this equality holds, as just proved, for any function in $L_2(0, \pi)$, condition 1 in Theorem 3.17 yields immediately the following theorem.

Theorem 4.4 *The system* (4.17) *is complete and orthogonal in* $L_2(0, \pi)$.

The proof of this theorem may also be carried out by using condition 2 in Theorem 3.17. Actually the Fourier series of a function $F(x) \in L_2(0, \pi)$ with respect to the system (4.17) coincides, due to the equalities $a_n = A_n$, with the Fourier series of the even function $f(x) \in L_2(-\pi, \pi)$ with respect to the system (4.8). The latter system is complete in $L_2(-\pi, \pi)$, and consequently the

Fourier series converges to f. Hence, the Fourier series of the function F converges to F, i.e. the system (4.17) is complete.

Analogously, we can prove the next proposition.

Theorem 4.5. *The system*

$$\sin x, \sin 2x, \sin 3x, \ldots, \sin nx, \ldots \qquad (4.20)$$

is complete and orthogonal in the space $L_2(0, \pi)$.

In the proof, properties of odd functions in $L_2(-\pi, \pi)$ are used. For, if $F(x) \in L_2(0, \pi)$, then its Fourier series with respect to the system (4.20) takes the form

$$F(x) = \sum_{n=1}^{\infty} B_n \sin nx \; ; \quad B_n = \frac{2}{\pi} \int_0^{\pi} F(x) \sin nx \, dx \; .$$

The function $f(x) \in L_2(-\pi, \pi)$, defined by

$$f(x) = \begin{cases} F(x) & \text{for } x \in (0, \pi), \\ -F(-x) & \text{for } x \in (-\pi, 0), \end{cases}$$

(the values at the points 0, π and $-\pi$ do not matter, for these points constitute a null set) is odd and formula (4.19) holds; according to (4.15) we have $b_n = B_n$. Substituting this into (4.16) we obtain Parseval's equality for the function $F(x)$ with respect to the system (4.20) which proves, due to condition 1 in Theorem 3.17, the completeness of the system (4.20). The orthogonality of the system (4.20) is a matter of routine.

In the space $L_2(0, \pi)$, two complete orthogonal systems are available for constructing the Fourier series, (4.17) and (4.20). If orthonormal systems are needed, we can use the systems

$$\frac{1}{\sqrt{\pi}}, \sqrt{\frac{2}{\pi}} \cos x, \sqrt{\frac{2}{\pi}} \cos 2x, \ldots, \sqrt{\frac{2}{\pi}} \cos nx; \ldots$$

and

$$\sqrt{\frac{2}{\pi}} \sin x, \sqrt{\frac{2}{\pi}} \sin 2x, \ldots, \sqrt{\frac{2}{\pi}} \sin nx; \ldots$$

Further complete orthogonal systems in spaces L_2 on intervals of length π can be obtained by using, for example, Theorem 4.3. Thus it is obvious that the system

$$1, \cos 2x, \sin 2x, \cos 4x, \sin 4x, \ldots, \cos 2nx, \sin 2nx, \ldots,$$

is complete and orthogonal in the space $L_2(-\pi/2, \pi/2)$, for it can be obtained from the system (4.8) by using Theorem 4.3, and the substitution $x = 2t$.

4.2.4 In the space $L_2(0, \pi/2)$ the following system is complete and orthogonal,

$$\sin x, \sin 3x, \sin 5x, \ldots, \sin(2n + 1)x, \ldots. \qquad (4.21)$$

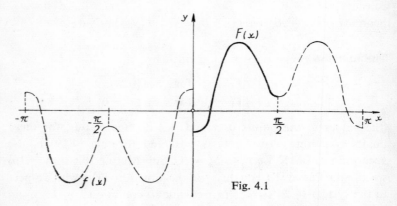

Fig. 4.1

The proof of this assertion is left to the reader. The orthogonality of the system (4.21) can be easily proved by calculation, the completeness follows, analogously as in the preceding section, from Parseval's equality for the function $f(x) \in L_2(-\pi, \pi)$ (with respect to the system (4.8)) obtained from $F(x) \in L_2(0, \pi/2)$ by extending it onto the interval $(-\pi, \pi)$ as follows,

$$f(x) = \begin{cases} F(\pi - x) & \text{for } x \in (\pi/2, \pi) \\ F(x) & \text{for } x \in (0, \pi/2) \\ -F(-x) & \text{for } x \in (-\pi/2, 0) \\ -F(x - \pi) & \text{for } x \in (-\pi, \pi/2) \end{cases}$$

(See Fig. 4.1.)

4.2.5 For the sake of completeness let us note that the system of complex functions

$$\frac{1}{\sqrt{(2\pi)}} e^{inx}, \quad n = 0, \pm 1, \pm 2, \ldots \tag{4.22}$$

is also complete and orthonormal in $L_2(a, a + 2\pi)$.

4.3 TRIGONOMETRIC FOURIER SERIES WITH A GENERAL PERIOD

Consider the space $L_2(-l, l)$, $l > 0$. We have then the following theorem.

Theorem 4.6. *The system*

$$1, \cos\frac{\pi x}{l}, \sin\frac{\pi x}{l}, \cos\frac{2\pi x}{l}, \sin\frac{2\pi x}{l}, \ldots, \cos\frac{n\pi x}{l}, \sin\frac{n\pi x}{l}, \ldots \tag{4.23}$$

is complete and orthogonal in $L_2(-l, l)$.

Proof: The substitution $t = \pi x/l$ transforms the interval $(-\pi, \pi)$ (of variable t) onto the interval $(-l, l)$ and the system $1, \cos t, \sin t, \cos 2t, \sin 2t, \ldots$ onto the system (4.23). The former system is complete and orthogonal in $L_2(-\pi, \pi)$; hence, as in Theorem 4.3, the same is true for the latter one.

Remark. The system (4.23) is not orthonormal. However, it can be easily normalized, if the first function is multiplied by $1/\sqrt{(2l)}$ and the others by $1/\sqrt{l}$.

Thus, a function $f(x) \in L_2(-l, l)$ can be expanded in the Fourier series

$$f(x) = \frac{a_0}{2} + \sum_{n=1}^{\infty} \left(a_n \cos\frac{n\pi x}{l} + b_n \sin\frac{n\pi x}{l} \right), \tag{4.24}$$

where

$$a_n = \frac{1}{l} \int_{-l}^{l} f(x) \cos \frac{n\pi x}{l} \, dx, \quad n = 0, 1, 2, \ldots;$$

$$b_n = \frac{1}{l} \int_{-l}^{l} f(x) \sin \frac{n\pi x}{l} \, dx, \quad n = 1, 2, \ldots.$$

The system (4.23) is complete and orthogonal also in the space $L_2(a, a + 2l)$, a being an arbitrary real number; moreover the Fourier series of a function $f(x)$, defined on $(-\infty, \infty)$ and periodic with period $2l$, coincides for any a with the series corresponding to the space $L_2(-l, l)$. This can be proved analogously as in Section 4.2.1, where the case $l = \pi$ was considered.

We stress again that the previous remark is true only for such functions, which are periodic with period $2l$ or are defined on an interval $(a, a + 2l)$ of length $2l$ and extended periodically (with period $2l$) onto $(-\infty, \infty)$. Expansions of a non-periodic function $f(x)$ in the space $L_2(a, a + 2l)$ may differ essentially for different values of a. This fact will be more apparent from the following examples.

Example 4.2. Consider the function $f(x) = x^2$ in the interval $(-l, l)$. Its Fourier series with respect to the system (4.23) has the form (4.24), where

$$a_0 = \frac{1}{l} \int_{-l}^{l} x^2 \, dx = \frac{2l^2}{3},$$

$$a_n = \frac{1}{l} \int_{-l}^{l} x^2 \cos \frac{n\pi x}{l} \, dx = \frac{4l^2}{n^2\pi^2}(-1)^n \quad (n = 1, 2, \ldots);$$

$$b_n = \frac{1}{l} \int_{-l}^{l} x^2 \sin \frac{n\pi x}{l} \, dx = 0 \quad (n = 1, 2, \ldots);$$

hence, we have

$$x^2 = \frac{l^2}{3} - \frac{4l^2}{\pi^2} \left[\cos \frac{\pi x}{l} - \frac{\cos(2\pi x/l)}{2^2} + \frac{\cos(3\pi x/l)}{3^2} - \ldots \right], \quad (4.25)$$

where the series on the right hand side converges to x^2 in the norm of $L_2(-l, l)$ and even uniformly in $[-l, l]$.

Example 4.3. Consider again the function $f(x) = x^2$ but now in the interval $(0, 2l)$. Its Fourier series with respect to the system (4.23) in the interval $(0, 2l)$ has again the form (4.24), but with the coefficients given by

$$a_0 = \frac{1}{l}\int_0^{2l} x^2 \, dx = \frac{8l^2}{3}; \quad a_n = \frac{1}{l}\int_0^{2l} x^2 \cos\frac{n\pi x}{l} \, dx = \frac{4l^2}{n^2\pi^2};$$

$$b_n = \frac{1}{l}\int_0^{2l} x^2 \sin\frac{n\pi x}{l} \, dx = -\frac{4l^2}{n\pi};$$

thus, we have

$$x^2 = \frac{4l^2}{3} + 4l^2 \left[\frac{\cos(\pi x/l)}{\pi^2} - \frac{\sin(\pi x/l)}{\pi} + \frac{\cos(2\pi x/l)}{2^2\pi^2} \right.$$
$$\left. - \frac{\sin(2\pi x/l)}{2\pi} + \frac{\cos(3\pi x/l)}{3^2\pi^2} - \frac{\sin(3\pi x/l)}{3\pi} + \ldots \right]. \quad (4.26)$$

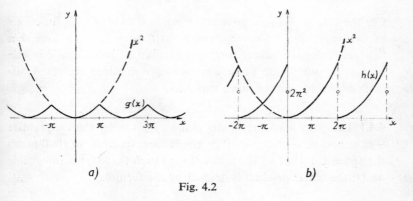

Fig. 4.2

The series (4.25) and (4.26) are different, although they correspond to "the same" function x^2; this difference has its origin in the fact that a *non-periodic* function has been expanded in different intervals.

Example 4.4. If the function $f(x) = x^2$ in $(-l, l)$ is periodically extended onto the entire real axis, the function illustrated in Fig. 4.2a is obtained. The Fourier series of this function, denoted by $g(x)$, in any interval $(a, a + 2l)$ has again the form (4.25), because

$$\int_a^{a+2l} g(x) \cos \frac{n\pi x}{l} \, dx = \int_{-l}^{l} x^2 \cos \frac{n\pi x}{l} \, dx \quad \text{etc.}$$

Analogously, the periodic extension of the function $f(x) = x^2$ in $(0, 2l)$ onto the entire real axis (see Fig. 4.2b) has in $(a, a + 2l)$ the Fourier series of the form (4.26).

Using Theorem 4.3 we can derive from (4.17) and (4.20) complete orthogonal systems in $L_2(0, l)$, etc. (see Problems 4.6a, b).

4.4 FOURIER SERIES OF FUNCTIONS OF SEVERAL VARIABLES

For the sake of simplicity we are going to deal only with functions of two variables. However, from what follows it is obvious that this restriction is merely formal and that all considerations can be carried out in the same way (with some minor technical difficulties) in the case of more variables.

4.4.1 Let Q be a square in the plane xy given by the inequalities $-\pi < x < \pi$, $-\pi < y < \pi$. By the Hilbert space H we shall mean the space $L_2(Q)$ of all functions $f(x, y)$ such that $|f|^2$ is integrable on Q; the scalar product is defined by the formula

$$(f, g) = \int_{-\pi}^{\pi} \int_{-\pi}^{\pi} f(x, y) \, \overline{g(x, y)} \, dx \, dy \, .$$

We have the following assertion.

SOME SPECIAL FOURIER SERIES

Theorem 4.7. *The system*

$$e_{mn}(x, y) = \frac{1}{2\pi} e^{i(mx+ny)} \quad (m, n = 0, \pm 1, \pm 2, \ldots) \quad (4.27)$$

is complete and orthonormal in $L_2(Q)$.

We shall prove the orthonormality of functions in the system (4.27). The completeness follows from the completeness of a somewhat more general system, which will be proved in Section 4.4.4. Thus, let us consider the scalar product of two functions from (4.27). We have

$$(e_{mn}, e_{jk}) = \frac{1}{4\pi^2} \int_{-\pi}^{\pi} \int_{-\pi}^{\pi} e^{i(mx+ny)} e^{-i(jx+ky)} \, dx \, dy$$

$$= \frac{1}{2\pi} \int_{-\pi}^{\pi} e^{i(m-j)x} \, dx \cdot \frac{1}{2\pi} \int_{-\pi}^{\pi} e^{i(n-k)y} \, dy \, .$$

Since the system $e^{ikx}/\sqrt{(2\pi)}$ $(k = 0, \pm 1, \pm 2, \ldots)$ is orthonormal in $L_2(-\pi, \pi)$, it follows that

$$\frac{1}{2\pi} \int_{-\pi}^{\pi} e^{i(k-j)x} \, dx = \begin{cases} 1 & \text{for} \quad k = j, \\ 0 & \text{for} \quad k \neq j. \end{cases} \quad (4.28)$$

The scalar product (e_{mn}, e_{jk}) is a product of two expressions of the form (4.28); hence we have

$$(e_{mn}, e_{jk}) = \begin{cases} 1, & \text{whenever} \quad m = j \quad \text{and} \quad n = k, \\ 0, & \text{whenever} \quad m \neq j \quad \text{or} \quad n \neq k, \end{cases}$$

which is what we wished to prove.

The Fourier series of a function $f(x, y) \in L_2(Q)$ with respect to the system (4.27) has the following form

$$f(x, y) = \sum_{m,n=-\infty}^{\infty} c_{mn} \frac{1}{2\pi} e^{i(mx+ny)}, \quad (4.29)$$

where

$$c_{mn} = (f, e_{mn}) = \frac{1}{2\pi} \int_{-\pi}^{\pi} \int_{-\pi}^{\pi} f(x, y) \, e^{-i(mx+ny)} \, dx \, dy \, .$$

Parseval's equality reads

$$\int_{-\pi}^{\pi} \int_{-\pi}^{\pi} |f(x, y)|^2 \, dx \, dy = \sum_{m,n=-\infty}^{\infty} |c_{mn}|^2 \, .$$

4.4.2 The expression (4.29) gives the Fourier series of a function $f(x, y)$ in the space $L_2(Q)$ in the so-called complex form. Using the formula $e^{ix} = \cos x + i \sin x$ we can derive the real form of the Fourier series, which, of course, is much more complicated. We get

$$f(x, y) = \sum_{m,n=0}^{\infty} \varepsilon_{mn} [\alpha_{mn} \cos mx \cos ny + \beta_{mn} \cos mx \sin ny$$
$$+ \gamma_{mn} \sin mx \cos ny + \delta_{mn} \sin mx \sin ny] \, , \qquad (4.30)$$

where

$$\alpha_{mn} = \frac{1}{\pi^2} \int_{-\pi}^{\pi} \int_{-\pi}^{\pi} f(x, y) \cos mx \cos ny \, dx \, dy \, ;$$

$$\beta_{mn} = \frac{1}{\pi^2} \int_{-\pi}^{\pi} \int_{-\pi}^{\pi} f(x, y) \cos mx \sin ny \, dx \, dy \, ;$$

$$\gamma_{mn} = \frac{1}{\pi^2} \int_{-\pi}^{\pi} \int_{-\pi}^{\pi} f(x, y) \sin mx \cos ny \, dx \, dy \, ;$$

$$\delta_{mn} = \frac{1}{\pi^2} \int_{-\pi}^{\pi} \int_{-\pi}^{\pi} f(x, y) \sin mx \sin ny \, dx \, dy \qquad (4.31)$$

$(m, n = 0, 1, 2, \ldots)$ and

$$\varepsilon_{mn} = \begin{cases} \frac{1}{4} & \text{for} \quad m = n = 0 \, , \\ \frac{1}{2} & \text{for} \quad m > 0, \, n = 0 \quad \text{and for} \quad m = 0, \, n > 0 \, , \\ 1 & \text{for} \quad m > 0, \, n > 0 \, . \end{cases}$$

This is the Fourier series with respect to the following system of functions:

$$\cos mx \cos ny; \quad \cos mx \sin ny; \quad \sin mx \cos ny;$$
$$\sin mx \sin ny \quad (m, n = 0, 1, 2, \ldots). \tag{4.32}$$

The system (4.32) is not orthonormal, but is orthogonal and complete in $L_2(Q)$, in fact in any $L_2(\tilde{Q})$ such that $\tilde{Q}$ is a square $a < x < a + 2\pi$, $b < y < b + 2\pi$.

Example 4.5. Consider the function $f(x, y) = xy$ in the square $Q = \{|x| < \pi, |y| < \pi\}$. Since this function is a product of a function of one variable x and a function of one other variable y, the coefficients (4.31) will be given by a product of two integrals; thus, we have immediately

$$\alpha_{mn} = \beta_{mn} = \gamma_{mn} = 0; \quad \delta_{mn} = (-1)^{m+n} 4/mn$$

for $m > 0$ and $n > 0$, and $\delta_{mn} = 0$ for $n = 0$ or $m = 0$. Hence, the Fourier series of the function xy in $L_2(Q)$ with respect to the system (4.32) takes the following form

$$xy = 4 \sum_{m,n=1}^{\infty} (-1)^{m+n} \frac{\sin mx \sin ny}{mn}. \tag{4.33}$$

Observe that the series (4.33) could also be obtained simply by multiplying the Fourier series of the function $f(x) = x$ in $(-\pi, \pi)$ with respect to the system (4.8) by the Fourier series of the function $f(y) = y$ with respect to the same system (see Problem 2.1a).

Example 4.6. Let us find the Fourier series of the function $f(x, y)$, defined in the square $\tilde{Q} = \{0 < x < 2\pi, 0 < y < 2\pi\}$ by

$$f(x, y) = \begin{cases} 1 & \text{for } 0 < x \leq y, \\ 0 & \text{for } y < x < 2\pi. \end{cases}$$

Using the formulas (4.31), where we integrate within the limits 0 and 2π, we have

$$\alpha_{mn} = \frac{1}{\pi^2} \int_0^{2\pi} \int_0^{2\pi} f(x, y) \cos mx \cos ny \, dx \, dy$$

$$= \frac{1}{\pi^2} \int_0^{2\pi} \cos ny \left(\int_0^y \cos mx \, dx \right) dy = \begin{cases} 2 & \text{for } m = n = 0, \\ 0 & \text{for other } m, n; \end{cases}$$

$$\beta_{0n} = \frac{1}{\pi^2} \int_0^{2\pi} \sin ny \left(\int_0^y dx \right) dy$$

$$= \frac{1}{\pi^2} \int_0^{2\pi} y \sin ny \, dy = -\frac{2}{n\pi} \quad \text{for } n > 0;$$

$$\beta_{mn} = \frac{1}{\pi^2} \int_0^{2\pi} \sin ny \left(\int_0^y \cos mx \, dx \right) dy$$

$$= \frac{1}{n\pi^2} \int_0^{2\pi} \sin ny \sin my \, dy = \begin{cases} 1/n\pi & \text{for } m = n > 0, \\ 0 & \text{for } m \neq n; \end{cases}$$

$$\gamma_{m0} = \frac{1}{\pi^2} \int_0^{2\pi} \left(\int_0^y \sin mx \, dx \right) dy$$

$$= \frac{1}{m\pi^2} \int_0^{2\pi} (1 - \cos my) \, dy = \frac{2}{m\pi} \quad \text{for } m > 0;$$

$$\gamma_{mn} = \frac{1}{\pi^2} \int_0^{2\pi} \cos ny \left(\int_0^y \sin mx \, dx \right) dy$$

$$= \frac{1}{m\pi} \int_0^{2\pi} \cos ny \, (1 - \cos my) \, dy = \begin{cases} -(1/m\pi) & \text{for } m = n > 0, \\ 0 & \text{for } m \neq n; \end{cases}$$

$$\delta_{mn} = \frac{1}{\pi^2} \int_0^{2\pi} \sin ny \left(\int_0^y \sin mx \, dx \right) dy$$

$$= \frac{1}{m\pi^2} \int_0^{2\pi} \sin ny \, (1 - \cos my) \, dy = 0 \quad \text{for all } m, n;$$

the remaining coefficients vanish ($\beta_{00} = 0$, $\gamma_{00} = 0$). Thus, from (4.30) we get the following Fourier series,

$$f(x, y) = \frac{1}{2} + \frac{1}{\pi} \sum_{n=1}^{\infty} \left(-\frac{\sin ny}{n} + \frac{\sin nx}{n} + \frac{\cos nx \sin ny}{n} - \frac{\sin nx \cos ny}{n} \right) = \frac{1}{2} + \frac{1}{\pi} \sum_{n=1}^{\infty} \frac{\sin nx - \sin ny - \sin n(x-y)}{n}. \quad (4.34)$$

This Fourier series could also be derived in a somewhat simpler way (see Example 4.7).

4.4.3 Consider now a rectangle K in the plane xy given by the inequalities $-l < x < l$, $-h < y < h$ ($h > 0, l > 0$). We have

Theorem 4.8. *The system*

$$\tilde{e}_{mn}(x, y) = \frac{1}{2\sqrt{(lh)}} e^{i\pi(mx/l + ny/h)} \quad (m, n = 0, \pm 1, \pm 2, \ldots) \quad (4.35)$$

is complete and orthonormal in the space $L_2(K)$. *To a function* $f(x, y) \in L_2(K)$ *there corresponds the Fourier series*

$$\sum_{m,n=-\infty}^{\infty} c_{mn} \frac{1}{2\sqrt{(lh)}} e^{i\pi(mx/l + ny/h)},$$

where

$$c_{mn} = (f, \tilde{e}_{mn}) = \frac{1}{2\sqrt{(lh)}} \int_{-l}^{l} \int_{-h}^{h} f(x, y) e^{-i\pi(mx/l + ny/h)} \, dx \, dy.$$

The proof is an analogue of Theorem 4.3. Actually, the system (4.35) can be obtained from the system (4.27) in $L_2(Q)$ by the substitutions $s = \pi x/l$, $t = \pi y/h$, which transform the square Q (the variables being s and t) onto the rectangle K (variables x and y).

4.4.4 The systems of orthogonal or orthonormal functions, considered so far, had the property that each function in a system was a product of two functions such that one depends only on the variable x and the other one on y. Such Fourier series are called *multiple series* and the theorems stated above are particular cases of the following general statement:

Theorem 4.9. *Let K be a rectangle in the plane xy, given by the inequalities $a < x < b$, $c < y < d$. Let $\{\varphi_m(x)\}$ be a complete orthonormal system of functions in the space $L_2(a, b)$ and let $\{\psi_n(y)\}$ be a complete orthonormal system of functions in the space $L_2(c, d)$. Then the system $\{\varphi_m(x)\psi_n(y)\}$ is complete and orthonormal in the space $L_2(K)$.*

Proof: 1. The *orthonormality* can be readily verified. Denoting $e_{mn}(x, y) = \varphi_m(x)\psi_n(y)$, then we have

$$(e_{mn}, e_{jk}) = \int_a^b \int_c^d \varphi_m(x)\psi_n(y)\overline{\varphi_j(x)}\,\overline{\psi_k(y)}\,dx\,dy$$

$$= \int_a^b \varphi_m(x)\overline{\varphi_j(x)}\,dx \int_c^d \psi_n(y)\overline{\psi_k(y)}\,dy = (\varphi_m, \varphi_j)(\psi_n, \psi_k).$$

Since the systems $\{\varphi_m\}$ and $\{\psi_n\}$ are orthonormal, the last equality implies that

$$(e_{mn}, e_{jk}) = \begin{cases} 1, & \text{whenever } m = j \text{ and } n = k, \\ 0, & \text{whenever } m \neq j \text{ or } n \neq k. \end{cases}$$

2. *Completeness*: Let $f(x, y) \in L_2(K)$. Keep y fixed and denote $g_y(x) = f(x, y)$. The functions $g_y(x)$ belong for almost all $y \in (c, d)$ to the space $L_2(a, b)$ (for otherwise, the function f would not be in $L_2(K)$, cf. Theorem 1.11); hence $g_y(x)$ can be expanded in a Fourier series with respect to the system $\{\varphi_m\}$ with coefficients depending on y, i.e.,

$$g_y(x) = \sum_{m=1}^{\infty} c_m(y)\varphi_m(x), \qquad (4.36)$$

where

$$c_m(y) = (g_y, \varphi_m) = \int_a^b f(x, y)\overline{\varphi_m(x)}\,dx.$$

Furthermore, Parseval's equality reads

$$F(y) = \int_a^b |g_y(x)|^2 \, dx = \int_a^b |f(x, y)|^2 \, dx = \sum_{m=1}^{\infty} |c_m(y)|^2 \, .$$

The function $F(y)$ is integrable, since

$$\int_c^d F(y) \, dy = \int_a^b \int_c^d |f(x, y)|^2 \, dx \, dy \, .$$

The function $c_m(y)$ is in $L_2(c, d)$, because

$$\int_c^d |c_n(y)|^2 \, dy \leqq \int_c^d \sum_{m=1}^{\infty} |c_m(y)|^2 \, dy = \int_c^d F(y) \, dy < \infty \, ;$$

thus, it can be expanded in a Fourier series with respect to the system $\{\psi_n\}$,

$$c_m(y) = \sum_{n=1}^{\infty} c_{mn} \psi_n(y) \, , \tag{4.37}$$

where

$$c_{mn} = (c_m, \psi_n) = \int_c^d c_m(y) \, \overline{\psi_n(y)} \, dy = \int_a^b \int_c^d f(x, y) \, \overline{\varphi_m(x)} \, \overline{\psi_n(y)} \, dx \, dy. \tag{4.38}$$

Parseval's equality reads

$$\int_c^d |c_m(y)|^2 \, dy = \sum_{n=1}^{\infty} |c_{mn}|^2 \, .$$

Substituting from (4.37) into (4.36), we obtain the Fourier series of the function $f(x, y)$ with respect to the system $\{e_{mn}\}$:

$$f(x, y) = \sum_{m=1}^{\infty} \sum_{n=1}^{\infty} c_{mn} \, \varphi_m(x) \, \psi_n(y)$$

with coefficients $c_{mn} = (f, e_{mn})$ determined by the formulas (4.38). From Parseval's equalities for the functions $g_y(x)$ and $c_m(y)$ we have

$$\int_a^b \int_c^d |f(x, y)|^2 \, dx \, dy = \int_c^d \sum_{m=1}^{\infty} |c_m(y)|^2 \, dy$$

$$= \sum_{m=1}^{\infty} \int_c^d |c_m(y)|^2 \, dy = \sum_{m=1}^{\infty} \sum_{n=1}^{\infty} |c_{mn}|^2$$

(interchanging the order of integration and summation is permissible, for the functions in question are non-negative — cf. Theorem 1.8), which is Parseval's equality for the system $\{e_{mn}\}$ in $L_2(K)$. This equality holds for every function $f \in L_2(K)$; hence, the system $\{\varphi_m(x)\,\psi_n(y)\}$ is complete by Theorem 3.17, condition 1.

The system (4.27) or (4.32), is a particular case of the system $\{\varphi_m(x)\,\psi_n(y)\}$ obtained by choosing $\varphi_k(x) = \psi_k(x) = e^{ikx}/\sqrt{(2\pi)}$ ($k = 0, \pm 1, \pm 2, \ldots$), or by taking the system (4.8) for both $\{\varphi_k\}$ and $\{\psi_k\}$. In addition, Theorem 4.9 yields many other possibilities of forming multiple Fourier series.

The method used in Theorem 4.9 for proving the completeness of a system can be sometimes applied for establishing double Fourier series. This procedure is demonstrated in the following example.

Example 4.7. Let us find again the Fourier series of the function $f(x, y)$ defined in Example 4.6 with respect to the system (4.32). For a fixed $y \in (0, 2\pi)$, the function $f(x, y) = g_y(x)$ takes the form

$$g_y(x) = \begin{cases} 1 & \text{for } 0 < x \leq y, \\ 0 & \text{for } y < x < 2\pi. \end{cases}$$

Let us expand the function $g_y(x)$ in a Fourier series in $L_2(0, 2\pi)$ with respect to the system (4.8). Since the Fourier coefficients of the function $g_y(x)$ are

$$a_0(y) = \frac{1}{\pi}\int_0^{2\pi} g_y(x)\,\mathrm{d}x = \frac{1}{\pi}\int_0^y \mathrm{d}x = \frac{y}{\pi};$$

$$a_n(y) = \frac{1}{\pi}\int_0^{2\pi} g_y(x)\cos nx\,\mathrm{d}x = \frac{1}{\pi}\int_0^y \cos nx\,\mathrm{d}x = \frac{1}{n\pi}\sin ny;$$

$$b_n(y) = \frac{1}{\pi}\int_0^{2\pi} g_y(x)\sin nx\,\mathrm{d}x = \frac{1}{\pi}\int_0^y \sin nx\,\mathrm{d}x = -\frac{1}{n\pi}[\cos ny - 1]$$

$(n = 1, 2, \ldots)$,

we have

$$f(x, y) = g_y(x)$$
$$= \frac{y}{2\pi} + \frac{1}{\pi} \sum_{n=1}^{\infty} \left(\frac{\sin ny}{n} \cos nx - \frac{\cos ny - 1}{n} \sin nx \right).$$

Replacing the function $y/2\pi$ by its Fourier series in $L_2(0, 2\pi)$ with respect to the system (4.8) (see Example 5.1) we obtain immediately the formula (4.34).

Hence, the multiple Fourier series are closely related to the Fourier series of functions of one variable. However, the theory of multiple Fourier series is not only a slight generalization of the theory of Fourier series of one variable. For example, the principle of localization, which will be treated in Chapter 7 for ordinary Fourier series, does not hold for double trigonometric Fourier series.

The multiple Fourier series do not yield, of course, the only method for expanding functions of several variables; there exists a number of other more general systems of orthogonal functions of several variables. However, we are not going to treat these problems more closely.

4.5 EXPANSIONS WITH RESPECT TO EIGENFUNCTIONS

Many complete orthogonal systems in the space $L_2(a, b)$ and in related spaces have a common feature: they are systems of the so-called *eigenfunctions* of certain differential equations. These systems play an important role particularly in applications (see Chapter 9); therefore, we are going to consider them more closely.

4.5.1 Let $p(x)$ be a non-negative function integrable in the interval (a, b), which vanishes at most at finitely many points. By the symbol $L_{2,p}(a, b)$ we denote the Hilbert space (for the sake of simplicity real) of functions $f(x)$ such that the integral

$$\int_a^b p(x) |f(x)|^2 \, \mathrm{d}x$$

is finite. The scalar product is, in the real case, defined for $f, g \in L_{2,p}(a, b)$ by

$$(f, g) = \int_a^b p(x) f(x) g(x) \, \mathrm{d}x \, . \tag{4.39}$$

The scalar product defined by (4.39) satisfies all required conditions (see Problem 3.14). The space $L_{2,p}(a, b)$ is a complete Hilbert space (see Problem 3.15).

The function $p(x)$ is called the *weight function* and the spaces $L_{2,p}(a, b)$ are sometimes called the *weight spaces*.

4.5.2 Consider a differential equation

$$\frac{\mathrm{d}}{\mathrm{d}x}\left(A(x) \frac{\mathrm{d}u}{\mathrm{d}x}\right) + B(x) u(x) = -\lambda \, p(x) u(x), \tag{4.40}$$

where $u(x)$ is the unknown function, λ is a parameter (number) and $A(x)$, $B(x)$ and $p(x)$ are given real functions satisfying the following conditions:

the function A is continuous and positive in $[a, b]$ and continuously differentiable in (a, b);
the function B is continuous in $[a, b]$;
the function p is positive, continuous and integrable in (a, b).

We are going to investigate the real solutions of equation (4.40) which, in addition, satisfy the boundary conditions

$$\alpha \, u(a) + \beta \, u'(a) = 0 \, ; \quad \gamma \, u(b) + \delta \, u'(b) = 0 \tag{4.41}$$

where α, β, γ and δ are real numbers satisfying the conditions $\alpha^2 + \beta^2 \neq 0$ and $\gamma^2 + \delta^2 \neq 0$.

The equation (4.40) along with the boundary conditions (4.41) will be called the *boundary value problem*. The considered boundary value problem obviously has the solution $u(x) \equiv 0$; however, we are not interested in this trivial solution.

Definition: A number λ, for which a non-trivial solution $u(x)$ of the boundary value problem (4.40) and (4.41) exists, is called an *eigenvalue* of this boundary value problem and the non-trivial solution $u(x)$ is called an *eigenfunction* corresponding to the eigenvalue λ.

If $u(x)$ is an eigenfunction, so is each function of the form $c \cdot u(x)$, c being an arbitrary constant. Therefore, the eigenfunctions will usually be normalized, i.e., the constant c will be chosen so that the norm of the eigenfunction in the weight space $L_{2,p}(a, b)$ is equal to one.

We state two important propositions without proof:

1. *Under the above assumptions on the boundary value problem there exist infinitely many eigenvalues. All these eigenvalues are real and can be arranged in the sequence*

$$\lambda_1 < \lambda_2 < \lambda_3 < \ldots < \lambda_n < \ldots,$$

such that $\lim_{n \to \infty} \lambda_n = +\infty$.

2. *For each eigenvalue λ_i there exists exactly one corresponding (normalized) eigenfunction $u_i(x)$.*

(For proof see [7], [12].)

Consider now in $L_{2,p}(a, b)$ a system of normalized eigenfunctions

$$u_1(x), u_2(x), \ldots, u_n(x), \ldots. \qquad (4.42)$$

We have the following important theorem.

Theorem 4.10. *The system* (4.42) *is complete and orthonormal in the space $L_{2,p}(a, b)$.*

We are not going to present the whole proof here; for the completeness of the system (4.42), the reader is referred to [7].

However, we shall prove the orthogonality of the functions in (4.42), i.e. we shall prove that $(u_i, u_k) = 0$ for $i \neq k$ (the scalar product being that defined in (4.39)):

To begin with, the identity

$$u\left[\frac{d}{dx}\left(A\frac{dv}{dx}\right) + Bv\right] - v\left[\frac{d}{dx}\left(A\frac{du}{dx}\right) + Bu\right]$$
$$= \frac{d}{dx}\left[A\left(u\frac{dv}{dx} - v\frac{du}{dx}\right)\right] \qquad (4.43)$$

is obviously true for any two differentiable functions u and v. Also for any two functions u and v satisfying the boundary conditions (4.41) we prove now that

$$u(a) v'(a) - u'(a) v(a) = u(b) v'(b) - u'(b) v(b) = 0. \qquad (4.44)$$

Since the functions u and v satisfy the boundary conditions we have

$$\alpha u(a) + \beta u'(a) = 0,$$
$$\alpha v(a) + \beta v'(a) = 0.$$

The numbers α and β cannot be simultaneously equal to zero, for $\alpha^2 + \beta^2 \neq 0$. Hence, both the last equations may hold simultaneously only if the determinant of the system of both equations (where α and β are considered as unknowns) is equal to zero, i.e. if

$$\begin{vmatrix} u(a), & u'(a) \\ v(a), & v'(a) \end{vmatrix} = u(a) v'(a) - u'(a) v(a) = 0.$$

In an analogous way, equality (4.44) for the point $x = b$ can be derived.

Let $u_i(x)$ and $u_k(x)$ be two eigenfunctions corresponding to different eigenvalues λ_i and λ_k. We have

$$\frac{d}{dx}\left(A\frac{du_i}{dx}\right) + Bu_i = -\lambda_i p u_i,$$

$$\frac{d}{dx}\left(A\frac{du_k}{dx}\right) + Bu_k = -\lambda_k p u_k.$$

Multiply the first equation by u_k, the second one by u_i and subtract one from the other; using the identity (4.43) we obtain

$$\frac{\mathrm{d}}{\mathrm{d}x}\left[A\left(u_i\frac{\mathrm{d}u_k}{\mathrm{d}x} - u_k\frac{\mathrm{d}u_i}{\mathrm{d}x}\right)\right] = (\lambda_i - \lambda_k)\, pu_iu_k\,.$$

Integrating this equality within the limits a and b we obtain

$$\left[A\left(u_i\frac{\mathrm{d}u_k}{\mathrm{d}x} - u_k\frac{\mathrm{d}u_i}{\mathrm{d}x}\right)\right]_a^b = (\lambda_i - \lambda_k)\int_a^b p(x)\, u_i(x)\, u_k(x)\, \mathrm{d}x\,.$$

The left hand side is equal to zero due to the equalities (4.44), because the functions u_i and u_k satisfy the boundary conditions (4.41). Hence, the right hand side vanishes as well, i.e.,

$$(\lambda_i - \lambda_k)(u_i, u_k) = 0\,.$$

Since $\lambda_i \neq \lambda_k$ by assumption, we have $(u_i, u_k) = 0$, and consequently, the functions in the system (4.42) are orthogonal. Since the system is normalized, (4.42) is orthonormal in $L_{2,p}(a, b)$.

Hence, Theorem 4.10 enables us to expand each function $f(x)$ from the space $L_{2,p}(a, b)$ in a Fourier series with respect to the eigenfunctions of our boundary value problem, i.e.,

$$f(x) = \sum_{n=1}^{\infty} c_n\, u_n(x)\,, \tag{4.45}$$

where

$$c_n = (f, u_n) = \int_a^b p(x)\, f(x)\, u_n(x)\, \mathrm{d}x\,.$$

Series (4.45) converges to the function $f(x)$ in the norm of the space $L_{2,p}(a, b)$.

Remark. Instead of the system $u_1(x), u_2(x), \ldots, u_n(x), \ldots$, complete and orthogonal in the weight space $L_{2,p}(a, b)$, we could investigate the system

$$\sqrt{[p(x)]}\, u_1(x),\ \sqrt{[p(x)]}\, u_2(x), \ldots, \sqrt{[p(x)]}\, u_n(x), \ldots,\quad (4.46)$$

which is complete and orthogonal in the space $L_2(a, b)$. The Fourier series of a function $f(x) \in L_{2,p}(a, b)$ with respect to the former system coincides, apart from the factor $\sqrt{[p(x)]}$, with the Fourier series of the function $\sqrt{[p(x)]} f(x) \in L_2(a, b)$ with respect to the system (4.46). Thus, the difference between the series in the space L_2 and that in the space $L_{2,p}$ is merely formal.

We are now going to present a number of examples demonstrating the diversity of the complete orthogonal systems obtained by means of the boundary value problem (4.40) and (4.41). The systems dealt with in the following examples are not orthonormal, but only orthogonal; however, as we know, this is hardly a defect.

Example 4.8. Consider, in the interval $(0, l)$, the differential equation $u'' = -\lambda u$ with the boundary conditions $u(0) = u(l) = 0$. This is a particular case of equation (4.40) with $A \equiv 1$, $B \equiv 0$ and $p \equiv 1$ and of the boundary conditions (4.41) with $\alpha = \gamma = 1$ and $\beta = \delta = 0$. Thus, $L_{2,p}(0, l)$ reduces to the usual space $L_2(0, l)$.

The differential equation $u'' = -\lambda u$ has the solution

$$u(x) = C_1 \cos \sqrt{(\lambda)}\, x + C_2 \sin \sqrt{(\lambda)}\, x, \qquad (4.47)$$

where the constants C_1 and C_2 are to be specified. Using the boundary conditions we have $C_1 = 0$ and $C_2 \sin \sqrt{(\lambda)}\, l = 0$. Since the trivial solution is of no interest for us, we must have $C_2 \neq 0$; hence, $\sin \sqrt{(\lambda)}\, l = 0$, i.e. $\sqrt{(\lambda)}\, l = n\pi$ or $\lambda_n = n^2\pi^2/l^2$. These numbers are the eigenvalues and

$$u_n(x) = \sin \frac{n\pi x}{l}, \quad n = 1, 2, \ldots, \qquad (4.48)$$

is the corresponding system of eigenfunctions.

This system is complete and orthogonal in $L_2(0, l)$ due to Theorem 4.10 and can be used for expanding functions in $L_2(0, l)$ in Fourier series.

Example 4.9. Consider again the equation $u'' = -\lambda u$ in the interval $(0, l)$, but this time with the boundary conditions $u(0) =$

= $u'(l) = 0$, i.e., we choose $\alpha = \delta = 1$, $\beta = \gamma = 0$ in (4.41). The solution of the equation has again the form (4.47); the boundary conditions imply $C_1 = 0$ and $\sqrt{(\lambda)}\, C_2 \cos \sqrt{(\lambda)}\, l = 0$. Thus, necessarily $\sqrt{(\lambda)}\, l = (2n + 1)\pi/2$ and we obtain the eigenvalues $\lambda_n = [(2n + 1)^2 \pi^2]/4l^2$, $n = 0, 1, 2, \ldots$ with the corresponding eigenfunctions

$$u_n(x) = \sin \frac{(2n + 1)\pi x}{2l}, \quad n = 0, 1, 2, \ldots. \qquad (4.49)$$

Also these functions can be used, by Theorem 4.10, for expanding functions in $L_2(0, l)$, because they constitute a complete and orthogonal system in $L_2(0, l)$.

The results obtained in the last examples are in essence not new. Actually, system (4.48) may also be obtained from the complete orthogonal system (4.20) using the substitution $x = \pi t/l$, and the system (4.49) from the complete orthogonal system (4.21) by using the substitution $x = \pi t/2l$ (see also Problem 4.6). Thus, the theory of eigenfunctions of boundary value problems is unnecessary for proving the completeness and orthogonality of the systems (4.48) and (4.49). However, these examples demonstrate that the methods used here enable us to include many familiar complete orthogonal systems in a general framework.

The following examples furnish systems which have not been dealt with yet.

Example 4.10. Consider again the equation $u'' = -\lambda u$ in the interval $(0, l)$, but now with the boundary conditions $u(0) = 0$, $u'(l) - u(l) = 0$, i.e., we choose $\alpha = \delta = -\gamma = 1$, $\beta = 0$ in (4.41). The solution of the system has again the form (4.47); the boundary conditions imply that $C_1 = 0$ and $C_2(\sqrt{(\lambda)} \cos \sqrt{(\lambda)}\, l - \sin \sqrt{(\lambda)}\, l) = 0$, i.e. $\sqrt{(\lambda)} = \tan \sqrt{(\lambda)}\, l$. Denoting by ξ_n the positive roots of the equation

$$\tan \xi = \frac{1}{l} \xi$$

(see Fig. 4.3), we have for the eigenvalues, $\lambda_n = \xi_n^2/l^2$. Then

$$\sin \frac{\xi_1}{l} x, \sin \frac{\xi_2}{l} x; \ldots; \sin \frac{\xi_n}{l} x; \ldots \qquad (4.50)$$

is the system of corresponding eigenfunctions. These functions constitute a complete and orthogonal system in $L_2(0, l)$ and can

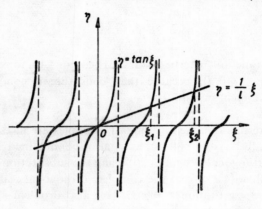

Fig. 4.3

therefore be used for constructing Fourier series of functions in $L_2(0, l)$ (witness the applications of system (4.50) in Chapter 9, Section 9.4.4).

Let us note that the orthogonality of the functions in (4.50) can be proved directly from their definition, without using the theory of eigenfunctions (see Problem 4.13). However, the proof of the completeness of the system (4.50) not using this theory would be much more difficult.

Example 4.11. Consider again the equation $u'' = -\lambda u$ in the interval $(0, l)$ with the boundary conditions $u'(0) - u(0) = 0$, $u'(l) + u(l) = 0$, i.e. we choose $\beta = -\alpha = \gamma = \delta = 1$ in (4.41). The first boundary condition implies that $C_1 - C_2 \sqrt{\lambda} = 0$; thus we have

$$u(x) = C_2[\sqrt{(\lambda)} \cos \sqrt{(\lambda)} \, x + \sin \sqrt{(\lambda)} \, x].$$

Excluding the trivial solution, we have necessarily $C_2 \neq 0$. The other boundary condition $u'(l) + u(l) = 0$ implies that

$$(1 - \lambda) \sin \sqrt{(\lambda)}\, l + 2\sqrt{(\lambda)} \cos \sqrt{(\lambda)}\, l = 0,$$

i.e.

$$\tan \sqrt{(\lambda)}\, l = \frac{2\sqrt{\lambda}}{\lambda - 1}.$$

Denoting by ξ_n the positive roots of the transcendental equation

$$\tan \xi = \frac{2l\xi}{\xi^2 - l^2}$$

we have $\lambda_n = \xi_n^2/l^2$ for the eigenvalues, and, for the corresponding system of eigenfunctions,

$$u_n(x) = \frac{\xi_n}{l} \cos \frac{\xi_n}{l} x + \sin \frac{\xi_n}{l} x, \quad n = 1, 2, \ldots. \quad (4.51)$$

Also this system is complete and orthogonal in $L_2(0, l)$.

Applying this procedure we could construct various other examples, e.g., by choosing another differential equation etc. The assumptions on a boundary value problem stated in Theorem 4.10, can be generalized and altered. However, these possibilities will not be tackled here; as an illustration we are going to present a boundary value problem furnishing a complete orthonormal system, which does not satisfy the assumptions of Theorem 4.10.

Example 4.12. Consider, in the interval $(0, l)$, the differential equation

$$\frac{d}{dx}\left(x \frac{du}{dx}\right) = -\lambda x\, u(x). \quad (4.52)$$

This is a special case of equation (4.40) with $A(x) = p(x) = x$ and $B(x) = 0$.* The boundary condition is given only at the

---

* The function A does not satisfy the condition that it is positive in $[0, l]$, for it vanishes at $x = 0$.

endpoint $x = l$, $u(l) = 0$. Instead of an exact boundary condition at the point $x = 0$ we shall require the solution $u(x)$ of the equation (4.52) to be bounded at points x close to zero. Since $p(x) = x$ in (4.52), the space under consideration is the weight space $L_{2,x}(0, l)$ with the weight function x.

The equation (4.52) can be written as

$$u'' + \frac{1}{x} u' + \lambda u = 0 ,$$

which is the *Bessel equation of zero index*. Its general solution has the form

$$u(x) = C_1 J_0(\sqrt{(\lambda)}\, x) + C_2 Y_0(\sqrt{(\lambda)}\, x) ,$$

where J_0 and Y_0 are *Bessel functions of the first and second kind*, respectively (of zero index).* C_1 and C_2 are constants not yet specified; since the function $Y_0(\sqrt{(\lambda)}\, x)$ is not bound for x close to zero, it follows from the requirement on the solution that $C_2 = 0$. Then the boundary condition $u(l) = 0$ implies that

$$C_1 J_0(\sqrt{(\lambda)}\, l) = 0 .$$

Excluding the trivial solution, we have necessarily $C_1 \neq 0$; hence, $\sqrt{(\lambda)}\, l$ is one of the positive roots ξ_n of the equation $J_0(\xi) = 0$. So we obtain the eigenvalues $\lambda_n = \xi_n^2/l^2$ and the eigenfunctions

$$u_n(x) = J_0\left(\frac{\xi_n}{l} x\right), \quad n = 1, 2, \ldots . \tag{4.53}$$

These functions are orthogonal in $L_{2,x}(0, l)$, i.e.

$$(u_m, u_n) = \int_0^l x J_0\left(\frac{\xi_n}{l} x\right) J_0\left(\frac{\xi_m}{l} x\right) \mathrm{d}x = 0 \quad \text{for} \quad m \neq n ,$$

---

* For more detailed information on Bessel functions see, e. g., WATSON, G. N.: *A Treatise on the Theory of Bessel Functions*, Cambridge 1944.

and constitute a complete system in this space. Hence, a function $f(x) \in L_{2,x}(0, l)$ can be expanded in a Fourier series

$$f(x) = \sum_{n=1}^{\infty} c_n J_0\left(\frac{\xi_n}{l} x\right),$$

where

$$c_n = \|u_n\|_{L_{2,x}(0,l)}^{-2} (f, u_n) = \frac{\int_0^l x f(x) J_0\left(\frac{\xi_n}{l} x\right) dx}{\int_0^l x J_0^2\left(\frac{\xi_n}{l} x\right) dx}.$$

Also in this case the orthogonality of functions $u_n(x)$ can be proved directly from their definition, i.e. by using (4.53) and various special properties of Bessel functions.

4.6 ORTHOGONAL POLYNOMIALS

Let (a, b) be a finite interval and consider the weight Hilbert space $L_{2,p}(a, b)$ (for the sake of simplicity assume it to be real) introduced in Section 4.5.1 with the weight function $p(x) \in L_1(a, b)$, $p(x) > 0$ except for a finite number of points.

Consider the system of powers

$$1, x, x^2, x^3, \ldots, x^n, \ldots . \tag{4.54}$$

We have the following important theorem.

Theorem 4.11. *If a function $f(x)$ in the space $L_{2,p}(a, b)$ is orthogonal to all functions in* (4.54) *(in the sense of the scalar product in the weight space $L_{2,p}(a, b)$), then $f(x)$ vanishes almost everywhere.*

Proof: 1. First, we prove the assertion for a special case, i.e. for f being continuous and $p(x) \equiv 1$. By the Weierstrass theorem (cf. Problem 3.21, where we transform the interval $[0, 1]$ onto the interval $[a, b]$ by a linear substitution), for each function $f(x)$ continuous in $[a, b]$ polynomials $p_n(x)$ exist such that $p_n \to f$ as $n \to \infty$ uniformly in the interval $[a, b]$; hence $p_n \to f$ in the norm of the space $L_2(a, b)$ (cf. Example 3.13). Since by assumption the function f is orthogonal to all functions in the system (4.54) with respect to the scalar product in the space $L_2(a, b)$, it is also orthogonal to the polynomials p_n, i.e.

$$(p_n, f) = \int_a^b p_n(x) f(x) \, \mathrm{d}x = 0 \, . \tag{4.55}$$

Since $p_n \to f$ in $L_2(a, b)$, Theorem 3.4 and (4.55) yield $0 = (p_n, f) \to (f, f)$. Thus, we have $\|f\|^2 = 0$, i.e. $f(x) = 0$ almost everywhere.

2. Let now $p(x)$ be a given weight function and let the function $f(x) \in L_{2,p}(a, b)$ be orthogonal to all powers of the variable x. Then $f(x)$ is also orthogonal to all polynomials $P(x)$,

$$\int_a^b p(x) f(x) P(x) \, \mathrm{d}x = 0 \, . \tag{4.56}$$

The function $p(x) f(x)$ is integrable, for by the Schwarz inequality (cf. Theorem 3.1 or Example 3.8) we have

$$\left| \int_a^b p(x) f(x) \, \mathrm{d}x \right| = \left| \int_a^b \sqrt{[p(x)]} \sqrt{[p(x)]} f(x) \, \mathrm{d}x \right|$$
$$\leq \left[\int_a^b p(x) \, \mathrm{d}x \right]^{1/2} \left[\int_a^b p(x) f^2(x) \, \mathrm{d}x \right]^{1/2} \, .$$

The function

$$F(x) = \int_a^x p(s) f(s) \, \mathrm{d}s$$

is thus continuous, moreover, $F(a) = 0$ and $F(b) = 0$ (the latter equality follows from the fact that the functions $f(x)$ and 1 are orthogonal with respect to the weight function $p(x)$). Integrating by parts, we obtain from (4.56)

$$[F(x) P(x)]_a^b - \int_a^b F(x) P'(x) \, dx = 0 ;$$

hence,

$$\int_a^b F(x) P'(x) \, dx = 0 .$$

Thus the continuous function $F(x)$ is orthogonal (in the sense of the scalar product in the space $L_2(a, b)$) to all functions which are derivatives of polynomials, i.e. to all polynomials. Such a function must vanish according to the first part of the present proof; since the weight function $p(x)$ vanishes at most at finitely many points, we have $f(x) = 0$ almost everywhere, and our theorem is proved.

Consider now an infinite interval (a, b) of the type (c, ∞) or $(-\infty, \infty)$. If the weight function is not appropriately chosen, the functions x^n need not belong to $L_{2,p}(a, b)$ at all (e.g. for $p(x) = (1 + x^2)^{-1}$). Thus, further restrictions have to be imposed on the weight function.

Theorem 4.12. *If the interval (a, b) is infinite and positive constants R, C and β exist such that*

$$|p(x)| \leq C e^{-\beta x} \quad \text{for} \quad |x| \geq R \tag{4.57}$$

and if $p(x) \in L_1(a, b)$, then the assertion of Theorem 4.11 holds.

We are not going to carry out the proof here. It uses some more profound properties of Fourier transforms which will not be mentioned also in Chapter 8. The reader is referred to [10], and, for special weight functions, to [14].

Orthogonalizing (or orthonormalizing) the system (4.54) in the weight space $L_{2,p}(a, b)$ by the procedure described in

Chapter 3, Section 3.6, we obtain a system of orthogonal polynomials

$$P_0(x), P_1(x), P_2(x), \ldots, P_n(x), \ldots ; \quad (4.58)$$

the orthogonalization is to be arranged so that $P_n(x)$ is a polynomial of n-th degree, i.e.,

$$P_n(x) = a_{nn}x^n + a_{n,n-1}x^{n-1} + \ldots + a_{n,1}x + a_{n,0}, \quad (4.59)$$

with $a_{nn} \neq 0$. The polynomials $P_n(x)$, in general, are determined uniquely up to a constant factor. This factor is usually chosen so that $a_{nn} = 1$, or that the polynomials $P_n(x)$ are normalized, i.e., $\|P_n\|_{L_{2,p}(a,b)} = 1$. The system obtained by orthonormalizing (4.58) is denoted by

$$Q_0(x), Q_1(x), Q_2(x), \ldots, Q_n(x), \ldots ; \quad (4.60)$$

thus, we have

$$Q_n(x) = \frac{1}{\varkappa_n} P_n(x), \quad \varkappa_n = \|P_n\|_{L_{2,p}(a,b)}. \quad (4.61)$$

The main proposition in the present section reads as follows:

Theorem 4.13. *Let the non-negative weight function $p(x)$ be positive for almost every $x \in (a, b)$ and let it belong to the space $L_1(a, b)$. If (a, b) is infinite, let $p(x)$ satisfy condition (4.57). Then the system (4.58) is complete and orthogonal in $L_{2p}(a, b)$, and the system (4.60) is complete and orthonormal in the same space.*

Proof. The orthogonality or orthonormality is obvious, because the functions under consideration were so defined. The completeness follows from Theorem 4.11 or Theorem 4.12. Actually, if a function $f(x) \in L_{2,p}(a, b)$ is orthogonal to all polynomials in (4.58), it is orthogonal to all powers $1, x, x^2, \ldots$ as well, and consequently it vanishes by the mentioned theorems. Since $L_{2,p}(a, b)$ is a complete Hilbert space (cf. Problem 3.15), the completeness of our systems follows from Theorem 3.17 (see also Problem 3.16).

Following the general theory of Hilbert spaces (Chapter 3) we can expand each function $f(x) \in L_{2,p}(a, b)$ in a Fourier series with respect to the orthonormal system (4.60),

$$f(x) = \sum_{n=0}^{\infty} c_n \, Q_n(x),$$

$$c_n = (f, Q_n) = \int_a^b p(x) f(x) \, Q_n(x) \, dx; \qquad (4.62)$$

or in a Fourier series with respect to the orthogonal system (4.58),

$$f(x) = \sum_{n=0}^{\infty} d_n \, P_n(x),$$

$$d_n = \|P_n\|^{-2} (f, P_n) = \frac{\int_a^b p(x) f(x) \, P_n(x) \, dx}{\int_a^b p(x) \, P_n^2(x) \, dx}. \qquad (4.63)$$

(Apparently, from (4.61), $d_n = (1/\varkappa_n) \, c_n$; since $Q_n(x) = (1/\varkappa_n) \, P_n(x)$, the series in (4.62) and (4.63) coincide. Thus the difference between the systems (4.60) and (4.58) is merely formal.) These Fourier series converge to $f(x)$ in the norm of the Hilbert space $L_{2,p}(a, b)$ and Parseval's equality holds,

$$\|f\|^2_{L_{2,p}(a,b)} = \int_a^b p(x) f^2(x) \, dx = \sum_{n=0}^{\infty} c_n^2 = \sum_{n=0}^{\infty} \varkappa_n^2 d_n^2. \qquad (4.64)$$

The theory of orthogonal polynomials has been developed to a large extent and finds wide application in practice. Here we state some important properties of orthogonal polynomials without proof. All the following statements are proved, for example, in [14].

4.6.1 Theorem on the roots of the polynomial $P_n(x)$

The polynomial $P_n(x)$ from (4.58) has n simple real roots. These roots lie inside the interval (a, b). The roots of $P_n(x)$ separate the roots of $P_{n+1}(x)$.

4.6.2 Recurrence formula

We have

$$P_{n+1}(x) = (A_n x + B_n) P_n(x) - C_n P_{n-1}(x) \quad (n = 1, 2, \ldots), \quad (4.65)$$

where

$$A_n = \frac{a_{n+1,n+1}}{a_{nn}}; \quad B_n = A_n \left(\frac{a_{n+1,n}}{a_{n+1,n+1}} - \frac{a_{n,n-1}}{a_{nn}} \right);$$

$$C_n = \frac{A_n \varkappa_n^2}{A_{n-1} \varkappa_{n-1}^2}. \quad (4.66)$$

(Let us note that a_{ij} is the coefficient of x^j in the polynomial $P_i(x)$ and that $a_{ii} \neq 0$. The number $\varkappa_n$ is the norm of the polynomial $P_n(x)$ in $L_{2,p}(a, b)$.)

4.6.3 Christoffel-Darboux summation formula

$$\sum_{k=0}^{n} \frac{1}{\varkappa_k^2} P_k(x) P_k(y)$$

$$= \frac{1}{\varkappa_n^2} \frac{a_{nn}}{a_{n+1,n+1}} \frac{P_{n+1}(x) P_n(y) - P_n(x) P_{n+1}(y)}{x - y}. \quad (4.67)$$

In what follows, we restrict our considerations to three types of weight functions $p(x)$ according to the character of the interval (a, b):

I. For the case of a finite interval (a, b) we choose

$$p(x) = (b - x)^\alpha (x - a)^\beta, \quad \text{where} \quad \alpha > -1 \quad \text{and} \quad \beta > -1. \quad (4.68)$$

II. For the case of an interval (a, ∞), a being finite, we choose

$$p(x) = e^{-x}(x - a)^\alpha, \quad \text{where} \quad \alpha > -1. \quad (4.69)$$

III. For the case of the interval $(-\infty, +\infty)$ we choose

$$p(x) = e^{-x^2/2}. \quad (4.70)$$

All these weight functions obviously satisfy the conditions of Theorems 4.11 and 4.12.

The orthogonal polynomials obtained by orthogonalizing the system (4.54) in spaces $L_{2,p}(a, b)$ with weights of the type (4.68), (4.69) and (4.70) are called the *classical orthogonal polynomials*.

For the classical orthogonal polynomials we have a number of further statements. Define the function $H(x)$ as follows:

$$H(x) = (b - x)(x - a) \quad \text{for the case I},$$
$$H(x) = x - a \quad \text{for the case II}, \qquad (4.71)$$
$$H(x) = 1 \quad \text{for the case III}.$$

Then we have

4.6.4 Rodriguez formula

$$P_n(x) = \frac{1}{\mu_n} \frac{1}{p(x)} \frac{d^n}{dx^n} [p(x) H^n(x)], \qquad (4.72)$$

where μ_n is a certain constant factor and $H(x)$ is the function defined in (4.71).

4.6.5 Differential equations for the classical orthogonal polynomials*

A classical orthogonal polynomial $P_n(x)$ satisfies the differential equation

$$H(x) y'' + \mu_1 P_1(x) y' = -\lambda_n y, \qquad (4.73)$$

---

* Using the Rodriguez formula for $P_1(x)$ we can rewrite equation (4.73) as

$$\frac{d}{dx}\left(pH \frac{dy}{dx}\right) = -\lambda py.$$

Thus, the situation here is closely related to the results of the preceding section, because the polynomials $P_n(x)$ may be considered as the eigenfunctions of the last equation corresponding to the eigenvalues λ_n in (4.74).

where
$$\lambda_n = -n\left(\mu_1 a_{11} + \frac{n-1}{2}\frac{d^2 H}{dx^2}\right). \qquad (4.74)$$

We are going to present a survey of several, most important cases of orthogonal polynomials. The proofs along with many further properties may be found in [14].

A. *Legendre polynomials*

Legendre polynomials are obtained if we take the weight function $p(x) \equiv 1$, i.e., if we orthogonalize the system (4.54) in $L_2(a, b)$ (See also Chapter 3, Section 3.6). Using the Rodriguez formula for the interval $(-1, +1)$ we get the following representation of Legendre polynomials

$$P_n(x) = \frac{1}{(-2)^n n!}\frac{d^n}{dx^n}[(1-x^2)^n] \quad (n = 0, 1, \ldots).$$

Since
$$\int_{-1}^{1} P_n(x)\,P_m(x)\,dx = \begin{cases} 0 & \text{for } n \neq m, \\ 1/(n+\tfrac{1}{2}) & \text{for } n = m, \end{cases}$$

the orthonormal polynomials are given by the formulas

$$Q_n(x) = \sqrt{(n+\tfrac{1}{2})}\,P_n(x) \quad (n = 0, 1, 2, \ldots).$$

B. *Tshebyshev polynomials of the first kind*

They are denoted by $T_n(x)$ and correspond to the weight function $(b-x)^{-1/2}(x-a)^{-1/2}$. Thus, for the interval $(-1, +1)$ we have $p(x) = (1-x^2)^{-1/2}$; orthogonalizing the system (4.54) in the corresponding weight space, we obtain

$$T_n(x) = \cos(n\cos^{-1} x)$$
$$= \frac{(-1)^n (1-x^2)^{1/2}}{1\cdot 3\cdot 5\cdot\ldots\cdot(2n-1)}\frac{d^n}{dx^n}[(1-x^2)^{n-1/2}].$$

Since
$$(T_n, T_m) = \int_{-1}^{1} (1 - x^2)^{-1/2} T_n(x) T_m(x) \, dx = \begin{cases} 0 & \text{for } n \neq m, \\ \pi/2 & \text{for } n = m, \end{cases}$$

the corresponding orthonormal system has the following form

$$Q_n(x) = \sqrt{\frac{2}{\pi}} T_n(x).$$

C. *Tshebyshev polynomials of the second kind*

They are denoted by $U_n(x)$ and correspond to the weight function $(b - x)^{1/2} (x - a)^{1/2}$. Thus, for the interval $(-1, +1)$ we have $p(x) = (1 - x^2)^{1/2}$; orthogonalizing the system (4.54) in the corresponding weight space, we get

$$U_n(x) = \frac{\sin[(n+1)\cos^{-1} x]}{\sqrt{(1 - x^2)}}$$

$$= \frac{(-1)^n (n+1)(1 - x^2)^{-1/2}}{1 \cdot 3 \cdot 5 \ldots (2n-1)(2n+1)} \frac{d^n}{dx^n}\left[(1 - x^2)^{n+1/2}\right].$$

Since
$$(U_n, U_m) = \int_{-1}^{1} (1 - x^2)^{1/2} U_n(x) U_m(x) \, dx = \begin{cases} 0 & \text{for } n \neq m, \\ \pi/2 & \text{for } n = m, \end{cases}$$

the functions

$$Q_n(x) = \sqrt{\frac{2}{\pi}} U_n(x)$$

constitute a complete orthonormal system in the space

$$L_{2, \sqrt{(1-x^2)}}(-1, +1).$$

D. *Gegenbauer polynomials*

They are denoted by $C_n^{(\lambda)}(x)$ and obtained for the weight function $[(b - x)(x - a)]^{\lambda - 1/2}$ ($\lambda > -\frac{1}{2}$). Both Tshebyshev and Legendre polynomials appear as particular cases of $C_n^{(\lambda)}(x)$, i.e.,

$$T_n(x) = C_n^{(0)}(x) \, ; \quad U_n(x) = C_n^{(1)}(x) \, ; \quad P_n(x) = C_n^{(1/2)}(x) \, .$$

E. *Jacobi polynomials*

Denoted by $G_n^{(\alpha,\beta)}(x)$, the Jacobi polynomials are obtained for the general weight function $p(x) = (b - x)^\alpha (x - a)^\beta$ defined in the case I. Gegenbauer polynomials are a particular case of $G_n^{(\alpha,\beta)}$, and are obtained for $\alpha = \beta = \lambda - \frac{1}{2}$. For the interval $(-1, +1)$ we have

$$G_n^{(\alpha,\beta)} = \frac{(1-x)^{-\alpha}(1+x)^{-\beta}}{(-2)^n \, n!} \frac{d^n}{dx^n} \left[(1-x)^{\alpha+n}(1+x)^{\beta+n} \right] .$$

F. *Laguerre polynomials*

Denoted by $L_n^{(\alpha)}(x)$, they correspond to the interval (a, ∞) and the weight function $p(x) = e^{-x}(x - a)^\alpha$ $(\alpha > -1)$. Thus, in the interval $(0, \infty)$ the system

$$L_n^{(\alpha)}(x) = \frac{1}{n!} e^x x^{-\alpha} \frac{d^n}{dx^n} \left[e^{-x} x^{n+\alpha} \right] \quad (n = 0, 1, 2, \ldots)$$

is complete and orthogonal in $L_{2,p}(0, \infty)$. The system $\{L_n^{(\alpha)}\}$ can be orthonormalized by using the formula

$$(L_n^{(\alpha)}, L_m^{(\alpha)}) = \int_0^\infty e^{-x} x^\alpha \, L_n^{(\alpha)}(x) \, L_m^{(\alpha)}(x) \, dx$$

$$= \begin{cases} 0 & \text{for } n \neq m, \\ (1/n!) \, \Gamma(\alpha + n + 1) & \text{for } n = m. \end{cases}$$

G. *Hermite polynomials*

Denoted by $H_n(x)$, they correspond to the interval $(-\infty, \infty)$ and to the weight function $p(x) = e^{-x^2/2}$. From the Rodriguez formula we get the following expression,

$$H_n(x) = (-1)^n \, e^{x^2/2} \frac{d^n}{dx^n} \left[e^{-x^2/2} \right] .$$

These functions are orthogonal and complete in $L_{2,p}(-\infty, \infty)$; the corresponding orthonormal system is given by the functions

$$Q_n(x) = (2\pi)^{-1/4} \frac{1}{\sqrt{n!}} H_n(x).$$

Using the substitution $x^2/2 = t$ we can express the Hermite polynomials in terms of the Laguerre polynomials $L_n^{(\alpha)}(x)$ with $\alpha = \pm \frac{1}{2}$.

The polynomials given above in **A** to **G** are the most important classical orthogonal polynomials. Many other polynomials can be obtained by a suitable choice of the weight function $p(x)$. We recommend the reader to establish an explicit form of the first orthogonal polynomials in the above examples, the corresponding differential equations, etc.

There are many other interesting properties of the general orthogonal polynomials $P_n(x)$; for example, they can be expressed in terms of various integrals and the corresponding *generating function* can be found. That is, a function $F(x, z)$ exists such that

$$F(x, z) = \sum_{n=0}^{\infty} P_n(x) z^n$$

or in other words, the polynomials $P_n(x)$ are the coefficients in the expansion of $F(x, z)$ in a power series in the variable z. All these problems are studied in the literature. Here, let us state only that, for Legendre polynomials,

$$F(x, z) = (1 - 2xz + z^2)^{-1/2} \, (\equiv R^{-1});$$

for Jacobi polynomials,

$$F(x, z) = 2^{\alpha+\beta} R^{-1}(1 - z + R)^{-\alpha}(1 + z + R)^{-\beta};$$

for Laguerre polynomials,

$$F(x, z) = e^{-xz/(1-z)}(1 - z)^{-\alpha-1};$$

for Hermite polynomials $H_n(x)/n!$,

$$F(x, z) = e^{xz - z^2/2}.$$

Problems

4.1. Prove that the linear substitution transforming the interval $(-\pi, \pi)$ onto the interval $(0, 2\pi)$ transforms the system (4.6) onto the same system except, possibly, for the sign.

4.2. Onto what system is the system (4.6) transformed by the linear substitution transforming the interval $(-\pi, \pi)$ onto $(a, a + 2\pi)$?

4.3. Prove that the systems (4.17) and (4.20) are orthogonal in $L_2(0, \pi)$.

4.4. Prove that the system (4.21) is complete and orthogonal in $L_2(0, \pi/2)$, and normalize it. (For a *hint* see Section 4.2.4.)

4.5. Find other complete orthogonal systems in the space $L_2(0, \pi/2)$.
(*Hint*: Use the substitution $x = 2t$ for the systems (4.17) and (4.20).)

4.6. Prove that the following systems are complete and orthogonal in $L_2(0, l)$:

(a) $1, \cos\dfrac{\pi x}{l}, \cos\dfrac{2\pi x}{l}, \ldots, \cos\dfrac{n\pi x}{l}, \ldots$;

(b) $\sin\dfrac{\pi x}{l}, \sin\dfrac{2\pi x}{l}, \ldots, \sin\dfrac{n\pi x}{l}, \ldots$;

(c) $\sin\dfrac{\pi x}{2l}, \sin\dfrac{3\pi x}{2l}, \ldots, \sin\dfrac{(2n + 1)\pi x}{2l}, \ldots$.

(*Hint*: Using appropriate substitutions we can get these systems from the systems (4.17), (4.20) and (4.21).)

4.7. Prove that the systems (4.27) and (4.32) are complete and orthogonal in $L_2(\tilde{Q})$, where $\tilde{Q}$ is a square defined by $a < x < a + 2\pi$, $b < y < b + 2\pi$ with arbitrary real numbers a and b.

(*Hint*: Proceed analogously as in the one-dimensional case in Section 4.2.1. Work out how to extend the function

SOME SPECIAL FOURIER SERIES

$f(x, y)$ belonging to $L_2(\tilde{Q})$ onto the entire plane so as to be periodic with period 2π in both variables x and y.)

4.8. Following the same pattern prove that the system (4.35) is complete and orthogonal in $L_2(\tilde{K})$, $\tilde{K}$ being a rectangle defined by $a < x < a + 2l$, $b < x < b + 2h$.

4.9. Establish the Fourier series of the function $f(x, y) = xy$ in the square $0 < x < 2\pi$, $0 < y < 2\pi$ and compare it with the Fourier series of the same function in the square $-\pi < x < \pi$, $-\pi < y < \pi$ given by formula (4.33).

4.10. Determine the Fourier series of the function $f(x, y) = \min(x, y)$ in the square $0 < x < 2\pi$, $0 < y < 2\pi$.

4.11. Find the real form of the double Fourier series corresponding to the complex form from Theorem 4.8.

4.12. Let A be the set of all functions $f(z)$ of a complex variable $z = x + iy$, which are analytic in the unit circle K defined by $|z| < 1$, and such that the integral $\|f\|^2 = \int_K |f(x + iy)|^2 \, dx \, dy$ is finite. Define a scalar product in A via the norm given by the formula (3.9). Prove that this space is complete and that the system of functions z^n ($n = 0, 1, \ldots$) is complete and orthogonal in A. (This means that the functions $r^n e^{in\varphi}$ are orthogonal in the circle K, where r and φ are polar coordinates in the plane.)

4.13. Let $c > 0$ and let ξ_i ($i = 1, 2, \ldots$) be the positive roots of the equation $\tan \xi = c\xi$. Prove that the system of functions $\sin \xi_i x$ ($i = 1, 2, \ldots$) is orthogonal in $L_2(0, 1)$.

4.14. Let $c > 0$ and let η_i ($i = 1, 2, \ldots$) be the positive roots of the equation $\cot \eta = c\eta$. Prove that the system of functions $\cos \eta_i x$ ($i = 1, 2, \ldots$) is orthogonal in $L_2(0, 1)$.

4.15. Derive an explicit form for the first five Legendre polynomials $P_n(x)$ in the interval $(-1, +1)$, the corresponding differential equation, recurrence formula etc.

4.16. Construct the differential equation for the orthogonal polynomials in **B** to **G**. Give the explicit forms of the first Tshebyshev, Laguerre and Hermite polynomials.

4.17. Prove that the polynomials $T_n(x)$ and $U_n(x)$ (in the interval $(-1, +1)$) and $H_n(x)$ are even functions for even n and odd functions for odd n.

4.18. For the Tshebyshev polynomials $T_n(x)$, prove the identity $T_n(T_m(x)) = T_{m \cdot n}(x)$.

CHAPTER 5 CALCULATION OF FOURIER SERIES

In the two preceding chapters we have seen that a function belonging to $L_2(a, b)$, for example, can be expanded in series of not only sines and cosines but of entirely different functions; from the examples given in Chapter 9 it will be apparent that these and other expansions are of considerable importance in practice.

For the time being, however, let us return to the original *trigonometric* Fourier series and show how various methods and particular properties of trigonometric functions can be applied to their calculation. As a matter of fact, we are going to give a survey of various methods which may be useful for the calculation of the Fourier series of an actual function, for the calculation of integrals, for the summation of series, etc.

Many results which are to be discussed here can, of course, be modified so as to be applicable for other types of Fourier series; this, however, will not be tackled. The trigonometric Fourier series are chosen as our topic primarily for the reason that special methods and applications can be demonstrated in the most transparent manner and because out of all the Fourier series, the trigonometric are used most frequently.

In the present chapter, we usually will not mention the convergence of the Fourier series under consideration. However, the reader can easily verify (by using the properties of the expanded function) which type of convergence is involved: theorems in Chapters 2 and 7 are sufficient for this purpose.

Another important remark: Referring to a periodic function (with period $2l$) in the following sections, we have one of the following two possibilities in mind:

1. The function is defined for almost all $x \in (-\infty, \infty)$ and is periodic, i.e. $f(x + 2kl) = f(x)$ for every integer k.

2. The function is defined at almost all points in an open interval $(a, a + 2l)$ of length $2l$ and is periodically extended, i.e. we define

$$f(x) = f(x - 2kl)$$

for $x \in (a + 2kl, a + 2(k + 1)l)$. At points $x = a + 2kl$ we put

$$f(a + 2kl) = \tfrac{1}{2}[f(a + 0) + f(a + 2l - 0)]$$

provided both limits — from the right at the point a and from the left at the point $a + 2l$ — exist and are finite; if this is not the case, the function $f(x)$ is not defined at the points $x = a + 2kl$.

It is quite natural to define the function at the points $x = a + 2kl$ just as the mean of limits from the right and from the left. Actually, we shall see in Chapter 7 that the Fourier series of a reasonably well-behaved (e.g. piece-wise smooth) function tends at a point of jump to the arithmetic mean of the limits from the right and from the left. Consequently, we are going to use this mean (provided it exists and is finite) for defining the value of the function at any point of jump (discontinuity of the first kind) interior to the interval $(a, a + 2l)$.

5.1 EXPANSIONS OF VARIOUS FUNCTIONS

Let the function $f(x)$ be periodic with period 2π. If

$$f(x) \in L_2(a, a + 2\pi),$$

it can be associated with the Fourier series

CALCULATION OF FOURIER SERIES

$$f(x) = \frac{a_0}{2} + \sum_{n=1}^{\infty} (a_n \cos nx + b_n \sin nx), \qquad (5.1)$$

where

$$a_n = \frac{1}{\pi} \int_a^{a+2\pi} f(x) \cos nx \, dx \quad (n = 0, 1, 2, \ldots);$$

$$b_n = \frac{1}{\pi} \int_a^{a+2\pi} f(x) \sin nx \, dx \quad (n = 1, 2, \ldots). \qquad (5.2)$$

The Fourier coefficients a_n and b_n may also be obtained by an integration over any interval of length 2π, since an integral of a periodic function over an interval of length equal to the period does not depend on the position of this interval.

In most cases, the intervals $(-\pi, \pi)$ or $(0, 2\pi)$ are considered, i.e., we consider a function defined on one of these intervals and periodically extended in the sense of the introductory remarks.

If the function $f(x)$ is periodic with period $2l$, the corresponding Fourier series reads

$$f(x) = \frac{a_0}{2} + \sum_{n=1}^{\infty} \left(a_n \cos \frac{n\pi x}{l} + b_n \sin \frac{n\pi x}{l} \right), \qquad (5.3)$$

where

$$a_n = \frac{1}{l} \int_a^{a+2l} f(x) \cos \frac{n\pi x}{l} \, dx \, ;$$

$$b_n = \frac{1}{l} \int_a^{a+2l} f(x) \sin \frac{n\pi x}{l} \, dx \quad (n = 0, 1, 2, \ldots). \qquad (5.4)$$

Analogously as in the preceding case, in (5.4) we may integrate over an arbitrary interval of length $2l$.

In most cases, the intervals $(-l, l)$ and $(0, 2l)$ are considered.

Example 5.1. Let $f(x) = x$ for $x \in (0, 2\pi)$, $f(0) = f(2\pi) = \pi$. Then

$$a_0 = \frac{1}{\pi} \int_0^{2\pi} x \, dx = 2\pi;$$

$$a_n = \frac{1}{\pi} \int_0^{2\pi} x \cos nx \, dx = 0 \quad (n = 1, 2, \ldots);$$

$$b_n = \frac{1}{\pi} \int_0^{2\pi} x \sin nx \, dx = -\frac{2}{n} \quad (n = 1, 2, \ldots).$$

Thus we have for $0 < x < 2\pi$,

$$x = \pi - 2\left[\sin x + \frac{\sin 2x}{2} + \frac{\sin 3x}{3} + \ldots\right] \qquad (5.5)$$

(see Fig. 5.1, where the function is periodically extended).

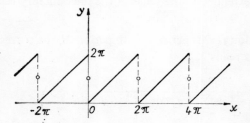

Fig. 5.1

Example 5.2. Let us find the Fourier series of the function $f(x) = |\sin x|$. This function is periodic with period 2π (even with period π) and is even; thus, we are going to calculate its Fourier coefficients for the interval $(-\pi, \pi)$. The coefficients b_n vanish (for f is even), whereas for the coefficients a_n we have

$$a_0 = \frac{1}{\pi}\left[\int_0^{\pi} \sin x \, dx - \int_{-\pi}^0 \sin x \, dx\right] = \frac{2}{\pi} \int_0^{\pi} \sin x \, dx = \frac{4}{\pi};$$

$$a_1 = \frac{2}{\pi} \int_0^{\pi} \sin x \cos x \, dx = 0;$$

$$a_n = \frac{2}{\pi}\int_0^\pi \sin x \cos nx \, dx = -\frac{2}{\pi} \cdot \frac{(-1)^n + 1}{n^2 - 1}$$

$$= \begin{cases} 0 & \text{for } n \text{ odd.,} \\ -(4/\pi)(1/(n^2-1)) & \text{for } n \text{ even..} \end{cases}$$

Thus, for all x,

$$|\sin x| = \frac{2}{\pi} - \frac{4}{\pi}\left[\frac{\cos 2x}{1 \cdot 3} + \frac{\cos 4x}{3 \cdot 5} + \frac{\cos 6x}{5 \cdot 7} + \ldots\right]. \quad (5.6)$$

(See Fig. 5.2.)

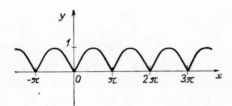

Fig. 5.2

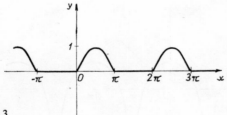

Fig. 5.3

Example 5.3. Let us find the Fourier series of the function defined as follows:

$$f(x) = \begin{cases} \sin x & \text{whenever } x \in [0, \pi], \\ 0 & \text{whenever } x \in [\pi, 2\pi]. \end{cases}$$

(See Fig. 5.3, where this function is periodically extended.) Using some calculations from the preceding example we obtain

$$a_0 = \frac{1}{\pi}\int_0^\pi \sin x \, dx = \frac{2}{\pi};$$

$$a_n = \frac{1}{\pi} \int_0^\pi \sin x \cos nx \, dx = \begin{cases} 0 & \text{for } n \text{ odd}, \\ -(2/\pi)(1/(n^2-1)) & \text{for } n \text{ even}. \end{cases}$$

$$b_1 = \frac{1}{\pi} \int_0^\pi \sin x \sin x \, dx = \tfrac{1}{2};$$

$$b_n = \frac{1}{\pi} \int_0^\pi \sin x \sin nx \, dx = 0 \quad \text{for} \quad n = 2, 3, \ldots.$$

Hence,

$$f(x) = \frac{1}{\pi} + \frac{1}{2} \sin x - \frac{2}{\pi} \left[\frac{\cos 2x}{1 \cdot 3} + \frac{\cos 4x}{3 \cdot 5} + \frac{\cos 6x}{5 \cdot 7} + \ldots \right]. \quad (5.7)$$

Example 5.4. Let $f(x) = \log(2 \cos x/2)$ for $x \in (-\pi, \pi)$. This function (along with its periodic extension onto the entire real line) is not defined at the points $x = \pm(2k+1)\pi$ $(k = 0, 1, 2, \ldots)$; both limits from the left and from the right are $-\infty$ (see Fig. 5.4).

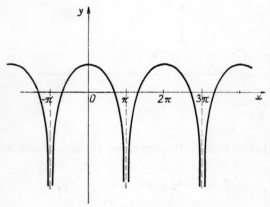

Fig. 5.4

However, it is continuous and integrable in the open interval $(-\pi, \pi)$ and consequently, its Fourier coefficients can be calculated. Since $f(x)$ is even, the coefficients b_n vanish. The coefficients a_n can be calculated directly; because this is rather laborious and we will obtain them later in a simpler way (cf. Example 5.26),

CALCULATION OF FOURIER SERIES

we state only the result: $a_0 = 0$, $a_n = (-1)^{n+1}/n$ $(n = 1, 2, \ldots)$. So we have for $-\pi < x < \pi$

$$\log\left(2\cos\frac{x}{2}\right) = \cos x - \frac{\cos 2x}{2} + \frac{\cos 3x}{3} - \ldots$$

$$= \sum_{n=1}^{\infty} (-1)^{n+1} \frac{\cos nx}{n}. \tag{5.8}$$

Example 5.5. Let the function $f(x)$ be defined on $[0, 2\pi]$ as follows:

$$f(x) = \begin{cases} 1 - x/h & \text{for } x \in (0, h], \\ 0 & \text{for } x \in [h, 2\pi), \\ \tfrac{1}{2} & \text{for } x = 0 \text{ and } x = 2\pi \end{cases}$$

(for the periodic extension see Fig. 5.5). Then

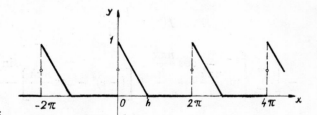

Fig. 5.5

$$a_0 = \frac{1}{\pi} \int_0^h \left(1 - \frac{x}{h}\right) dx = \frac{h}{2\pi};$$

$$a_n = \frac{1}{\pi} \int_0^h \left(1 - \frac{x}{h}\right) \cos nx \, dx = \frac{1}{n^2 \pi h} [1 - \cos nh] \quad (n \geq 1);$$

$$b_n = \frac{1}{\pi} \int_0^h \left(1 - \frac{x}{h}\right) \sin nx \, dx = \frac{1}{n^2 \pi h} [nh - \sin nh] \quad (n \geq 1).$$

Denoting the extended function again by $f(x)$, we have for all x,

$$f(x) = \frac{h}{4\pi} + \frac{1}{\pi h} \sum_{n=1}^{\infty} \left[(1 - \cos nh) \frac{\cos nx}{n^2} + (nh - \sin nh) \frac{\sin nx}{n^2}\right].$$

Example 5.6. Consider the function

$$f(x) = \begin{cases} 0 & \text{for} \quad x \in (-l, h), \\ \tfrac{1}{2} & \text{for} \quad x = -l, h, l, \\ 1 & \text{for} \quad x \in (h, l). \end{cases}$$

Then

$$a_0 = \frac{1}{l}\int_h^l dx = \frac{1}{l}(l - h);$$

$$a_n = \frac{1}{l}\int_h^l \cos\frac{n\pi x}{l} dx = -\frac{1}{n\pi}\sin\frac{n\pi h}{l} \quad (n \geq 1);$$

$$b_n = \frac{1}{l}\int_h^l \sin\frac{n\pi x}{l} dx = -\frac{1}{n\pi}\left[(-1)^n - \cos\frac{n\pi h}{l}\right] \quad (n \geq 1),$$

and thus we have for $x \in [-l, l]$

$$f(x) = \frac{l - h}{2l}$$

$$-\frac{1}{\pi}\sum_{n=1}^{\infty}\left\{\sin\frac{n\pi h}{l}\frac{\cos\dfrac{n\pi x}{l}}{n} + \left[(-1)^n - \cos\frac{n\pi h}{l}\right]\frac{\sin\dfrac{n\pi x}{l}}{n}\right\}.$$

For example, if $h = 0$, then $f(x) = \tfrac{1}{2}\,\text{sgn}\,x + \tfrac{1}{2}$; compare this (for $l = \pi$) with the result of Example 2.1.

5.2. EXPANSIONS OF FUNCTIONS DEFINED ON AN INTERVAL OF HALF LENGTH

As mentioned in preceding chapters, the coefficients b_n from (5.2) corresponding to an even function (considered, for example, in $(-\pi, \pi)$) are equal to zero, whereas the coefficients a_n vanish for an odd function.

Hence, having a function $f(x)$, defined in the interval $(0, \pi)$ we can extend it onto the interval $(-\pi, 0)$ either so that the extended function is even and the cosine expansion is then obtained, or so that it is odd and the sine expansion is obtained.

Thus, there are two possibilities for the interval $(0, \pi)$: either

$$f(x) = \frac{a_0}{2} + \sum_{n=1}^{\infty} a_n \cos nx \quad \text{where} \quad a_n = \frac{2}{\pi} \int_0^\pi f(x) \cos nx \, dx$$
$$(n = 0, 1, 2, \ldots), \tag{5.9}$$

or

$$f(x) = \sum_{n=1}^{\infty} b_n \sin nx \quad \text{where} \quad b_n = \frac{2}{\pi} \int_0^\pi f(x) \sin nx \, dx$$
$$(n = 1, 2, \ldots). \tag{5.10}$$

Fig. 5.6 clarifies the meaning of these formulas. Fig. 5.6a represents the original function in $(0, \pi)$, Fig. 5.6b represents the function with the expansion (5.9), which is in fact expansion (5.1) for the corresponding even function, and Fig. 5.6c represents the function with the expansion (5.10), which is again expansion (5.1) but for the corresponding odd function.

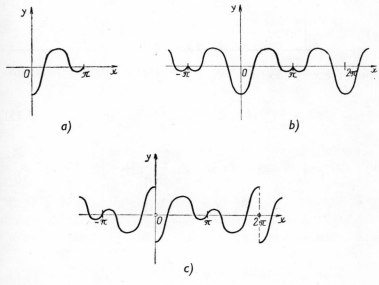

Fig. 5.6

Analogously, for functions defined in the interval $(0, l)$ we have either

$$f(x) = \frac{a_0}{2} + \sum_{n=1}^{\infty} a_n \cos \frac{n\pi x}{l} \quad \text{where} \quad a_n = \frac{2}{l} \int_0^l f(x) \cos \frac{n\pi x}{l} \, dx$$
$$(n = 0, 1, \ldots) \tag{5.11}$$

or

$$f(x) = \sum_{n=1}^{\infty} b_n \sin \frac{n\pi x}{l} \quad \text{where} \quad b_n = \frac{2}{l} \int_0^l f(x) \sin \frac{n\pi x}{l} \, dx$$
$$(n = 1, 2, \ldots). \tag{5.12}$$

All this will become obvious after going through the following examples.

Example 5.7. For (a) the sine-expansion and (b) the cosine-expansion of the function $f(x) = e^{\alpha x}$ in $(0, \pi)$ (α being a real number $\neq 0$) we have

(a) $\quad b_n = \dfrac{2}{\pi} \displaystyle\int_0^\pi e^{\alpha x} \sin nx \, dx = \dfrac{2}{\pi} \dfrac{n}{\alpha^2 + n^2} [1 - (-1)^n e^{\alpha \pi}],$

hence, for $x \in (0, \pi)$,

$$e^{\alpha x} = \frac{2}{\pi} \sum_{n=1}^{\infty} [1 - (-1)^n e^{\alpha \pi}] \frac{n}{\alpha^2 + n^2} \sin nx. \tag{5.13}$$

(b) $\quad a_n = \dfrac{2}{\pi} \displaystyle\int_0^\pi e^{\alpha x} \cos nx \, dx = \dfrac{2}{\pi} \cdot \dfrac{\alpha}{\alpha^2 + n^2} [(-1)^n e^{\alpha \pi} - 1]$

$$\text{for} \quad n = 0, 1, 2, \ldots;$$

hence, even for $x \in [0, \pi]$

$$e^{\alpha x} = \frac{e^{\alpha \pi} - 1}{\alpha \pi} + \frac{2\alpha}{\pi} \sum_{n=1}^{\infty} \frac{(-1)^n e^{\alpha \pi} - 1}{\alpha^2 + n^2} \cos nx. \tag{5.14}$$

CALCULATION OF FOURIER SERIES

The reader is advised to realize the difference between these two cases and the example in Problem 2.1d, and draw the graphs of the periodically extended functions in all three cases.

Example 5.8. For the cosine-expansion of the function $f(x) = x$ in $[0, \pi]$ we obtain,

$$a_0 = \frac{2}{\pi} \int_0^\pi x \, dx = \pi \, ; \quad a_n = \frac{2}{\pi} \int_0^\pi x \cos nx \, dx = \frac{2}{\pi n^2}[(-1)^n - 1]$$

$$= \begin{cases} -(4/\pi n^2) & \text{for } n \text{ odd}, \\ 0 & \text{for } n \text{ even}. \end{cases}$$

Thus, in $[0, \pi]$,

$$x = \frac{\pi}{2} - \frac{4}{\pi}\left[\cos x + \frac{\cos 3x}{3^2} + \frac{\cos 5x}{5^2} + \ldots\right]. \quad (5.15)$$

Observe that the function $f(x) = x$ in $[0, \pi]$ extended onto $[-\pi, 0]$ so as to get an even function is in fact the function $|x|$ in $[-\pi, \pi]$ and thus the expansion (5.15) is also the expansion of $f(x) = |x|$ in $[-\pi, \pi]$ (compare with Example 2.2).

Example 5.9. The sine-expansion of the function $f(x) = \cos x$ in $(0, \pi)$, $f(0) = f(\pi) = 0$, reads

$$b_1 = \frac{2}{\pi} \int_0^\pi \cos x \sin x \, dx = 0 \, ;$$

$$b_n = \frac{2}{\pi} \int_0^\pi \cos x \sin nx \, dx = \begin{cases} 4n/(n^2 - 1)\pi & \text{for } n \text{ even}, \\ 0 & \text{for } n \text{ odd}. \end{cases}$$

Thus, we have in $(0, \pi)$

$$\cos x = \frac{4}{\pi}\left[\frac{2}{3}\sin 2x + \frac{4}{15}\sin 4x + \frac{6}{35}\sin 6x + \ldots\right]$$

$$= \frac{8}{\pi} \sum_{n=1}^\infty \frac{n}{4n^2 - 1} \sin 2nx \, .$$

The graph of the function corresponding to the preceding series, is plotted in Fig. 10.14.

Example 5.10. The cosine-expansion of the function $f(x) = x^2$ in $[0, 2\pi]$. Let us recall that the function $f(x) = x^2$ in $[0, 2\pi]$ has already been expanded in the Fourier series in Example 4.3 for $l = \pi$, and that a Fourier series containing both sine and cosine terms was obtained. Looking for the cosine-expansion we have to consider the interval $(0, 2\pi)$ as the interval $(0, l)$ with $l = 2\pi$; then we obtain the expansion with respect to the functions $\cos(n\pi x/l) = \cos(nx/2)$:

$$a_0 = \frac{2}{l} \int_0^l x^2 \, dx = \frac{1}{\pi} \int_0^{2\pi} x^2 \, dx = \frac{8\pi^2}{3};$$

$$a_n = \frac{1}{\pi} \int_0^{2\pi} x^2 \cos\frac{n}{2} x \, dx = (-1)^n \frac{16}{n^2} \quad (n = 1, 2, \ldots).$$

Thus, for $x \in [0, 2\pi]$

$$x^2 = \frac{4\pi^2}{3} + 16 \sum_{n=1}^{\infty} \frac{(-1)^n}{n^2} \cos\frac{n}{2} x. \tag{5.16}$$

It should be noted that the expansion (5.16) corresponds also to the even function $f(x) = x^2$ in $(-2\pi, 2\pi)$. Hence, expansion (5.16) may be obtained by using (4.25) with $l = 2\pi$.

Remark. In an analogous way we could expand functions defined on an interval of length equal to one fourth, one eighth, etc., of the length of the original interval. Referring to results in Chapter 4, Section 4.2.4, we leave a detailed treatment to the reader.

5.3 EXPANSIONS OF A NON-PERIODIC FUNCTION

If a function $f(x)$ is not periodic and is defined, for example, on the entire interval $(-\infty, \infty)$ we have to realize that the Fourier series of this function in $(a, a + 2l)$, given by formulas (5.3) and (5.4), describes the function $f(x)$ in general only in this interval.*

---

* Even then not always; for if the function has jumps inside this interval, the sum of the corresponding Fourier series is not $f(x)$, but $\frac{1}{2}[f(x + 0) + f(x - 0)]$ (cf. Chapter 7).

When extending periodically non-periodic functions in some preceding examples, we did not consider the values at the end-points. Actually, if we considered a closed interval $[a, a + 2l]$, a periodic extension with period $2l$ would be impossible unless $f(a) = f(a + 2l)$. Therefore, we usually choose the function in $(a, a + 2l)$ and define it at the end-points according to the introductory remark by

$$f(a) = f(a + 2l) = \tfrac{1}{2}[f(a + 0) + f(a + 2l - 0)] \quad (5.17)$$

(provided this is possible — cf. Example 5.4).

Hence, the Fourier series (5.3) of a non-periodic function $f(x)$ describes the function $f(x)$ in the interval $(a, a + 2l)$. At the end-points, the Fourier series of a sufficiently well-behaved function assumes the value (5.17) and outside the interval it is the periodic extension of its values in $[a, a + 2l)$ (with period $2l$).

If we wish to describe the behaviour of a function $f(x)$ by its Fourier series as fully as possible we have to choose the interval $(a, a + 2l)$, i.e., the number l, as large as possible. The larger the number l is, the better the Fourier series describes the original function $f(x)$.*

Thus, it is quite natural to ask what happens if, e.g., for the interval $(-l, l)$, l tends to infinity. The analogue of the Fourier series for $l = \infty$ is the Fourier transform which is to be dealt with in Chapter 8.

Example 5.11. Consider the function $f(x) = x$. In the interval $(-\pi, \pi)$ it is expressed by the series

$$x = 2\left(\sin x - \frac{\sin 2x}{2} + \frac{\sin 3x}{3} - \frac{\sin 4x}{4} + \ldots\right)$$

(cf. Problem 2.1a). In the interval $(-10\pi, 10\pi)$ it is given by the series

$$x = 20\left(\sin \frac{x}{10} - \frac{\sin(2x/10)}{2} + \frac{\sin(3x/10)}{3} - \frac{\sin(4x/10)}{4} + \ldots\right).$$

---

* This, of course, depends also on the position of the point a which may be chosen in different ways.

The first series corresponds to the periodic function plotted in Fig. 5.7a, the second one to the periodic function in Fig. 5.7b.

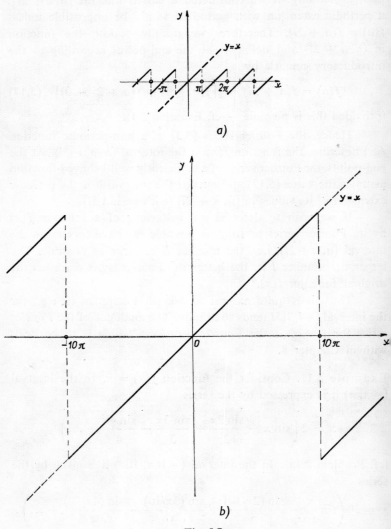

Fig. 5.7

5.4 COMPLEX AND PHASE FORMS OF THE FOURIER SERIES

For the sake of completeness, let us recall two further forms of the Fourier series, i.e., (5.1) and (5.3) derived in Chapter 2.

For a function periodic with period 2π we have

$$f(x) = \sum_{n=-\infty}^{\infty} c_n e^{inx} \quad \text{where} \quad c_n = \frac{1}{2\pi} \int_a^{a+2\pi} f(x) e^{-inx} \, dx$$

($n = 0, \pm 1, \pm 2, \ldots$), whereas for a function periodic with period $2l$ we have

$$f(x) = \sum_{n=-\infty}^{\infty} c_n e^{i(n\pi x/l)}, \quad \text{where} \quad c_n = \frac{1}{2l} \int_a^{a+2l} f(x) e^{-i(n\pi x/l)} \, dx.$$

The complex form is more advantageous, because it is simpler, all the Fourier coefficients are given by a single formula, and is better suited for theoretical considerations.

The phase form (for functions periodic with period 2π, cf. p. 31),

$$f(x) = \sum_{n=0}^{\infty} h_n \sin(nx + \varphi_n) \quad \text{or} \quad f(x) = \sum_{n=0}^{\infty} k_n \cos(nx + \psi_n)$$

is particularly suitable for some engineering applications.

5.5 APPLICATION OF VARIOUS OPERATIONS WITH SERIES

For the sake of simplicity we are going to deal mostly with functions periodic with period 2π. However, the reader will certainly realize that this limitation is by no means essential.

A. *Substitution*

The function $f(x)$ defined on a finite interval (a, b) may be transformed by a linear substitution

$$x = At + B \quad (A \text{ and } B \text{ being real numbers}, A \neq 0)$$

into a function $g(t)$ which is defined on some of the intervals discussed in preceding sections, e.g. $(-\pi, \pi)$, $(0, 2\pi)$, $(0, \pi)$ etc. The function $g(t) = f(At + B)$ is expanded in the corresponding Fourier series

$$g(t) = \frac{a_0}{2} + \sum_{n=1}^{\infty} (a_n \cos nt + b_n \sin nt) ;$$

using then the inverse substitution $t = (1/A)x - B/A$, the series for $f(x)$ in (a, b) is obtained,

$$f(x) = \frac{a_0}{2} + \sum_{n=1}^{\infty} \left[a_n \cos\left(\frac{nx}{A} - \frac{nB}{A}\right) + b_n \sin\left(\frac{nx}{A} - \frac{nB}{A}\right) \right].$$

Applying various substitutions to series already known, we obtain expansions of various other functions, of course, on intervals other than the original ones.

Example 5.12. By the substitution $x = t - \pi$, the interval $(-\pi, \pi)$ (for the variable x) is mapped onto $(0, 2\pi)$ (for the variable t). Since $\cos nx = \cos n(t - \pi) = (-1)^n \cos nt$ and $\cos x/2 = \cos(t - \pi)/2 = \sin t/2$, the function $\log(2 \cos x/2)$ from Example 5.4 is transformed to $\log(2 \sin t/2)$; using formula (5.8), we get for $t \in (0, 2\pi)$ the expansion,

$$\log\left(2 \sin \frac{t}{2}\right) = -\cos t - \frac{\cos 2t}{2} - \frac{\cos 3t}{3} - \ldots = -\sum_{n=1}^{\infty} \frac{\cos nt}{n}.$$

Example 5.13. Denote by $f_0(x)$ the function from Example 5.3 extended periodically onto $(-\infty, \infty)$, i.e.

$$f_0(x) = \begin{cases} \sin x & \text{for } x \in [k\pi, (k+1)\pi] \\ 0 & \text{for } x \in [(k+1)\pi, (k+2)\pi] \end{cases} \quad (k \text{ even}).$$

CALCULATION OF FOURIER SERIES

Denote

$$f_1(x) = f_0\left(x - \frac{2\pi}{3}\right) \quad \text{and} \quad f_2(x) = f_0\left(x + \frac{2\pi}{3}\right)$$

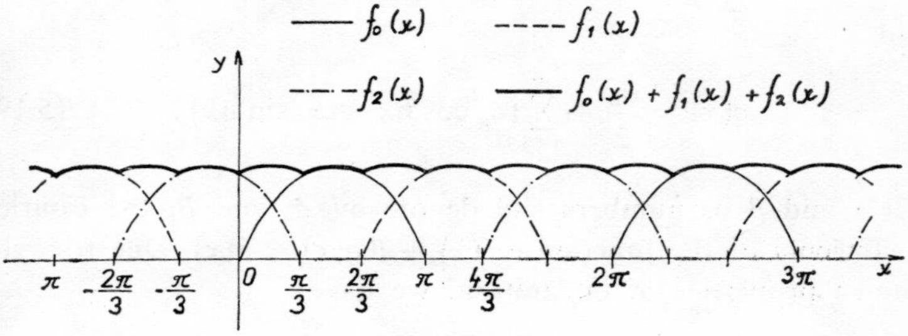

Fig. 5.8

(see Fig. 5.8). Using these substitutions in formula (5.7) we obtain the expansions of the functions f_1 and f_2. Since

$$f_0(x) = \frac{1}{\pi} + \frac{1}{2}\sin x$$

$$- \frac{2}{\pi}\sum_{k=0}^{\infty}\left[\frac{\cos(6k+2)x}{(6k+2)^2 - 1} + \frac{\cos(6k+4)x}{(6k+4)^2 - 1} + \frac{\cos(6k+6)x}{(6k+6)^2 - 1}\right],$$

we finally have (the upper signs on the right hand side correspond to the upper signs on the left hand side and similarly for the lower signs)

$$f_0\left(x \pm \frac{2\pi}{3}\right) = \frac{1}{\pi} - \frac{1}{4}\sin x \pm \frac{\sqrt{3}}{4}\cos x$$

$$- \frac{2}{\pi}\sum_{k=0}^{\infty}\left[\frac{-\frac{1}{2}\cos(6k+2)x \pm \frac{\sqrt{3}}{2}\sin(6k+2)x}{(6k+6)^2 - 1}\right.$$

$$\left. + \frac{-\frac{1}{2}\cos(6k+4)x \mp \frac{\sqrt{3}}{2}\sin(6k+4)x}{(6k+4)^2 - 1} + \frac{\cos(6k+6)x}{(6k+6)^2 - 1}\right].$$

B. *Addition and multiplication*

Let
$$f(x) = \frac{a_0}{2} + \sum_{n=1}^{\infty} (a_n \cos nx + b_n \sin nx) \tag{5.18}$$
and
$$g(x) = \frac{\tilde{a}_0}{2} + \sum_{n=1}^{\infty} (\tilde{a}_n \cos nx + \tilde{b}_n \sin nx). \tag{5.19}$$

Let α and β be numbers and denote by A_n and B_n the Fourier coefficients of the function $\alpha f(x) + \beta g(x) = h(x)$. Due to well-known properties of integration, we have

$$\int_a^b [\alpha p(x) + \beta q(x)] \, dx = \alpha \int_a^b p(x) \, dx + \beta \int_a^b q(x) \, dx \ ;$$

using this formula for calculation of the Fourier coefficients of the function h, we obtain

$$A_n = \alpha a_n + \beta \tilde{a}_n \ ; \quad B_n = \alpha b_n + \beta \tilde{b}_n \ .$$

Thus, Fourier series of the function $\alpha f(x) + \beta g(x)$ may be found by multiplying the series (5.18) by α, the series (5.19) by β and by adding formally the two series obtained. However, the equalities in (5.18) and (5.19) must hold on the same intervals.

Example 5.14. The Fourier series of the function $Ax^2 + Bx + C$ may be obtained by using the Fourier series of the functions x^2 and x.

(a) Using the results of Example 4.3 for $l = \pi$ and Example 5.1, we have in $(0, 2\pi)$,

$$Ax^2 + Bx + C = \frac{4A\pi^2}{3} + B\pi + C$$
$$+ \sum_{n=1}^{\infty} \left[4A \frac{\cos nx}{n^2} - (4A\pi + 2B) \frac{\sin nx}{n} \right]. \tag{5.20}$$

(b) In $(-\pi, \pi)$, we have by Problem 2.1.a and b

$$Ax^2 + Bx + C = \frac{A\pi^2}{3} + C$$

$$+ \sum_{n=1}^{\infty} \left[4A(-1)^n \frac{\cos nx}{n^2} - 2B(-1)^n \frac{\sin nx}{n} \right]. \quad (5.21)$$

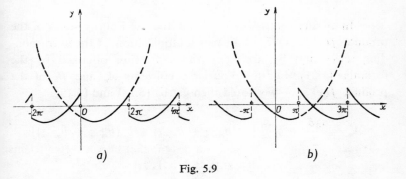

Fig. 5.9

In Fig. 5.9a and b there are plotted the functions corresponding to the series on the right hand side of (5.20) and (5.21), respectively, i.e. the functions $Ax^2 + Bx + C$ extended periodically from the basic intervals and defined as the arithmetic means of the limits from the right and from the left at 0 and 2π and at $-\pi$ and π respectively.

Example 5.15. Summing up the functions f_0, f_1 and f_2 from Example 5.13 we obtain a function plotted in Fig. 5.8. The Fourier series of the function

$$F(x) = f_0(x) + f_0\left(x - \frac{2\pi}{3}\right) + f_0\left(x + \frac{2\pi}{3}\right)$$

thus has the form

$$F(x) = \frac{3}{\pi} - \frac{6}{\pi} \left[\frac{\cos 6x}{35} + \frac{\cos 12x}{143} + \ldots \right] = \frac{3}{\pi} - \frac{6}{\pi} \sum_{k=1}^{\infty} \frac{\cos 6kx}{36k^2 - 1}$$

and this formula holds for all x. The function $F(x)$ has applications, e.g., in electrical engineering. It has the period $\pi/3$.

Example 5.16. The Fourier series of the function $f(x)$ from Example 5.3 may be obtained from the Fourier series of the function $|\sin x|$ (cf. Example 5.2). Since $f(x) = \frac{1}{2}|\sin x| + \frac{1}{2}\sin x$, we get (5.7) from (5.6) without calculating the Fourier coefficients.

In an analogous way we could find the Fourier series of the product $f(x) \cdot g(x)$ by a formal multiplication of the corresponding series and by arranging the expression obtained by the formulas (1.21). For the Fourier coefficients A_n and B_n of the product $f(x) g(x)$ we would then get by (5.18) and (5.19),

$$A_n = \frac{a_0 \tilde{a}_n}{2} + \frac{1}{2} \sum_{m=1}^{\infty} \left[a_m(\tilde{a}_{m+n} + \tilde{a}_{m-n}) + b_m(\tilde{b}_{m+n} + \tilde{b}_{m-n}) \right]$$

$$(n = 0, 1, 2, \ldots),$$

$$B_n = \frac{a_0 \tilde{b}_n}{2} + \frac{1}{2} \sum_{m=1}^{\infty} \left[a_m(\tilde{b}_{m+n} - \tilde{b}_{m-n}) + b_m(\tilde{a}_{m+n} - \tilde{a}_{m-n}) \right]$$

$$(n = 1, 2, \ldots), \qquad (5.22)$$

where $\tilde{a}_{-k} = \tilde{a}_k$ and $\tilde{b}_{-k} = -\tilde{b}_k$ for $k > 0$.

These formulas were obtained by a formal procedure; however, it can be easily verified that for f and g from $L_2(a, a + 2\pi)$ the series (5.22) are meaningful (cf. Problems 5.8, 5.9 and 5.10).

The formulas (5.22) are, of course, rather complicated; consequently, this procedure is applicable in practice only if the Fourier series of one of the functions is finite, i.e. if one of the functions f and g is a trigonometric polynomial of a not very high degree.

Example 5.17. The Fourier series of the function $f(x) = x \sin x$ in $[-\pi, \pi]$ may be found by using the result of Problem 2.1a:

Since $x = 2 \sum_{n=1}^{\infty} (-1)^{n+1} (\sin nx)/n$ in $(-\pi, \pi)$ and $\sin nx \sin x = \frac{1}{2}[\cos(n-1)x - \cos(n+1)x]$, we have after several rearrangements,

$$x \sin x = 1 - \tfrac{1}{2} \cos x + 2 \sum_{n=2}^{\infty} (-1)^n \frac{\cos nx}{n^2 - 1}.$$

C. Term by term integration of Fourier series

Let $f(x)$ be a function periodic with period 2π (absolutely) integrable in $(-\pi, \pi)$. Then it can be associated with the Fourier series (which need not converge even in the mean),

$$f(x) \sim \frac{a_0}{2} + \sum_{n=1}^{\infty} (a_n \cos nx + b_n \sin nx).$$

The function

$$F(x) = \int_0^x \left[f(t) - \frac{a_0}{2} \right] dt \qquad (5.23)$$

is continuous and periodic with period 2π, for

$$F(x) - F(x - 2\pi) = \int_{x-2\pi}^{x} \left[f(t) - \frac{a_0}{2} \right] dt$$

$$= \int_{-\pi}^{\pi} \left[f(t) - \frac{a_0}{2} \right] dt = \int_{-\pi}^{\pi} f(t) \, dt - \pi a_0 = 0.$$

The function $F(x)$ is absolutely continuous and differentiable almost everywhere, $F'(x) = f(x) - a_0/2$, and consequently, has in $(-\infty, \infty)$ the Fourier series

$$F(x) = \frac{A_0}{2} + \sum_{n=1}^{\infty} (A_n \cos nx + B_n \sin nx), \qquad (5.24)$$

where the right hand side series is uniformly convergent for all x (cf. Chapter 7, Theorem 7.10).

For $n \geq 1$, there is a very simple relation between the coefficients A_n, B_n and a_n, b_n. Actually,

$$A_n = \frac{1}{\pi} \int_{-\pi}^{\pi} F(x) \cos nx \, dx$$

$$= \frac{1}{\pi n} [F(x) \sin nx]_{-\pi}^{\pi} - \frac{1}{n\pi} \int_{-\pi}^{\pi} F'(x) \sin nx \, dx$$

$$= -\frac{1}{n\pi} \int_{-\pi}^{\pi} \left[f(x) - \frac{a_0}{2} \right] \sin nx \, dx$$

$$= -\frac{1}{n\pi} \int_{-\pi}^{\pi} f(x) \sin nx \, dx = -\frac{b_n}{n};$$

similarly, we can show that $B_n = a_n/n$. The coefficient A_0 may be derived from (5.24) for $x = 0$ (because $F(0) = 0$),

$$\frac{A_0}{2} = -\sum_{n=1}^{\infty} A_n = \sum_{n=1}^{\infty} \frac{b_n}{n} \tag{5.25}$$

or directly from the definition of the coefficient A_0 by integrating expression (5.23).

Thus, we have finally

$$F(x) = \sum_{n=1}^{\infty} \frac{a_n \sin nx + b_n(1 - \cos nx)}{n} \tag{5.26}$$

and the following proposition holds.

Theorem 5.1. *The Fourier series of the function $\int_0^x f(t) \, dt$ may be obtained from the Fourier series of the function $f(t)$ by a term by term integration, i.e.,*

$$\int_0^x f(t) \, dt = \int_0^x \frac{a_0}{2} \, dt + \sum_{n=1}^{\infty} \int_0^x (a_n \cos nt + b_n \sin nt) \, dt.$$

The proof follows from (5.23) and (5.26), since

$$\int_0^x (a_n \cos nt + b_n \sin nt) \, dt = \frac{1}{n} [a_n \sin nx + b_n(1 - \cos nx)].$$

CALCULATION OF FOURIER SERIES

According to (5.26) we have

$$\int_0^x f(t)\,dt = \int_0^x \frac{a_0}{2}\,dt + \sum_{n=1}^\infty \frac{b_n}{n} + \sum_{n=1}^\infty \frac{-b_n \cos nx + a_n \sin nx}{n},$$
(5.27)

which is a formula useful for practical applications. Since

$$\int_0^x \frac{a_0}{2}\,dt = a_0 \frac{x}{2} = a_0 \sum_{n=1}^\infty (-1)^{n+1} \frac{\sin nx}{n} \quad \text{for} \quad -\pi < x < \pi$$

by Problem 2.1a, we have another formula: for $x \in (-\pi, \pi)$,

$$\int_0^x f(t)\,dt = \sum_{n=1}^\infty \frac{b_n}{n} + \sum_{n=1}^\infty \frac{-b_n \cos nx + (a_n + (-1)^{n+1} a_0) \sin nx}{n}.$$
(5.28)

Formula (5.27) holds for all x, provided the function $f(x)$ is periodic; if $f(x)$ is not periodic and is defined only in $(-\pi, \pi)$, the validity of the formula (5.27) as well as of the formula (5.28) is confined to the interval $(-\pi, \pi)$.

It is necessary to emphasize that the term by term integration of the Fourier series of a function f is permissible without any assumption on the convergence of this series; in other words, the bare fact that the Fourier series has a sense, i.e. that the function is integrable, is sufficient.

In addition to permitting us to find the Fourier series of various functions by a term by term integration, Theorem 5.1 provides another method. Indeed, it is apparent that

$$\int_a^b f(x)\,dx = \frac{a_0}{2}(b - a) + \sum_{n=1}^\infty \frac{a_n(\sin nb - \sin na) - b_n(\cos nb - \cos na)}{n}, \quad (5.29)$$

where a and b may be arbitrary real numbers, provided the function $f(x)$ is periodic with period 2π, and where $-\pi \leq a < b \leq \pi$

in the case of a function $f(x)$ defined in $(-\pi, \pi)$. Thus formula (5.29) may be used for the calculation of definite integrals, if the Fourier coefficients of the integrated function are known and if we are able to sum up the series on the right hand side.

Remark. As a by-product of the preceding considerations we proved that the series

$$\sum_{n=1}^{\infty} \frac{b_n}{n}$$

is convergent for any function $f(x)$, (absolutely) integrable in $(-\pi, \pi)$, even if the corresponding Fourier series does not converge. The sum of this series is equal to the integral $(1/\pi) \int_{-\pi}^{\pi} F(x) \, dx$ which is finite. Let us apply this result.

Example 5.18. The trigonometric series

$$\sum_{n=2}^{\infty} \frac{\sin nx}{\log n}$$

(cf. Example 2.4) is not a Fourier series of any integrable function. For, if it were a Fourier series of a function $f(x)$, the numbers $(1/\log n) = b_n$ would be the Fourier coefficients of an (odd) function $f(x)$ and, by the preceding remark, the series

$$\sum_{n=2}^{\infty} \frac{1}{n \log n} \tag{5.30}$$

would be convergent. However, the series (5.30) is divergent.

As already mentioned, Theorem 5.1 may be used for obtaining Fourier series of various primitive functions. We are going to present two examples:

Example 5.19. We have

$$f(x) = \frac{x}{2} = \sin x - \frac{\sin 2x}{2} + \frac{\sin 3x}{3} - \ldots$$

CALCULATION OF FOURIER SERIES

in the interval $(-\pi, \pi)$. A term by term integration and formula (5.27) or (5.28) yield

$$\frac{x^2}{4} = \int_0^x \frac{t}{2}\,dt = \left(1 - \frac{1}{2^2} + \frac{1}{3^2} - \frac{1}{4^2} + \cdots\right)$$
$$-\cos x + \frac{\cos 2x}{2^2} - \frac{\cos 3x}{3^2} + \cdots = C - \sum_{n=1}^{\infty} \frac{(-1)^{n+1}}{n^2} \cos nx,$$

where $C = \sum_{n=1}^{\infty} (-1)^{n+1} (1/n^2)$. Since we do not know the sum of this series, we shall find it by using the definition of the coefficient A_0, i.e.

$$\int_0^x \left[f(t) - \frac{a_0}{2}\right] dt = \frac{x^2}{4} = F(x)$$

and

$$C = \frac{A_0}{2} = \frac{1}{2\pi} \int_{-\pi}^{\pi} \frac{x^2}{4}\,dx = \frac{\pi^2}{12},$$

thus we have finally,

$$\frac{\pi^2}{12} - \frac{x^2}{4} = \sum_{n=1}^{\infty} (-1)^{n+1} \frac{\cos nx}{n^2} \quad \text{for} \quad x \in (-\pi, \pi). \quad (5.31)$$

Compare this result with that from Example 5.14b for $A = -\frac{1}{4}$, $B = 0$ and $C = \pi^2/12$.

In this example we have also proved that

$$\sum_{n=1}^{\infty} (-1)^{n+1} \frac{1}{n^2} = \frac{\pi^2}{12}$$

(cf. also Section 5.8).

Example 5.20. The Fourier coefficients b_n are equal to zero in the series (5.31). Using term by term integration again, we obtain,

$$\int_0^x \left(\frac{\pi^2}{12} - \frac{t^2}{4}\right) dt = \frac{\pi^2}{12} x - \frac{x^3}{12} = \sum_{n=1}^{\infty} (-1)^{n+1} \frac{\sin nx}{n^3} \quad (5.32)$$

for $x \in (-\pi, \pi)$ and even for $x \in [-\pi, \pi]$.

All the preceding considerations were carried out for the interval $(-\pi, \pi)$; this fact is again not essential, since appropriate rearrangements permit us to use an arbitrary interval of length 2π, or $2l$.

Example 5.21. From Example 5.14a we have for $A = 0$, $B = -\frac{1}{2}$ and $C = \pi/2$ the relation

$$\frac{\pi - x}{2} = \sum_{n=1}^{\infty} \frac{\sin nx}{n} \quad \text{for} \quad x \in (0, 2\pi).$$

Integrating term by term within the limits 0 and x, we obtain

$$\frac{\pi x}{2} - \frac{x^2}{4} = \sum_{n=1}^{\infty} \frac{1}{n^2} - \sum_{n=1}^{\infty} \frac{\cos nx}{n^2}.$$

Since we already know that $\sum_{n=1}^{\infty} 1/n^2 = \pi^2/6$ (cf. Example 4.1; this result could be derived directly from the definition of A_0, i.e.

$$\sum_{n=1}^{\infty} \frac{1}{n^2} = \frac{A_0}{2} = \frac{1}{2\pi} \int_0^{2\pi} \left(\frac{\pi x}{2} - \frac{x^2}{4} \right) dx)$$

we obtain

$$\frac{x^2}{4} - \frac{\pi}{2} x + \frac{\pi^2}{6} = \sum_{n=1}^{\infty} \frac{\cos nx}{n^2} \quad (5.33)$$

for $x \in (0, 2\pi)$ and even for $x \in [0, 2\pi]$. Compare (5.33) with the result of Example 5.14a, choosing properly A, B and C.

D. *Differentiation of Fourier series*

Theorem 5.2. *Let $f(x)$ be continuous in $[-\pi, \pi]$ and let $f(x)$ have a derivative $f'(x)$ there (in the sense of absolutely continuous function),* which is integrable in $(-\pi, \pi)$. Then*

$$f'(x) \sim \frac{c}{2} + \sum_{n=1}^{\infty} \left[(nb_n + (-1)^n c) \cos nx - na_n \sin nx \right], \quad (5.34)$$

---

* Cf. Chapter 1, Sect. 1.4.

CALCULATION OF FOURIER SERIES

where a_n and b_n are the Fourier coefficients of the function $f(x)$ and c is a constant defined by

$$c = \frac{1}{\pi}\left[f(\pi) - f(-\pi)\right]. \;*$$

Proof: Let

$$f'(x) \sim \frac{a_0'}{2} + \sum_{n=1}^{\infty}\left(a_n' \cos nx + b_n' \sin nx\right).$$

The Fourier series is meaningful, since f' is integrable; however, we cannot put the equality sign between f' and its Fourier series in general. Then, by definition

$$a_0' = \frac{1}{\pi}\int_{-\pi}^{\pi} f'(x)\, \mathrm{d}x = \frac{1}{\pi}\left[f(\pi) - f(-\pi)\right] = c\,.$$

Obviously,

$$f'(x) - \frac{c}{2} \sim \sum_{n=1}^{\infty}\left(a_n' \cos nx + b_n' \sin nx\right). \qquad (5.35)$$

By Theorem 5.1, the Fourier series of the function

$$\int_0^x \left[f'(t) - \frac{c}{2}\right] \mathrm{d}t = f(x) - \frac{cx}{2} - f(0)$$

may be found by integrating the series (5.35) term by term; in other words, the series (5.35) may be obtained by a term by term differentiation of the series for $f(x) - cx/2 - f(0)$. Since

$$f(x) = \frac{a_0}{2} + \sum_{n=1}^{\infty}\left(a_n \cos nx + b_n \sin nx\right),$$

---

* The formula (5.34) holds, of course, only for $x \in (-\pi, \pi)$. If considering the function $f(x)$ in $[0, 2\pi]$, we would have to expand the function $x/2$ in $(0, 2\pi)$ and a somewhat different series would be obtained for $f'(x)$.

for $x \in (-\pi, \pi)$, we obtain for $x \in (-\pi, \pi)$ by using the expansion

$$\frac{x}{2} = \sum_{n=1}^{\infty} (-1)^{n+1} \frac{\sin nx}{n}$$

the relation

$$f(x) - \frac{cx}{2} - f(0) = \frac{a_0}{2} - f(0)$$
$$+ \sum_{n=1}^{\infty} \left[a_n \cos nx + \left(b_n + \frac{(-1)^n}{n} c \right) \sin nx \right];$$

differentiating term by term we get

$$f'(x) - \frac{c}{2} \sim \sum_{n=1}^{\infty} \left[-n a_n \sin nx + (n b_n + (-1)^n c) \cos nx \right]$$

which is in fact (5.34).

Corollary: If, under the assumptions of Theorem 5.2, $f(\pi) = f(-\pi)$, then $c = 0$ and hence

$$f'(x) \sim \sum_{n=1}^{\infty} n(b_n \cos nx - a_n \sin nx) \tag{5.36}$$

i.e. the series for $f(x)$ may be formally differentiated term by term. (Compare with Theorem 2.9 which states the same under somewhat more special assumptions.)

Remark. If $f(\pi) = f(-\pi)$, then function $f(x)$ from Theorem 5.2 may be extended continuously and periodically onto $(-\infty, \infty)$; thus the preceding corollary asserts in fact that *if $f(x)$ is continuous and periodic with period 2π in $(-\infty, \infty)$ and if it has a derivative (in the sense of absolutely continuous function) $f'(x) \in L_1(-\pi, \pi)$, then its Fourier series may be differentiated term by term and the Fourier series of the derivative $f'(x)$ is obtained.* In addition, Theorem 5.2 asserts that for non-periodic functions a term by term differentiation of the Fourier series is not permissible in general. (The following fact indicates this as well: the series in (5.34) contains a constant term which should vanish after the differentiation is performed.)

The constant c is determined by the relation $c = \frac{1}{2}[f(\pi) - f(-\pi)]$. If we do not know the function f explicitly, but only its Fourier series, the calculation of the number c is difficult. However, we have

$$c = \lim_{n \to \infty} [(-1)^{n+1} nb_n]. \tag{5.37}$$

This formula follows from the fact that the Fourier coefficients of an absolutely integrable function tend to zero as $n \to \infty$ (cf. Chapter 3, Appendix). Thus, the limit of the Fourier coefficients of $\cos nx$ in (5.34) must be zero, i.e. $\lim_{n \to \infty} [nb_n + (-1)^n c] = 0$, which implies (5.37).

Theorem 5.2 has a disadvantage: it assumes that the function $f(x)$ is continuous and has an integrable derivative — a fact usually difficult to verify, if $f(x)$ is given merely by its Fourier series. Thus, the following theorem is very useful:

Theorem 5.3. *Let us have a series*

$$\frac{a_0}{2} + \sum_{n=1}^{\infty} (a_n \cos nx + b_n \sin nx) \tag{5.38}$$

and let c be the number defined by formula (5.37). If the series

$$\frac{c}{2} + \sum_{n=1}^{\infty} ([nb_n + (-1)^n c] \cos nx - na_n \sin nx) \tag{5.39}$$

is the Fourier series of an integrable function $g(x)$, then the series (5.38) is the Fourier series of the function

$$f(x) = \int_0^x g(t) \, dt + \frac{a_0}{2} + \sum_{n=1}^{\infty} a_n \tag{5.40}$$

and converges to $f(x)$ in the interval $(-\pi, \pi)$. The function $f(x)$ is continuous for $x \in (-\pi, \pi)$ and $f'(x) = g(x)$ almost everywhere in $(-\pi, \pi)$.

Proof: Applying Theorem 5.1 to the series (5.39), formula (5.28) gives the relation

$$\int_0^x g(t)\,dt = -\sum_{n=1}^\infty a_n$$
$$+ \sum_{n=1}^\infty \frac{na_n \cos nx + \left[nb_n + (-1)^n c + (-1)^{n+1} c\right] \sin nx}{n}$$
$$= -\sum_{n=1}^\infty a_n + \sum_{n=1}^\infty (a_n \cos nx + b_n \sin nx);$$

this yields

$$\int_0^x g(t)\,dt + \frac{a_0}{2} + \sum_{n=1}^\infty a_n = \frac{a_0}{2} + \sum_{n=1}^\infty (a_n \cos nx + b_n \sin nx).$$

The properties of the function $f(x)$ follow from this expression.

For differentiating a Fourier series the general theorems on term by term differentiation of a series are also applicable. An immediate analogue of Theorem 1.2 is the following proposition.

Theorem 5.4. *If the trigonometric series*

$$\frac{a_0}{2} + \sum_{n=1}^\infty (a_n \cos nx + b_n \sin nx) \tag{5.41}$$

is convergent at least at one point $x \in (a, b)$ *and if the series of derivatives*

$$\sum_{n=1}^\infty n(b_n \cos nx - a_n \sin nx) \tag{5.42}$$

is uniformly convergent in (a, b), *then the series* (5.41) *is also uniformly convergent and has the sum* $f(x)$, *and the series* (5.42) *converges to the function* $f'(x)$.

Many analogous theorems concerning the second and higher derivatives can be stated; their formulation is left to the reader.

CALCULATION OF FOURIER SERIES

The theorems on differentiating the Fourier series are again applicable to finding the Fourier series of new functions and to the summation of trigonometric series.

Example 5.22. The function $|\sin x|$ is continuous and periodic in $(-\infty, \infty)$ and has there, except for the points $x = k\pi$, a derivative defined by $f'(x) = \operatorname{sgn} x \cdot \cos x$ in $(-\pi, \pi)$ and periodically extended outside $(-\pi, \pi)$. Using the result of Example 5.2 we obtain by (5.36),

$$\operatorname{sgn} x \cos x \sim \frac{8}{\pi}(\tfrac{1}{3}\sin 2x + \tfrac{2}{15}\sin 4x + \tfrac{3}{35}\sin 6x + \ldots) \quad (5.43)$$

in the interval $(-\pi, \pi)$. Theorem 5.2 (or its Corollary) does not ensure the equality but only the sign $\sim$ here; the equality is obtained by using Theorem 5.4: the series for $f'(x)$ has the form

$$\frac{8}{\pi} \sum_{n=1}^{\infty} \frac{n}{4n^2 - 1} \sin 2nx$$

and, by Problem 2.6 is uniformly convergent in each interval not containing the points $x = k\pi$. (Note that the sequence $n/(4n^2 - 1)$ is of bounded variation.) Hence,

$$f'(x) = \frac{8}{\pi} \sum_{n=1}^{\infty} \frac{n}{4n^2 - 1} \sin 2nx$$

for all $x \neq k\pi$ and the equality holds for $x \in (-\pi, \pi)$ in (5.43) (for $x = 0$ the equality is evident).

Example 5.23. Let

$$g(x) = -\tfrac{1}{2} + \cos x + \sum_{n=2}^{\infty} (-1)^n \frac{\cos nx}{n^2 - 1}.$$

This series converges uniformly, and consequently, the function $g(x)$ is periodic and continuous on $(-\infty, \infty)$; moreover, the function $g(x)$ has a derivative in $(-\pi, \pi)$, because, by Theorem 2.6, the series

$$-\sin x - \sum_{n=2}^{\infty} (-1)^n \frac{n}{n^2 - 1} \sin nx$$

converges even uniformly in every closed interval contained in $(-\pi, \pi)$, and thus is a derivative of $g(x)$ there due to Theorem 5.4 (i.e., to Theorem 1.2). According to the Corollary of Theorem 5.2 or to Theorem 5.4 we can differentiate term by term and obtain on $(-\pi, \pi)$

$$g'(x) = -\sin x - \sum_{n=2}^{\infty} (-1)^n \frac{n}{n^2 - 1} \sin nx \;.$$

Example 5.24. Let us find the sum of the series

$$\sum_{n=2}^{\infty} (-1)^n \frac{n}{n^2 - 1} \sin nx = f(x)$$

by using Theorem 5.3. Formula (5.37) yields

$$c = \lim_{n \to \infty} \left(- \frac{n^2}{n^2 - 1} \right) = -1 \;.$$

In our case, the series (5.39) has the form

$$-\tfrac{1}{2} + \cos x + \sum_{n=2}^{\infty} \left[(-1)^n \frac{n^2}{n^2 - 1} + (-1)^{n+1} \right] \cos nx$$

$$= -\tfrac{1}{2} + \cos x + \sum_{n=2}^{\infty} (-1)^n \frac{\cos nx}{n^2 - 1}$$

and converges (even uniformly) by Theorem 5.3 to the sum $g(x) = f'(x)$ in $(-\pi, \pi)$.

Using now the result of Example 5.23, we have $f''(x) = g'(x) = -\sin x - f(x)$, i.e. the function $f(x)$ satisfies the differential equation

$$f'' + f = -\sin x \;.$$

It can be easily ascertained that the solution of this equation has the form

$$f(x) = \alpha \sin x + \beta \cos x + \frac{x \cos x}{2}.$$

Thus, it remains to find the constants α and β. Since $f(0) = 0$, then necessarily $\beta = 0$. By differentiating we obtain

$$f'(x) = \alpha \cos x + \frac{\cos x}{2} - \frac{x \sin x}{2}$$

$$= -\tfrac{1}{2} + \cos x + \sum_{n=2}^{\infty} (-1)^n \frac{\cos nx}{n^2 - 1},$$

which gives

$$\alpha = \frac{1}{4} \quad \left(= \sum_{n=2}^{\infty} (-1)^n \frac{1}{n^2 - 1} \right)$$

for $x = 0$. Hence, in $(-\pi, \pi)$ we have

$$f(x) = \frac{\sin x}{4} + \frac{x \cos x}{2}.$$

5.6 APPLICATION OF FUNCTIONS OF A COMPLEX VARIABLE

Let us have a function $F(z)$ of a complex variable $z = z_1 + iz_2 = re^{ix}$ ($r \geq 0$, $0 \leq x < 2\pi$), which is analytic (holomorphic) in the disc $|z| < 1$. Then $F(z)$ can be expanded in a power series

$$F(z) = c_0 + c_1 z + c_2 z^2 + \ldots + c_n z^n + \ldots \quad (5.44)$$

convergent for $|z| < 1$. Suppose that the series on the right hand side of (5.44) is convergent also at the points $z = e^{ix}$ of the boundary except for finitely many points (i.e., except for a finite number of arguments $x \in (0, 2\pi)$). Suppose further that the coefficients c_n are real. Since

$$c_n z^n = c_n e^{inx} = c_n \cos nx + i c_n \sin nx \quad \text{for} \quad z = e^{ix},$$

we have

$$c_0 + c_1 z + c_2 z^2 + \ldots + c_n z^n + \ldots = f(x) + ig(x),$$

where

$$f(x) = c_0 + c_1 \cos x + c_2 \cos 2x + \ldots + c_n \cos nx + \ldots;$$
$$g(x) = c_1 \sin x + c_2 \sin 2x + \ldots + c_n \sin nx + \ldots. \quad (5.45)$$

Now, the following important ABEL's theorem is true. *If the series* (5.45) *are convergent at a point* x, *then*

$$f(x) + i\, g(x) = \lim_{r \to 1-} F(re^{ix}).$$

Denoting this limit by $F(e^{ix})$, we have

$$F(e^{ix}) = f(x) + ig(x).$$

This relation is useful primarily for the following two reasons:

1. Having an analytic function $F(z)$ we can use it for finding the trigonometric (Fourier) series corresponding to the functions $f(x)$ and $g(x)$, i.e. to the real and imaginary parts of the function $F(z)$ at the point $z = e^{ix}$, respectively.

2. We can find the sum of some trigonometric series. Indeed, having two series of the form (5.45) we can try to find the corresponding function $F(z)$ and thus establish the sums of the series (5.45) as the real and imaginary parts of the function $F(e^{ix})$ (considered in general as $\lim_{r \to 1-} F(re^{ix})$).

Example 5.25. Let

$$F(z) = e^z = 1 + z + \frac{z^2}{2!} + \ldots + \frac{z^n}{n!} + \ldots;$$

this series is convergent for all z and consequently, for $z = e^{ix}$ in particular, x being arbitrary. Since

$$F(e^{ix}) = e^{e^{ix}} = e^{\cos x + i\sin x} = e^{\cos x} e^{i\sin x}$$
$$= e^{\cos x}[\cos(\sin x) + i\sin(\sin x)],$$

by (5.45) we get the Fourier series of the functions $e^{\cos x}\cos(\sin x)$ and $e^{\cos x}\sin(\sin x)$, i.e.,

$$e^{\cos x}\cos(\sin x) = 1 + \cos x + \frac{\cos 2x}{2!} + \frac{\cos 3x}{3!} + \ldots$$
$$+ \frac{\cos nx}{n!} + \ldots;$$

$$e^{\cos x}\sin(\sin x) = \sin x + \frac{\sin 2x}{2!} + \frac{\sin 3x}{3!} + \ldots + \frac{\sin nx}{n!} + \ldots.$$

These series are convergent for all x.

Example 5.26. Let

$$F(z) = \log(1+z) = z - \frac{z^2}{2} + \frac{z^3}{3} - \frac{z^4}{4} + \ldots + (-1)^{n+1}\frac{z^n}{n} + \ldots$$

for $|z| < 1$; the series is convergent and equal to $\log(1 + z)$ even at the boundary points z except for the point $z = -1$, i.e. at the points $z = e^{ix}$, where $x \neq (2k + 1)\pi$. Consider the interval $(-\pi, \pi)$. Since for $\zeta = \varrho e^{i\tau}$ we have

$$\log \zeta = \log \varrho e^{i\tau} = \log \varrho + i\tau \quad (\tau \in (-\pi, \pi))$$

and

$$\zeta = 1 + e^{ix} = 2\cos\frac{x}{2} e^{ix/2}$$

in our case, i.e.

$$\varrho = 2\cos\frac{x}{2}, \quad \tau = \frac{x}{2},$$

we obtain

$$F(e^{ix}) = \log(1 + e^{ix}) = \log\left(2\cos\frac{x}{2}\right) + i\frac{x}{2};$$

hence, by (5.45),

$$\log\left(2\cos\frac{x}{2}\right) = \cos x - \frac{\cos 2x}{2} + \frac{\cos 3x}{3} - \cdots$$

$$+ (-1)^{n+1}\frac{\cos nx}{n} + \cdots;$$

$$\frac{x}{2} = \sin x - \frac{\sin 2x}{2} + \frac{\cos 3x}{3} - \cdots + (-1)^{n+1}\frac{\sin nx}{n} + \cdots.$$

Due to Abel's theorem the equalities hold for $x \in (-\pi, \pi)$. The Fourier series in question correspond to functions which have already been discussed. For the first one we refer the reader to Example 5.4, where the result for the Fourier series was quoted, cf. (5.8), for the second one to Problem 2.1a.

Example 5.27. Let α be real, $|\alpha| > 1$. Let us find the sums of the series

$$1 + \frac{\cos x}{\alpha} + \frac{\cos 2x}{\alpha^2} + \frac{\cos 3x}{\alpha^3} + \cdots + \frac{\cos nx}{\alpha^n} + \cdots = f(x)$$

and

$$\frac{\sin x}{\alpha} + \frac{\sin 2x}{\alpha^2} + \frac{\sin 3x}{\alpha^3} + \cdots + \frac{\sin nx}{\alpha^n} + \cdots = g(x).$$

These series are the real and imaginary parts, respectively, of the series $1 + e^{ix}/\alpha + e^{2ix}/\alpha^2 + \cdots$. The series $F(z) = 1 + z/\alpha + z^2/\alpha^2 + \cdots$ is a geometric series with the quotient z/α and thus we have for $|z| < |\alpha|$,

$$F(z) = \frac{1}{1 - z/\alpha} = \frac{\alpha}{\alpha - z}.$$

CALCULATION OF FOURIER SERIES

The series for $F(z)$ is convergent at all points with $|z| < |\alpha|$, in particular at $z = e^{ix}$. Since

$$F(e^{ix}) = \frac{\alpha}{\alpha - e^{ix}} = \alpha \frac{(\alpha - \cos x) + i \sin x}{\alpha^2 - 2\alpha \cos x + 1},$$

we have

$$f(x) = \frac{\alpha(\alpha - \cos x)}{\alpha^2 - 2\alpha \cos x + 1} \quad \text{and} \quad g(x) = \frac{\alpha \sin x}{\alpha^2 - 2\alpha \cos x + 1}.$$

In all our considerations, the coefficients c_n in (5.44) were supposed to be real. However, the same could be done for complex coefficients. Actually, if $c_n = (a_n - ib_n)$ for $n \geq 1$, $c_0 = a_0/2 - i(b_0/2)$, then for $z = e^{ix}$ we have

$$c_n z^n = (a_n - ib_n)(\cos nx + i \sin nx)$$
$$= (a_n \cos nx + b_n \sin nx) + i(-b_n \cos nx + a_n \sin nx)$$

so that

$$F(e^{ix}) = f(x) + ig(x)$$

where

$$f(x) = \frac{a_0}{2} + \sum_{n=1}^{\infty} (a_n \cos nx + b_n \sin nx),$$

$$g(x) = -\frac{b_0}{2} + \sum_{n=1}^{\infty} (-b_n \cos nx + a_n \sin nx). \quad (5.46)$$

Here again, for a given function $F(z)$ we can find the Fourier series f and g by using (5.46) and conversely, the sums of the latter two series can be found by establishing the corresponding function $F(z)$. This procedure, however, is slightly more complicated than in the case of the real coefficients c_n.

5.7 CONJUGATE SERIES

A trigonometric series

$$\frac{a_0}{2} + \sum_{n=1}^{\infty} (a_n \cos nx + b_n \sin nx) \tag{5.47}$$

with real coefficients a_n and b_n can formally be considered, according to the results of the preceding section, as the real part of a power series

$$\frac{a_0}{2} + \sum_{n=1}^{\infty} (a_n - ib_n) z^n \tag{5.48}$$

for $z = e^{ix}$. By (5.46), the imaginary part of the series (5.48) is given (again formally) by the series

$$\sum_{n=1}^{\infty} (-b_n \cos nx + a_n \sin nx). \tag{5.49}$$

The series (5.49) is called the *conjugate series* of (5.47).

Remark. Some authors (cf. e.g. [6]) understand by the conjugate series the series $\sum_{n=1}^{\infty} (b_n \cos nx - a_n \sin nx)$, which is the negative of (5.49).

If the series (5.47) is the Fourier series of a function $f(x)$ periodic with period 2π and integrable in $(-\pi, \pi)$, then the behaviour of the conjugate series (5.49) depends to a large extent on the behaviour of the function $f(x)$, for a_n and b_n are the Fourier coefficients of this function. If, for example, the function $f(x)$ satisfies the assumptions of Theorem 2.11, then not only the series (5.47), but also the series (5.49) converges uniformly to a periodic function $g(x)$, for which

$$g(x) = -\frac{1}{2\pi} \int_0^{\pi} [f(x + t) - f(x - t)] \cot \frac{t}{2} dt \tag{5.50}$$

holds (cf. Problem 5.17). Note that this formula also holds for more general functions $f(x)$; then the integral on the right hand side is understood as a limit of the integral over the interval (ε, π) as $\varepsilon \to 0+$ (cf. [4]).

The conjugate series is divergent, e.g., at any point of jump of the function $f(x)$. However, we are not going to investigate the properties of conjugate series here. Let us present only the following statement.

If the series (5.47) *is the Fourier series of a function* $f(x)$, *the series* (5.49) *need not be the Fourier series of any function* $g(x)$, *i.e. there need not exist an integrable function* $g(x)$ *satisfying the integral equations*

$$-b_n = \frac{1}{\pi} \int_{-\pi}^{\pi} g(x) \cos nx \, dx \quad \text{and} \quad a_n = \frac{1}{\pi} \int_{-\pi}^{\pi} g(x) \sin nx \, dx$$

$(n = 1, 2, \ldots)$. This fact can be demonstrated by the following example.

Example 5.28. The series $\sum_{n=2}^{\infty} (\cos nx)/\log n$ is the Fourier series of an integrable function (this is proved in [6] and in [4]). The conjugate series takes the form $\sum_{n=2}^{\infty} (\sin nx)/\log n$; from example 5.18 however, the latter is not the Fourier series of any integrable function.

5.8 EVALUATION OF VARIOUS NUMBER SERIES AND INTEGRALS BY MEANS OF FOURIER SERIES

Since the system $1, \cos x, \sin x, \cos 2x, \sin 2x, \ldots, \cos nx, \sin nx, \ldots$ is complete in $L_2(a, a + 2\pi)$, we have, for a function in this space, Parseval's equality

$$\int_a^{a+2\pi} |f(x)|^2 \, dx = \pi \left[\frac{|a_0|^2}{2} + \sum_{n=1}^{\infty} (|a_n|^2 + |b_n|^2) \right] \quad (5.51)$$

(cf. (3.18)). In addition to the theoretical importance, this equality has a practical significance as well:

1. If the Fourier coefficients of a function f and the sum of the series on the right hand side of (5.51) are known, Parseval's equality yields a formula for a definite integral. Thus, we can use Parseval's equality for the calculation of some definite integrals.
2. If the value of the integral on the left hand side of (5.51) is known, then this equality yields a formula for the sum of the series on the right-hand side. Thus, we can use Parseval's equality for the summation of series with constant terms.

For the same purpose we can use also other equalities, such as the equality (5.29) or the equality (∗) from Problem 5.9.

Example 5.29. Let us calculate the integral

$$J = \int_0^\pi \log^2\left(2\cos\frac{x}{2}\right)dx.$$

Since the function $\log^2(2\cos x/2)$ is even, we have

$$J = \frac{1}{2}\int_{-\pi}^\pi \log^2\left(2\cos\frac{x}{2}\right)dx.$$

In Example 5.26 we have shown that

$$\log\left(2\cos\frac{x}{2}\right) = \sum_{n=1}^\infty (-1)^{n+1}\frac{\cos nx}{n}$$

in $(-\pi, \pi)$; hence,

$$\frac{|a_0|^2}{2} + \sum_{n=1}^\infty (|a_n|^2 + |b_n|^2) = \sum_{n=1}^\infty \frac{1}{n^2} = \frac{\pi^2}{6}$$

(cf. Example 4.1) and by (5.51),

$$J = \tfrac{1}{2}\pi \sum_{n=1}^\infty \frac{1}{n^2} = \frac{\pi^3}{12}.$$

CALCULATION OF FOURIER SERIES

Example 5.30. Let us find the sum of the series

$$\sum_{n=0}^{\infty} \frac{1}{(2n+1)^2}.$$

By Example 2.1 and Problem 2.8,

$$\operatorname{sgn} x = \frac{4}{\pi} \sum_{n=0}^{\infty} \frac{\sin(2n+1)x}{2n+1} \quad \text{in} \quad (-\pi, \pi).$$

Parseval's equality gives

$$\int_{-\pi}^{\pi} (\operatorname{sgn} x)^2 \, dx = \frac{16}{\pi} \sum_{n=0}^{\infty} \frac{1}{(2n+1)^2};$$

thus, it follows that

$$\sum_{n=0}^{\infty} \frac{1}{(2n+1)^2} = \frac{\pi^2}{8}.$$

For the summation of particular number series we can analogously use the Fourier series directly. Indeed, if the equality

$$f(x) = \frac{a_0}{2} + \sum_{n=1}^{\infty} (a_n \cos nx + b_n \sin nx),$$

holds in an interval, then choosing arbitrarily the argument x we obtain sums of various particular number series.

Example 5.31. In $(0, \pi)$

$$\frac{\pi}{4} = \sum_{n=0}^{\infty} \frac{\sin(2n+1)x}{2n+1}$$

(cf. the preceding example). So we have for $x = \pi/2$ and $x = \pi/3$, respectively,

$$\frac{\pi}{4} = 1 - \tfrac{1}{3} + \tfrac{1}{5} - \tfrac{1}{7} + \ldots = \sum_{n=0}^{\infty} \frac{(-1)^n}{2n+1},$$

$$\frac{\pi}{2\sqrt{3}} = 1 - \tfrac{1}{5} + \tfrac{1}{7} - \tfrac{1}{11} + \tfrac{1}{13} - \tfrac{1}{17} + \tfrac{1}{19} - \ldots.$$

Example 5.32. Using the series

$$\frac{\pi^2}{3} + 4\sum_{n=1}^{\infty}(-1)^n \frac{\cos nx}{n^2} = x^2 \quad \text{in} \quad [-\pi, \pi]$$

we obtain

(a) for $x = 0$

$$\frac{\pi^2}{12} = \sum_{n=1}^{\infty}\frac{(-1)^{n+1}}{n^2} = 1 - \frac{1}{2^2} + \frac{1}{3^2} - \frac{1}{4^2} + \ldots;$$

(b) for $x = \pi$, the well known formula,

$$\frac{\pi^2}{6} = \sum_{n=1}^{\infty}\frac{1}{n^2} = 1 + \frac{1}{2^2} + \frac{1}{3^2} + \frac{1}{4^2} + \ldots.$$

This, however, is only a fragment of various methods using Fourier series. Let us only mention Example 4.1 in Chapter 4, where Dalzell's test for completeness has been used, or Example 5.19, where the sum of the series $\sum_{n=1}^{\infty}[(-1)^{n+1}/n^2]$ has been found by using the remark on p. 172. The reader himself will certainly realize which other relations could be used for various practical purposes.

Problems

5.1. Find the Fourier series of the following functions (draw the graph of the corresponding periodic extension)

(a) $f(x) = \begin{cases} x - [x] & \text{if } x \text{ is not integral,} \\ \frac{1}{2} & \text{if } x \text{ is integral.} \end{cases}$

$[x]$ denotes the integral part of x.

Hint: for $(-l, l)$ choose $(-1, +1)$.

(b) $f(x) = \sinh \alpha x$ in $(-\pi, \pi)$ (α is a real number);

(c) $f(x)$ is given by the graph in Fig. 5.10.

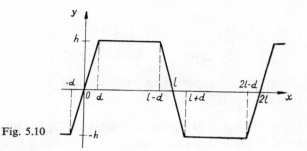

Fig. 5.10

5.2. Find (a) the sine-expansion, (b) the cosine-expansion of the function $Ax^2 + Bx + C$ in $(0, \pi)$.

5.3. Find the cosine-expansion in $(0, l)$ of the function plotted in Fig. 5.11.

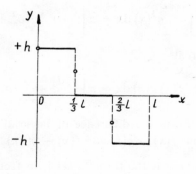

Fig. 5.11

5.4. Find the cosine-expansion of the function $f(x) = \sin \alpha x$ in $[0, \pi]$.

5.5. Find the Fourier series of the function $x \cos x$ in $(-\pi, \pi)$ by using the Fourier series of the function x. Compare the result with Example 5.24.

5.6. Find the Fourier series of the function $\log(\tan(x/2))$ in $(0, \pi)$. (*Hint*: Use the result of Examples 5.12 and 5.4 and subtract the corresponding series.)

5.7. Prove that
$$\sum_{n=0}^{\infty} (-1)^n \frac{\sin(2n+1)x}{2n+1} = \tfrac{1}{2} \log \tan\left(\frac{\pi}{4} - \frac{x}{2}\right)$$
for $x \in (-\pi/2, \pi/2)$.

(*Hint*: Use the preceding problem and the substitution $x = \pi/2 - t$.)

5.8. Prove formulas (5.22) by a formal multiplication of the series (5.18) and (5.19).

5.9. Consider, for simplicity, the real functions f and g. Then Parseval's equalities hold (we use the notation from (5.18) and (5.19)),
$$\int_a^{a+2\pi} f^2(x)\, dx = \pi \left[\frac{a_0^2}{2} + \sum_{n=1}^{\infty} (a_n^2 + b_n^2)\right];$$
$$\int_a^{a+2\pi} g^2(x)\, dx = \pi \left[\frac{\tilde{a}_0^2}{2} + \sum_{n=1}^{\infty} (\tilde{a}_n^2 + \tilde{b}_n^2)\right].$$

Prove that
$$\int_a^{a+2\pi} f(x) g(x)\, dx = \pi \left[\frac{a_0 \tilde{a}_0}{2} + \sum_{n=1}^{\infty} (a_n \tilde{a}_n + b_n \tilde{b}_n)\right]. \quad (*)$$

(*Hint*: Form the difference of Parseval's equalities for the functions $f(x) + g(x)$ and $f(x) - g(x)$.)

5.10. Realizing that the formula $(*)$ is in fact formula (5.22) for $n = 0$, prove that the formulas (5.22) are meaningful also for $n > 0$.

(*Hint*: Use the fact that the formulas (5.22) are analogues of formula $(*)$ from Problem 5.9, where, of course, the function $g(x) \cos nx$ instead of $g(x)$ for A_n and the function $g(x) \sin nx$ instead of $g(x)$ for B_n are considered.)

CALCULATION OF FOURIER SERIES

5.11. Prove that

(a) $\sum_{n=1}^{\infty} \dfrac{\sin nx}{n^2} = -\int_0^x \log\left(2\sin\dfrac{t}{2}\right) dt$ for $x \in (0, 2\pi)$,

(b) $\sum_{n=1}^{\infty} \dfrac{\cos nx}{n^3} = \int_0^x \left[\int_0^y \log\left(2\sin\dfrac{t}{2}\right) dt\right] dy + \sum_{n=1}^{\infty} \dfrac{1}{n^3}$

for $x \in (0, 2\pi)$.

5.12. Find the Fourier series of the derivative of the function

$$f(x) = \sum_{n=1}^{\infty} \left(\dfrac{\cos nx}{n^3} + (-1)^n \dfrac{\sin nx}{n+1}\right).$$

5.13. Find the sum of the series

$$\sum_{n=2}^{\infty} \dfrac{\sin nx}{n(n^2-1)} = f(x) \quad \text{in} \quad (0, 2\pi).$$

(*Hint*: Prove that the sum $f(x)$ satisfies the differential equation $f'' + f = \sin x + (x - \pi)/2$, i.e. $f(x) = (x - \pi)/2 - (x/2)\cos x + A\cos x + B\sin x$. For finding the constant A use the condition $f(\pi) = 0$, for B use $f'(\pi) = \frac{1}{4} = \sum_{n=2}^{\infty} (-1)^n/(n^2 - 1)$. Result: $A = \pi/2$, $B = \frac{3}{4}$.)

5.14. Find the Fourier series of the functions $\cos(\alpha \sin x)$ and $\sin(\alpha \sin x)$ for real α.

(*Hint*: Proceed analogously as in Section 5.6. Use the fact that

$$e^{-(\alpha/2)(z - 1/z)} = \sum_{n=-\infty}^{\infty} J_n(\alpha) z^n \quad \text{for all complex numbers } z \neq 0,$$

J_n being Bessel functions; then put $z = e^{ix}$ and use the formula $J_{-m}(\alpha) = (-1)^m J_m(\alpha)$.)

5.15. What are the sums of the series

$$\sum_{n=2}^{\infty} \dfrac{\cos nx}{n(n-1)} \quad \text{and} \quad \sum_{n=2}^{\infty} \dfrac{\sin nx}{n(n-1)}$$

in $(0, 2\pi)$?

(*Hint*: Use the formula

$$\sum_{n=2}^{\infty} \frac{z^n}{n(n-1)} = z + (1-z)\log(1-z).)$$

5.16. What are the sums of the series

$$\sum_{n=2}^{\infty} (-1)^n \frac{n}{n^2-1} \cos nx \quad \text{and} \quad \sum_{n=2}^{\infty} (-1)^n \frac{n}{n^2-1} \sin nx$$

in $(-\pi, \pi)$? Compare the result for the second series with Example 5.24.

(*Hint*: Use the equality

$$\sum_{n=2}^{\infty} \frac{n}{n^2-1}(-z)^n = \frac{1}{2}\left[\left(z + \frac{1}{z}\right)\log(1+z) - 1 + \frac{z}{2}\right].)$$

5.17. Denote by $s_n(x)$ the partial sum of the conjugate series (5.49). Prove that

$$s_n(x) = -\frac{1}{2\pi}\int_0^\pi [f(x+t) - f(x-t)] \frac{\cos\frac{t}{2} - \cos\frac{2n+1}{2}t}{\sin\frac{t}{2}}\, dt.$$

(*Hint*: Substitute the integral expressions for the Fourier coefficients a_n and b_n into $s_n(x)$ and use the formulas (1.25).) A comparison with (5.50) yields

$$s_n(x) - g(x) = \frac{1}{\pi}\int_0^\pi [f(x+t) - f(x-t)] \frac{\cos\frac{(2n+1)t}{2}}{2\sin\frac{t}{2}}\, dt. \quad (**)$$

To show that the function $g(x)$ in (5.50) is really the sum of the conjugate series (5.49), we have to prove that the difference (**) converges to zero for $n \to \infty$.

CALCULATION OF FOURIER SERIES

5.18. Prove the following proposition. If f is periodic and $|f'| \leq M$ then the sum $g(x)$ of the conjugate series (which exists) is bounded by the constant $M \cdot N$.

(*Hint*: Since $f(x + t) - f(x - t) = f'(\xi)\, 2t$, we have by (5.50), $|g(x)| \leq (M/\pi) \int_0^\pi t \cot(t/2)\, dt$; the last integral is finite.)

5.19. Calculate the integral $\displaystyle\int_0^\pi \log^2 \tan(x/2)\, dx$.

5.20. Sum the following series: (a) $\displaystyle\sum_{n=1}^\infty 1/n^4$; (b) $\displaystyle\sum_{n=1}^\infty (-1)^n/n^2$; (c) $\displaystyle\sum_{n=1}^\infty 1/n^6$; (d) $\displaystyle\sum_{n=1}^\infty (-1)^n/(2n+1)^3$.

(*Hint*: (a) use Parseval's equality in (5.33); (b) use (5.33) for $x = \pi$; (c) use Parseval's equality for (5.32); (d) in (5.32) choose $x = \pi/2$.)

CHAPTER 6 APPROXIMATE HARMONIC ANALYSIS

6.1 INTRODUCTION

The present chapter deals with some methods which enable us to find the Fourier series of a given function and, conversely, to find the sum of a given Fourier series. At the first glance, both these problems seem to be quite simple: in the first case, certain integrals are to be evaluated and in the other one a series is to be summed. However, the convergence of a Fourier series is rather slow. If a function is continuous and periodic and has piece-wise continuous derivatives of higher orders, then its Fourier series converges approximately as rapidly as the series $\sum_{n=1}^{\infty} 1/n^2$ (cf. Theorem 2.11); if a function is smooth and non-periodic, then its Fourier coefficients tend to zero as fast as $1/n$, i.e. the corresponding Fourier series cannot be absolutely convergent. Thus, if we wished to sum a Fourier series directly with a given accuracy, it would be necessary to sum up an inadequately large number of terms. As for the first problem, in the majority of cases it is hopeless to find an explicit formula for the Fourier coefficients; either the corresponding integrals cannot be expressed in terms of elementary functions, or their calculation is complicated. In practice we encounter also functions defined by their (approximate) values at discrete points; in such cases we are compelled to use a method of approximation (cf. Section 6.4). Hence, methods speeding up the convergence of a series or facilitating the calculation of Fourier coefficients are very valuable and important.

6.2 KRYLOV'S METHOD

Consider a periodic function f (with period 2π), which is continuous and has continuous derivatives of sufficiently high orders in the interval $[0, 2\pi)$, except for a finite number of points; assume that the function or its derivatives are not necessarily continuous at those points but have a jump there, i.e. finite limits of the functions $f^{(k)}$ both from the left and from the right exist there. Krylov's method rests on a decomposition of f into two functions

$$f(x) = g(x) + h(x),$$

where $g(x)$ is a sufficiently smooth periodic function (its Fourier series thus converges quite rapidly) and $h(x)$ is a function with already known Fourier series. Thus, first of all, the function $h(x)$ must be chosen such that the difference $g(x) = f(x) - h(x)$ is smooth. In other words, for all x_0 we must have

$$g(x_0 + 0) = g(x_0 - 0); \quad g'(x_0 + 0) = g'(x_0 - 0);$$
$$g''(x_0 + 0) = g''(x_0 - 0); \ldots *$$

etc. for a sufficiently large number of derivatives, i.e., equations

$$f(x_0 + 0) - f(x_0 - 0) = h(x_0 + 0) - h(x_0 - 0);$$
$$f'(x_0 + 0) - f'(x_0 - 0) = h'(x_0 + 0) - h'(x_0 - 0);$$
$$f''(x_0 + 0) - f''(x_0 - 0) = h''(x_0 + 0) - h''(x_0 - 0);$$
$$\dots\dots\dots\dots\dots\dots\dots\dots\dots\dots\dots\dots\dots\dots\dots\dots\dots\dots\dots$$
$$f^{(l)}(x_0 + 0) - f^{(l)}(x_0 - 0) = h^{(l)}(x_0 + 0) - h^{(l)}(x_0 - 0) \quad (6.1)$$

have to hold for a sufficiently large l.

Stated differently, at points of continuity of the function f and all its derivatives of orders up to the l-th, the function h along with all derivatives of orders up to the l-th has to be also continuous.

---

* Note that $g(c + 0)$, as usual, denotes the limit of g from the right at c, i.e., $g(c + 0) = \lim\limits_{x \to c+} g(x)$; $g(c - 0)$ denotes the limit from the left.

On the other hand, the discontinuities of the function f and its derivatives up to the l-th order have to coincide with the discontinuities of h and its derivatives.

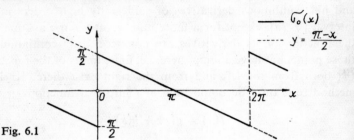

Fig. 6.1

Now, we are going to construct the function h. Denote

$$\sigma_0(x) = \frac{\pi - x}{2} \quad \text{in} \quad (0, 2\pi), \tag{6.2}$$

and extend this function periodically with period 2π (see Fig. 6.1); let the extended function be denoted again by $\sigma_0(x)$. From Example 5.1 we have

$$\sigma_0(x) = \sum_{n=1}^{\infty} \frac{\sin nx}{n}.$$

Due to Theorem 5.1, each primitive function to the function σ_0 has in the interval $(0, 2\pi)$ the form

$$\sigma_1(x) = c_1 + \int_0^x \sigma_0(t)\,dt = c_1 + \sum_{n=1}^{\infty} \frac{1}{n^2} - \sum_{n=1}^{\infty} \frac{\cos nx}{n^2},$$

where c_1 is an arbitrary constant. Choose this constant such that

$$\int_0^{2\pi} \sigma_1(x)\,dx = 0,$$

i.e. such that $c_1 + \sum_{n=1}^{\infty} 1/n^2 = 0$; since

$$0 = \int_0^{2\pi} \sigma_1(x)\,dx = \int_0^{2\pi}\left(c_1 + \int_0^x \frac{\pi - t}{2}\,dt\right)dx$$

we have $c_1 = -\pi^2/6$ (cf. Example 4.1).

Thus, we have finally,

$$\sigma_1(x) = \frac{\pi^2}{12} - \frac{(\pi - x)^2}{4} \quad \text{for} \quad x \in (0, 2\pi). \tag{6.3}$$

Analogously, we define the other functions ($x \in (0, 2\pi)$),

$$\sigma_2(x) = c_2 + \int_0^x \sigma_1(t)\,dt,$$

$$\sigma_3(x) = c_3 + \int_0^x \sigma_2(t)\,dt,$$

$$\cdots\cdots\cdots\cdots\cdots\cdots\cdots\cdots\cdots$$

$$\sigma_l(x) = c_l + \int_0^x \sigma_{l-1}(t)\,dt,$$

where the constants c_i ($i = 2, 3, ..., l$) are chosen such that

$$\int_0^{2\pi} \left(c_i + \int_0^x \sigma_{i-1}(t)\,dt \right) dx = 0,$$

i.e. that

$$\int_0^{2\pi} \sigma_i(t)\,dt = 0.$$

Theorem 6.1. *The functions $\sigma_i(x)$ periodically extended from $[0, 2\pi)$ onto the entire real axis with period 2π have the following important properties:*

1. *Each $\sigma_i(x)$ has derivatives of all orders in the open interval $(0, 2\pi)$.*
2. *For $i \neq j$, $\sigma_i^{(j)}(0 + 0) = \sigma_i^{(j)}(0 - 0) = \sigma_i^{(j)}(2\pi - 0) = \sigma_i^{(j)}(2\pi + 0)$.*
3. $\sigma_i^{(i)}(0 + 0) - \sigma_i^{(i)}(0 - 0) = \sigma_i^{(i)}(0 + 0) - \sigma_i^{(i)}(2\pi - 0) = \pi$.
4. *The constant term of the Fourier series of $\sigma_i(x)$ is equal to zero.*

The proof will be carried out by induction. The assertion holds obviously for $\sigma_0(x)$. Suppose the assertion is true for $\sigma_{i-1}(x)$.

The function $\sigma_i(x)$ has been constructed so that $\sigma_i^{(j)}(x) = \sigma_{i-1}^{(j-1)}(x)$ for $x \in (0, 2\pi)$ and $j \geq 1$. By induction hypothesis,

$$\sigma_i^{(j)}(0+0) - \sigma_i^{(j)}(2\pi - 0) = \sigma_{i-1}^{(j-1)}(0+0) - \sigma_{i-1}^{(j-1)}(2\pi - 0)$$

$$= \begin{cases} 0 & \text{for } j \neq i, \\ \pi & \text{for } j = i. \end{cases}$$

Next, for $j = 0$, i.e. $\sigma_i^{(0)}(x) = \sigma_i(x)$, we have

$$\sigma_i(2\pi - 0) - \sigma_i(0+0) = \int_0^{2\pi} \sigma_{i-1}(x)\,\mathrm{d}x = 0$$

for $i \geq 1$ which proves properties 1 to 3. Property 4 follows from the fact that

$$\int_0^{2\pi} \sigma_i(x)\,\mathrm{d}x = 0.$$

Remark 6.1. The Fourier series of the function $\sigma_i(x)$ may thus be obtained by a formal integration of the Fourier series of $\sigma_{i-1}(x)$.

Let now y be a fixed real number and define the function $h_y(x)$ by

$$h_y(x) = \sum_{i=0}^{l} \frac{1}{\pi} \left[f^{(i)}(y+0) - f^{(i)}(y-0) \right] \sigma_i(x-y).$$

According to Theorem 6.1, we have for $j \leq l$,

$$h_y^{(j)}(y+0) - h_y^{(j)}(y-0)$$

$$= \sum_{i=0}^{l} \frac{1}{\pi} \left[f^{(i)}(y+0) - f^{(i)}(y-0) \right] \left[\sigma_i^{(j)}(0+0) - \sigma_i^{(j)}(0-0) \right]$$

$$= \frac{1}{\pi} \left[f^{(j)}(y+0) - f^{(j)}(y-0) \right] \pi = f^{(j)}(y+0) - f^{(j)}(y-0);$$

$$\tag{6.4}$$

since $\sigma_i^{(j)}(0+0) = \sigma_i^{(j)}(0-0)$ for $i \neq j$. It follows that the function $h_y(x)$ has continuous derivatives of all orders in each open interval not containing the points $y + 2k\pi$ (k being an integer).

Consider now a function f of the type discussed at the beginning of the present section and denote by $y_1 < y_2 < \ldots < y_{m-1} < y_m$ all points of jump of the function f and its derivatives up to the l-th order in $[0, 2\pi)$. Define the function $h(x)$ as follows,

$$h(x) = \sum_{i=1}^{m} h_{y_i}(x).$$

Except for $y_1, \ldots, y_m$, the function $h(x)$ is smooth in $[0, 2\pi)$ and by (6.4) we have (6.1). Also, the function $h(x)$ is a combination of the functions $\sigma_i(x - y_j)$ ($i = 0, 1, \ldots, l; j = 1, 2, \ldots, m$) which have Fourier series that can be easily calculated by using the Fourier series of $\sigma_i(x)$.

The function $h(x)$ satisfies all the required conditions: the function $g(x) = f(x) - h(x)$ has continuous derivatives up to the l-th order and its Fourier series converges sufficiently fast. (For an estimate of the rapidity of convergence depending on l cf. Theorem 7.3.)

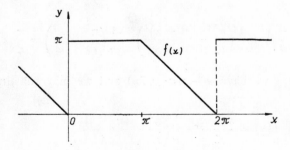

Fig. 6.2

Since the Fourier coefficients of the function h are known we can find the Fourier coefficients of f by calculating the Fourier coefficients of g. These coefficients tend to zero as $n \to \infty$ sufficiently rapidly and hence it suffices to calculate far less coefficients of g than the function f itself necessitates.

It is necessary to note that if the function $\sigma_i(x - y_j)$ is calculated, we use the *periodic extension* of the function σ_i, i.e. if the number $x - y_j$ does not belong to $[0, 2\pi)$, we have to substitute in (6.2), (6.3) etc. the number $x - y_j + 2k\pi$ instead, k being an integer such that $0 \leq x - y_j + 2k\pi < 2\pi$.

Example 6.1. Define a function $f(x)$ on $[0, 2\pi)$ as follows

$$f(x) = \begin{cases} \pi & \text{for } x \in [0, \pi), \\ 2\pi - x & \text{for } x \in [\pi, 2\pi), \end{cases}$$

(see Fig. 6.2) and extend it periodically with period 2π outside $[0, 2\pi)$. The extended function and its derivatives have in $[0, 2\pi)$ jumps at $y_1 = 0$ and $y_2 = \pi$: $f(0 + 0) - f(0 - 0) = \pi$; $f'(0 + 0) - f'(0 - 0) = 1$; $f'(\pi + 0) - f'(\pi - 0) = -1$. Hence

$$h_{y_1}(x) = \sigma_0(x) + \frac{1}{\pi} \sigma_1(x); \quad h_{y_2}(x) = -\frac{1}{\pi} \sigma_1(x - \pi).$$

For simplicity, calculate the function $h(x)$ in $(0, \pi)$ and $[\pi, 2\pi)$ separately. For $x \in (0, \pi)$ we have (*caution*: if calculating $h_{y_2}(x)$ we must substitute $x + \pi = x - \pi + 2\pi \in [0, 2\pi)$ instead of x into (6.3)!)

$$h(x) = \frac{\pi - x}{2} + \frac{1}{\pi}\left(\frac{\pi^2}{12} - \frac{(\pi - x)^2}{4}\right) - \frac{1}{\pi}\left(\frac{\pi^2}{12} - \frac{x^2}{4}\right) = \frac{\pi}{4}.$$

For $x \in [\pi, 2\pi)$ we have $x - \pi \in [0, 2\pi)$ and we may substitute directly,

$$h(x) = \frac{\pi - x}{2} + \frac{1}{\pi}\left(\frac{\pi^2}{12} - \frac{(\pi - x)^2}{4}\right)$$
$$- \frac{1}{\pi}\left(\frac{\pi^2}{12} - \frac{(2\pi - x)^2}{4}\right) = \frac{5}{4}\pi - x.$$

So we have in $(0, \pi)$

$$f(x) - h(x) = \pi - \tfrac{1}{4}\pi = \tfrac{3}{4}\pi$$

and in $[\pi, 2\pi)$

$$f(x) - h(x) = 2\pi - x - \tfrac{5}{4}\pi + x = \tfrac{3}{4}\pi,$$

i.e., the function $g(x) = f(x) - h(x)$ is equal to the constant $\tfrac{3}{4}\pi$.

Example 6.2. As a matter of fact, we have already expanded the function from Example 6.1 in the Fourier series,

$$h(x) = \sigma_0(x) + \frac{1}{\pi}\sigma_1(x) - \frac{1}{\pi}\sigma_1(x - \pi)$$

$$= \sum_{n=1}^{\infty} \frac{\sin nx}{n} - \frac{1}{\pi}\sum_{n=1}^{\infty} \frac{\cos nx}{n^2} + \frac{1}{\pi}\sum_{n=1}^{\infty} \frac{\cos n(x - \pi)}{n^2}$$

$$= \sum_{n=1}^{\infty} \frac{\sin nx}{n} - \frac{1}{\pi}\sum_{n=1}^{\infty} \frac{\cos nx}{n^2} + \frac{1}{\pi}\sum_{n=1}^{\infty} (-1)^n \frac{\cos nx}{n^2}$$

$$= \sum_{n=1}^{\infty} \left(\frac{(-1)^n - 1}{\pi n^2} \cos nx + \frac{1}{n}\sin nx \right).$$

Since the corresponding function g has a simple form $g(x) = \tfrac{3}{4}\pi$ here, we have for $x \in (0, 2\pi)$,

$$f(x) = \tfrac{3}{4}\pi + \sum_{n=1}^{\infty} \left(\frac{(-1)^n - 1}{\pi n^2} \cos nx + \frac{1}{n}\sin nx \right).$$

Example 6.3. Let us extend the function f periodically with period 2π, provided

$$f(x) = \begin{cases} \pi + \sin^3 x & \text{for } x \in [0, \pi), \\ 2\pi - x & \text{for } x \in [\pi, 2\pi). \end{cases}$$

The reader can easily verify that here $h(x)$ is the same function as in Example 6.1 with $l = 1$; the function $g(x) = f(x) - h(x)$, of course, will be different,

$$g(x) = \begin{cases} \tfrac{3}{4}\pi + \sin^3 x & \text{for } x \in [0, \pi), \\ \tfrac{3}{4}\pi & \text{for } x \in [\pi, 2\pi). \end{cases}$$

If we wished now to calculate the Fourier coefficients giving f with an error less than 10^{-4}, we would have to calculate about 10 000 coefficients, because the Fourier coefficients of f tend to zero as rapidly as $1/n$. However, we have the decomposition $f = g + h$, and the periodic extension of g has a bounded third derivative, i.e.,

$$|g'''(x)| \leq 27.$$

Denoting now A_n and B_n the Fourier coefficients of g and A_n''' and B_n''' the Fourier coefficients of g''', we get (by using Theorem 2.9 three times or by a triple integration by parts) $A_n = (1/n^3) B_n'''$ and $B_n = -(1/n^3) A_n'''$, and consequently

$$\max\left(|A_n|, |B_n|\right) \leq \frac{1}{\pi n^3} \int_0^{2\pi} |g'''(x)| \, \mathrm{d}x \leq \frac{1}{\pi n^3} \int_0^{\pi} 27 \, \mathrm{d}x = \frac{27}{n^3}.$$

To achieve the required degree of accuracy, it suffices here to calculate only those coefficients A_n, B_n, for which

$$\frac{27}{n^3} > 10^{-4}$$

i.e.,

$$n^3 < 270\,000, \quad n \leq 64.$$

The number obtained $n = 64$ is only a rough estimate, because a calculation of less coefficients would clearly yield the desired accuracy. The Fourier coefficients of f are then the sum of the known Fourier coefficients of h and of the Fourier coefficients of g (calculated numerically, e.g.).

6.3 APPROXIMATE SUMMATION OF SERIES

Krylov's method consists in decomposing a function f into two functions such that one of them can easily be expanded in a Fourier series and the Fourier series of the other converges rapidly. A similar method can be used, if the sum of a Fourier series is to be established in terms of the coefficients.

Consider, for example, the sine-series

$$\sum_{n=1}^{\infty} b_n \sin nx \qquad (6.5)$$

(the case of a cosine-series is analogous) and suppose that the coefficients b_n are defined by a certain function $g(x)$ which can be expanded in a power series in a neighbourhood of zero, i.e.

$$b_n = g\left(\frac{1}{n}\right), \quad g(0) = 0.$$

Let $p(x) = \alpha_1 x + \alpha_2 x^2 + \ldots + \alpha_m x^m$ be a partial sum of the Taylor expansion of $g(x)$ in a neighbourhood of zero and assume that $|g(x) - p(x)| < K x^{m+1}$ for $|x| < \varepsilon$. Let us define the function $q(x)$ by $g(x) = p(x) + q(x)$; then $b_n = p(1/n) + q(1/n)$. If $1/n < \varepsilon$, then

$$\left| q\left(\frac{1}{n}\right) \right| < K \frac{1}{n^{m+1}}. \qquad (6.6)$$

Thus, the series (6.5) can be written as follows,

$$\sum_{n=1}^{\infty} b_n \sin nx = \alpha_1 \sum_{n=1}^{\infty} \frac{\sin nx}{n} + \alpha_2 \sum_{n=1}^{\infty} \frac{\sin nx}{n^2} + \ldots$$

$$+ \alpha_m \sum_{n=1}^{\infty} \frac{\sin nx}{n^m} + \sum_{n=1}^{\infty} q\left(\frac{1}{n}\right) \sin nx.$$

By (6.6), the last series converges very rapidly and consequently, it suffices to sum up a relatively small number of terms of the series to achieve an extreme accuracy. Then, for summing up the series (6.5), it suffices to know the sums

$$\sum_{n=1}^{\infty} \frac{\sin nx}{n^k}$$

for $k = 1, 2, \ldots, m$.

We are going to present the sums of certain series which may be useful both in the method just discussed and elsewhere in practice. Some of them have already been derived in preceding chapters, others can be found in Chapter 10.

$$\sigma_0(x) = \sum_{n=1}^{\infty} \frac{\sin nx}{n} = \frac{\pi - x}{2} \qquad (0 < x < 2\pi);$$

$$-\sigma_1(x) = \sum_{n=1}^{\infty} \frac{\cos nx}{n^2} = \frac{(\pi - x)^2}{4} - \frac{\pi^2}{12} \qquad (0 \leq x \leq 2\pi);$$

$$-\sigma_2(x) = \sum_{n=1}^{\infty} \frac{\sin nx}{n^3} = \frac{2\pi^2 x - 3\pi x^2 + x^3}{12} \qquad (0 \leq x \leq 2\pi)$$

(cf. Section 6.2 of this chapter);

$$\sum_{n=1}^{\infty} \frac{\cos nx}{n} = -\log\left(2 \sin \frac{x}{2}\right) \qquad (0 < x < 2\pi);$$

$$\sum_{n=1}^{\infty} \frac{\sin nx}{n^2} = -\int_0^x \log\left(2 \sin \frac{t}{2}\right) dt \qquad (0 \leq x \leq 2\pi);$$

$$\sum_{n=1}^{\infty} \frac{\cos nx}{n^3} = \int_0^x \left[\int_0^y \log\left(2 \sin \frac{t}{2}\right) dt\right] dy + \sum_{n=1}^{\infty} \frac{1}{n^3},$$

where $\sum_{n=1}^{\infty} 1/n^3 = 1 \cdot 202\ 056\ 902 \ldots$; unfortunately, the sums of these last series are not expressible by elementary functions.

Sometimes, the following more general approach is more appropriate. Let the coefficients b_n of the series (6.5) be very close to the coefficients b_n^* of another Fourier series with a known sum $h(x)$, i.e. let $b_n = b_n^* + \gamma_n$; then

$$\sum_{n=1}^{\infty} b_n \sin nx = \sum_{n=1}^{\infty} b_n^* \sin nx + \sum_{n=1}^{\infty} \gamma_n \sin nx$$

$$= h(x) + \sum_{n=1}^{\infty} \gamma_n \sin nx\ .$$

APPROXIMATE HARMONIC ANALYSIS

If our choice was lucky enough, the series

$$\sum_{n=1}^{\infty} \gamma_n \sin nx$$

converges faster than the original one.

Example 6.4. We are to find the sum of the series

$$S(x) = \sum_{n=2}^{\infty} \frac{n^3}{n^4 - 1} \sin nx \qquad (6.7)$$

with an accuracy of 10^{-8}. Proceeding in the first way, we obtain

$$\frac{n^3}{n^4 - 1} = g\left(\frac{1}{n}\right),$$

where

$$g(x) = \frac{x^{-3}}{x^{-4} - 1} = \frac{x}{1 - x^4}.$$

For $|x| < 1$ we have $g(x) = x(1 + x^4 + x^8 + \ldots)$; thus,

$$g(x) - (x + x^5) = \frac{x^9}{1 - x^4}$$

and

$$S(x) = \sum_{n=2}^{\infty} \frac{\sin nx}{n} + \sum_{n=2}^{\infty} \frac{\sin nx}{n^5} + \sum_{n=2}^{\infty} \frac{n^{-9}}{1 - n^{-4}} \sin nx .$$

Since $1/(1 - n^{-4}) < 2$ for $n \geq 2$, we have

$$\left| \sum_{n=m+1}^{\infty} \frac{n^{-9}}{1 - n^{-4}} \sin nx \right| < 2 \sum_{n=m+1}^{\infty} \frac{1}{n^9} \leq 2 \int_{m}^{\infty} x^{-9} \, dx = \tfrac{2}{8} m^{-8} .$$

For $m = 10$, $\tfrac{2}{8} m^{-8} < 10^{-8}$. Thus, using the following formula, the error does not exceed 10^{-8},

$$S(x) = \sum_{n=1}^{\infty} \frac{\sin nx}{n} + \sum_{n=1}^{\infty} \frac{\sin nx}{n^5} - 2 \sin x + \sum_{n=2}^{10} \frac{1}{n^5(n^4 - 1)} \sin nx ;$$

the first two series on the right hand side have sums $\sigma_0(x)$ and $\sigma_4(x)$, respectively (cf. the preceding section).

Observe that if we used the partial sums of series (6.7) for calculating $S(x)$, then, for the required degree of accuracy, we would have to sum up approximately 10^8 terms of the series.

Example 6.5. Let us find the sum of the series

$$S(x) = \sum_{n=1}^{\infty} \frac{n}{n^2+1} (-1)^{n+1} \sin nx.$$

In Example 5.11 it was shown that

$$\sum_{n=1}^{\infty} \frac{(-1)^{n+1}}{n} \sin nx = 2x \quad \text{for} \quad x \in (-\pi, \pi).$$

Thus, putting $b_n = (-1)^{n+1} n/(n^2+1)$ and $b_n^* = (-1)^{n+1}/n$, we have

$$\gamma_n = \frac{(-1)^n}{n(n^2+1)} \quad \text{and} \quad |\gamma_n| \leq \frac{1}{n^3}.$$

Then the series

$$S(x) - 2x = \sum_{n=1}^{\infty} \gamma_n \sin nx$$

converges much faster than the original one.

Both the methods described above have disadvantages of their own. The first method may be used only if the sum of the Fourier series and its derivatives have no discontinuities except for points $x = 2k\pi$, k being an integer. The second method can be applied only if a close series with a known sum can be found.

Therefore, we are going to present here another method, which is, in a certain sense, a converse to Krylov's method of increasing the convergence. This method is applicable to any Fourier series whose sum along with its derivatives up to a certain order have only a finite number of jumps in $[0, 2\pi]$. For the sake of simplicity, consider a Fourier series in the complex form

$$\sum_{n=-\infty}^{\infty} c_n e^{inx}. \tag{6.8}$$

Suppose that the coefficients c_n ($n \neq 0$) are expressible as follows,

$$c_n = g_1\left(\frac{1}{n}\right) e^{inx_1} + g_2\left(\frac{1}{n}\right) e^{inx_2} + \ldots + g_k\left(\frac{1}{n}\right) e^{inx_k},$$

where the functions $g_1(x), g_2(x), \ldots, g_k(x)$ satisfy the condition

$$\left| g_i(x) - \sum_{j=1}^{m} \alpha_j^{(i)} x^j \right| \leq K |x|^{m+1}$$

whenever x is in a neighbourhood of zero ($\alpha_j^{(i)}$ and K are constants). Then

$$\sum_{\substack{n=-\infty \\ n \neq 0}}^{\infty} c_n e^{inx} = \sum_{\substack{n=-\infty \\ n \neq 0}}^{\infty} \sum_{p=1}^{k} \sum_{q=1}^{m} \alpha_q^{(p)} \frac{1}{n^q} e^{inx_p} e^{inx} + \sum_{\substack{n=-\infty \\ n \neq 0}}^{\infty} \gamma_n e^{inx},$$

where $|\gamma_n| \leq kKn^{-m-1}$ for n sufficiently large. Interchanging the order of summation of the triple series, we have

$$\sum_{\substack{n=-\infty \\ n \neq 0}}^{\infty} c_n e^{inx} = \sum_{p=1}^{k} \sum_{q=1}^{m} \alpha_q^{(p)} S_q(x + x_p) + \sum_{\substack{n=-\infty \\ n \neq 0}}^{\infty} \gamma_n e^{inx}.$$

The last series converges fast enough and the sums

$$S_q(x + x_p) = \sum_{\substack{n=-\infty \\ n \neq 0}}^{\infty} \frac{1}{n^q} e^{in(x+x_p)}$$

may be obtained by using the functions

$$S_q(x) = \sum_{\substack{n=-\infty \\ n \neq 0}}^{\infty} \frac{1}{n^q} e^{inx}$$

(cf. Problem 6.5).

Let us note that the points $-x_p$ are then the discontinuities of the sum (6.8) and its derivatives up to the $(m-1)$-th order.

Example 6.6. Let us increase the convergence of the series

$$S(x) = \sum_{n=2}^{\infty} (-1)^n \frac{n}{n^2 - 1} \sin nx.$$

Using the identity

$$\frac{n}{n^2 - 1} = \frac{1}{n} + \frac{1}{n(n^2 - 1)}$$

we have

$$S(x) = \sum_{n=2}^{\infty} (-1)^n \frac{\sin nx}{n} + \sum_{n=2}^{\infty} \frac{(-1)^n}{n(n^2 - 1)} \sin nx.$$

Find the sum of the first series; since $\sin nx = (1/2i)[e^{inx} - e^{-inx}]$, we have

$$\sum_{n=1}^{\infty} \frac{(-1)^n}{n} \sin nx = \frac{1}{2i} \sum_{\substack{n=-\infty \\ n \neq 0}}^{\infty} \frac{1}{n} e^{in(x-\pi)} = \frac{1}{2i} S_1(x - \pi).$$

Hence,

$$S(x) = \frac{1}{2i} S_1(x - \pi) + \sin x + \sum_{n=2}^{\infty} \frac{(-1)^n}{n(n^2 - 1)} \sin nx,$$

where the series on the right hand side converges sufficiently rapidly.

6.4 NUMERICAL CALCULATION OF FOURIER COEFFICIENTS

If an explicit expression for the Fourier coefficients cannot be found or if it is too complicated for practical uses, then it is necessary to calculate the Fourier coefficients approximately. The same is true if the function is defined only by its values at a finite number of points.

APPROXIMATE HARMONIC ANALYSIS

There are many methods for numerical calculation of Fourier coefficients and their applicability depends on the computing device available (various graphical aids, analogue and digital computers, special devices for calculation of Fourier coefficients etc.). Almost in every scientific discipline some special methods are used which fit the phenomena investigated there and yield more or less reliable results; the description of these special methods, however, exceeds the scope of this book.

In the present section, one of the most widely used methods for calculation of Fourier coefficients is explained. Its advantage consists in its simplicity and the fact that no special device is needed; hence, the method is quite universal. A disadvantage is that the method gives sometimes only a crude picture of the Fourier coefficients, because the error of calculation may be rather large.

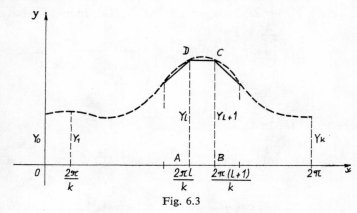

Fig. 6.3

To begin with, let us derive the *trapezoid formula* for the calculation of an integral. Consider a function $F(x)$ defined by its values at the points

$$0;\ \frac{2\pi}{k};\ 2\frac{2\pi}{k};\ 3\frac{2\pi}{k};\ \ldots;\ (k-1)\frac{2\pi}{k};\ k\frac{2\pi}{k} = 2\pi.$$

Denote the values of F at those points by

$$Y_0,\ Y_1,\ Y_2,\ Y_3,\ \ldots,\ Y_{k-1},\ Y_k$$

(see Fig. 6.3) and suppose that $Y_k = Y_0$. The integral

$$\int_{2\pi l/k}^{2\pi(l+1)/k} F(x)\,dx$$

expresses approximately the area of the trapezoid with vertices $ABCD$, i.e. the number $\frac{1}{2}(2\pi/k)(Y_l + Y_{l+1})$. Thus, the integral

$$\int_0^{2\pi} F(x)\,dx$$

is approximately equal to

$$\frac{2\pi}{k}\sum_{l=0}^{k-1}\frac{1}{2}(Y_l + Y_{l+1}) = \frac{2\pi}{k}(Y_0 + Y_1 + \ldots + Y_{k-1}).$$

Using this formula for calculating the Fourier coefficients, i.e., setting $F(x) = f(x)\cos mx$ or $F(x) = f(x)\sin mx$, we have*

$$a_m = \frac{1}{\pi}\frac{2\pi}{k}\left[y_0 + y_1\cos m\frac{2\pi}{k} + y_2\cos m\frac{2\cdot 2\pi}{k} + \ldots\right.$$
$$\left. + y_{k-1}\cos m\frac{(k-1)2\pi}{k}\right],$$

where $y_i = f(i\cdot 2\pi/k)$, i.e.

$$\frac{k}{2}a_m = y_0 + y_1\cos m\frac{2\pi}{k} + y_2\cos m\frac{2\cdot 2\pi}{k} + \ldots$$
$$+ y_{k-1}\cos m\frac{2(k-1)\pi}{k};\quad m = 0, 1, 2, \ldots. \quad (6.9)$$

Similarly,

$$\frac{k}{2}b_m = y_1\sin m\frac{2\pi}{k} + y_2\sin m\frac{2\cdot 2\pi}{k} + \ldots$$
$$+ y_{k-1}\sin m\frac{2(k-1)\pi}{k};\quad m = 1, 2, \ldots. \quad (6.10)$$

* Assuming always that $f(0) = f(2\pi)$.

Twelve points-scheme

Suppose that the values of $f(x)$ are known at the points

$$0, \frac{\pi}{6}, \frac{\pi}{3}, \frac{\pi}{2}, \frac{2\pi}{3}, \frac{5\pi}{6}, \pi, \frac{7\pi}{6}, \frac{4\pi}{3}, \frac{3\pi}{2}, \frac{5\pi}{3}, \frac{11\pi}{6}$$

i.e., for the angles

0°, 30°, 60°, 90°, 120°, 150°, 180°, 210°, 240°, 270°, 300°, 330°;

then $k = 12$. Using the trigonometric formulas, the values of $\cos m(2\pi/12)$ and $\sin m(2\pi/12)$ may be expressed by numbers

$$\pm 1; \pm \sin 30° = \pm 0.5; \quad \pm \sin 60° \cong \pm 0.866.$$

In this case, denote by a_0 the constant term of the Fourier series of f (not by $a_0/2$ as in Chapter 2); then

$$12a_0 = y_0 + y_1 + y_2 + y_3 + y_4 + y_5 + y_6 + y_7 + y_8 + y_9 \\ + y_{10} + y_{11};$$

$$6a_1 = (y_2 + y_{10} - y_4 - y_8) \sin 30° \\ + (y_1 + y_{11} - y_5 - y_7) \sin 60° + (y_0 - y_6);$$

$$6a_2 = (y_1 + y_5 + y_7 + y_{11} - y_2 - y_4 - y_8 - y_{10}) \sin 30° \\ + (y_0 + y_6 - y_3 - y_9);$$

$$6a_3 = y_0 + y_4 + y_8 - y_2 - y_6 - y_{10};$$

$$6b_1 = (y_1 + y_5 - y_7 - y_{11}) \sin 30° \\ + (y_2 + y_4 - y_8 - y_{10}) \sin 60° + (y_3 - y_9);$$

$$6b_2 = (y_1 + y_2 + y_7 + y_8 - y_4 - y_5 - y_{10} - y_{11}) \sin 60°;$$

$$6b_3 = y_1 + y_5 + y_9 - y_3 - y_7 - y_{11}$$

etc.

The calculation can be simplified by using the following *Runge's scheme*:

	y_0	y_1	y_2	y_3	y_4	y_5	y_6
		y_{11}	y_{10}	y_9	y_8	y_7	
sum	u_0	u_1	u_2	u_3	u_4	u_5	u_6
difference		v_1	v_2	v_3	v_4	v_5	

	u_0	u_1	u_2	u_3
	u_6	u_5	u_4	
sum	s_0	s_1	s_2	s_3
difference	d_0	d_1	d_2	

	v_1	v_2	v_3
	v_5	v_4	
sum	σ_1	σ_2	σ_3
difference	δ_1	δ_2	

and by the following formulas,

$$12a_0 = s_0 + s_1 + s_2 + s_3 ;$$
$$6a_1 = d_0 + 0{\cdot}866 d_1 + 0{\cdot}5 d_2 ;$$
$$6a_2 = (s_0 - s_3) + 0{\cdot}5(s_1 - s_2) ;$$
$$6a_3 = d_0 - d_2 ;$$
$$6b_1 = 0{\cdot}5\sigma_1 + 0{\cdot}866\sigma_2 + \sigma_3 ;$$
$$6b_2 = 0{\cdot}866(\delta_1 + \delta_2) ;$$
$$6b_3 = \sigma_1 - \sigma_3 . \tag{6.11}$$

Theoretically, we could calculate also further a_m and b_m with $m > 3$ by (6.9) and (6.10). However, here we have confined ourselves only to the calculation of the first coefficients, because

1. either the Fourier coefficients of f tend to zero so fast that the sum

$$a_0 + a_1 \cos x + b_1 \sin x + a_2 \cos 2x + b_2 \sin 2x$$
$$+ a_3 \cos 3x + b_3 \sin 3x$$

is a satisfactory approximation to the function f,

APPROXIMATE HARMONIC ANALYSIS

2. or the Fourier coefficients decrease so slowly that the further a_m and b_m are not obtained from these formulas with satisfactory accuracy.

In order to illustrate the accuracy of these calculations, let us do two elementary examples and compare the exact values of Fourier coefficients with the approximate ones obtained by the twelve points-scheme.

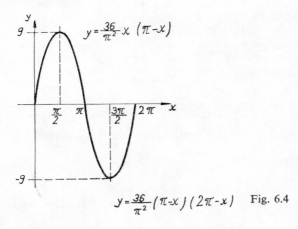

Fig. 6.4

Example 6.7. Consider the function f defined by

$$f(x) = \begin{cases} \dfrac{36}{\pi^2} x(\pi - x) & \text{for } 0 \leq x \leq \pi, \\ \dfrac{36}{\pi^2} (\pi - x)(2\pi - x) & \text{for } \pi < x \leq 2\pi \end{cases}$$

(see Fig. 6.4) and set up a table of its values at the subdividing points.

x	0°	30°	60°	90°	120°	150°	180°	210°	240°	270°	300°	330°
y	0	5	8	9	8	5	0	-5	-8	-9	-8	-5

Runge's scheme reads here as follows,

y	0	5 −5	8 −8	9 −9	8 −8	5 −5	0
u v	0	0 10	0 16	0 18	0 16	0 10	0

u	0	0	0	0	v	10	16	18
	0	0	0	0		10	16	
s	0	0	0	0	σ	20	32	18
d	0	0			δ	0	0	

Thus, we have

$$12a_0 = 0\,;\quad 6a_1 = 0\,;\quad 6a_2 = 0\,;\quad 6a_3 = 0\,;$$
$$6b_1 = 0{\cdot}5 \,.\, 20 + 0{\cdot}866 \,.\, 32 + 18\,;\quad 6b_2 = 0;\quad 6b_3 = 2\,,$$

i.e.

$$b_1 \cong 9{\cdot}285\,;\quad b_3 \cong 0{\cdot}333\,.$$

Let us compare this result with the exact Fourier expansion

$$f(x) = \frac{288}{\pi^3}\left(\sin x + \tfrac{1}{27}\sin 3x + \ldots\right),$$

hence $(\pi^{-3} \cong 0{\cdot}032\,251)$,

$$b_1 = \frac{288}{\pi^3} \cong 9{\cdot}288\,;\quad b_3 = \frac{288}{27\pi^3} \cong 0{\cdot}344\,.$$

Example 6.8. Consider the function f defined by

$$f(x) = \begin{cases} \dfrac{12}{\pi}x & \text{for } 0 \leqq x \leqq \dfrac{\pi}{2}, \\[2mm] \dfrac{12}{\pi}(\pi - x) & \text{for } \dfrac{\pi}{2} < x \leqq \dfrac{3\pi}{2}, \\[2mm] \dfrac{12}{\pi}(x - 2\pi) & \text{for } \dfrac{3\pi}{2} < x \leqq 2\pi \end{cases}$$

APPROXIMATE HARMONIC ANALYSIS 217

(see Fig. 6.5). If the reader sets up the table of values and Runge's scheme he obtains,

$$s_i = d_i = \delta_i = 0; \quad \sigma_1 = 8, \quad \sigma_2 = 16, \quad \sigma_3 = 12$$

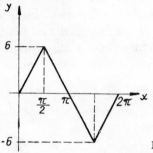

Fig. 6.5

and consequently,

$$a_0 = a_1 = a_2 = a_3 = b_2 = 0$$

and

$$6b_1 = 0\cdot 5 \,.\, 8 + 0\cdot 866 \,.\, 16 + 12; \quad 6b_3 = 8 - 12 = -4,$$

i.e.

$$b_1 \cong 4\cdot 976; \quad b_3 \cong -0\cdot 666.$$

The first terms in the expansion of the function f are

$$f(x) = \frac{48}{\pi^2} (\sin x - \tfrac{1}{9} \sin 3x + \ldots),$$

hence ($\pi^{-2} \cong 0\cdot 101\ 321$)

$$b_1 = \frac{48}{\pi^2} \cong 4\cdot 863; \quad b_3 = -\frac{48}{9\pi^2} \cong -0\cdot 540.$$

Comparing the results for both examples we see that the relative error in Example 6.8 was larger than in Example 6.7. This is due to the fact that the accuracy of calculation is influenced by the differential properties of the function f: The smoother the function is (and consequently the better the estimates of derivatives are), the more accurate the calculation is.

Twenty four points-scheme

In those cases where the twelve points-scheme fails to yield sufficiently accurate results (e.g., if the function does not have a continuous derivative or the derivative assumes too large values), we can use schemes having more subdividing points for the approximate calculation of Fourier coefficients.

Denote by

$$y_0, y_1, y_2, y_3, \ldots, y_{21}, y_{22}, y_{23}$$

the values of f at the points

$$0, \frac{\pi}{12}, \frac{\pi}{6}, \frac{\pi}{4}, \ldots, \frac{21\pi}{12}, \frac{11\pi}{6}, \frac{23\pi}{12}$$

i.e., for the angles

$$0°, 15°, 30°, 45°, 60°, 75°, \ldots, 285°, 300°, 315°, 330°, 345°.$$

The coefficients $\cos m(2\pi/24)$ and $\sin m(2\pi/24)$ can be given in terms of the values

$$\pm 1; \pm \sin 30°; \pm \sin 45°; \pm \sin 60°; \pm \sin 75°;$$

finally, we get Runge's twenty four points-scheme, analogous to the twelve points-scheme.

	y_0	y_1	y_2	y_3	y_4	y_5	y_6	y_7	y_8	y_9	y_{10}	y_{11}	y_{12}
	y_{23}	y_{22}	y_{21}	y_{20}	y_{19}	y_{18}	y_{17}	y_{16}	y_{15}	y_{14}	y_{13}		
sum	u_0	u_1	u_2	u_3	u_4	u_5	u_6	u_7	u_8	u_9	u_{10}	u_{11}	u_{12}
difference		v_1	v_2	v_3	v_4	v_5	v_6	v_7	v_8	v_9	v_{10}	v_{11}	

	u_0	u_1	u_2	u_3	u_4	u_5	u_6
	u_{12}	u_{11}	u_{10}	u_9	u_8	u_7	
sum	p_0	p_1	p_2	p_3	p_4	p_5	p_6
difference	q_0	q_1	q_2	q_3	q_4	q_5	

APPROXIMATE HARMONIC ANALYSIS

	v_1 v_2 v_3 v_4 v_5 v_6
	v_{11} v_{10} v_9 v_8 v_7
sum	r_1 r_2 r_3 r_4 r_5 r_6
difference	s_1 s_2 s_3 s_4 s_5

	p_0 p_1 p_2 p_3		s_1 s_2 s_3
	p_6 p_5 p_4		s_5 s_4
sum	k_0 k_1 k_2 k_3	sum	m_1 m_2 m_3
difference	l_0 l_1 l_2	difference	n_1 n_2

Then we have

$$24a_0 = k_0 + k_1 + k_2 + k_3$$

$$12a_1 = [q_0 + 0{\cdot}5q_4 + 0{\cdot}6124(q_1 + q_5)]$$
$$\quad + [0{\cdot}8660q_2 + 0{\cdot}7071q_3 + 0{\cdot}3536(q_1 - q_5)]$$

$$12a_2 = l_0 + 0{\cdot}8660l_1 + 0{\cdot}5l_2$$

$$12a_3 = (q_0 - q_4) + 0{\cdot}7071(q_1 - q_3 - q_5)$$

$$12a_4 = (k_0 - k_3) + 0{\cdot}5(k_1 - k_2)$$

$$12a_5 = [q_0 + 0{\cdot}5q_4 - 0{\cdot}6124(q_1 + q_5)]$$
$$\quad - [0{\cdot}8660q_2 + 0{\cdot}7071q_3 + 0{\cdot}3536(q_1 - q_5)]$$

$$12a_6 = l_0 - l_2$$
$$\ldots\ldots$$

$$12b_1 = [0{\cdot}5r_2 + r_6 + 0{\cdot}6124(r_1 + r_5)]$$
$$\quad + [0{\cdot}7071r_3 + 0{\cdot}8660r_4 - 0{\cdot}3536(r_1 - r_5)]$$

$$12b_2 = 0{\cdot}5m_1 + 0{\cdot}8660m_2 + m_3$$

$$12b_3 = (r_2 - r_6) + 0{\cdot}7071(r_1 + r_3 - r_5)$$

$$12b_4 = 0{\cdot}8660(n_1 + n_2)$$

$$12b_5 = [0{\cdot}5r_2 + r_6 + 0{\cdot}6124(r_1 + r_5)]$$
$$\quad - [0{\cdot}7071r_3 + 0{\cdot}8660r_4 - 0{\cdot}3536(r_1 - r_5)]$$

$$12b_6 = m_1 - m_3$$
$$\ldots\ldots$$

(6.12)

Example 6.9. Let us find the Fourier coefficients of the function f from Example 6.7 by using the twenty four points-scheme. First, we set up the table of values,

x	0°	15°	30°	45°	60°	75°	90°	105°	120°	135°	150°	165°
y	0	2·75	5	6·75	8	8·75	9	8·75	8	6·75	5	2·75
x	180°	195°	210°	225°	240°	255°	270°	285°	300°	315°	330°	345°
y	0	−2·75	−5	−6·75	−8	−8·75	−9	−8·75	−8	−6·75	−5	−2·75

Then we form the first part of the corresponding Runge's scheme,

```
y | 0   2·75    5   6·75    8   8·75    9   8·75    8   6·75    5   2·75  0
  |    −2·75  −5  −6·75   −8  −8·75  −9  −8·75  −8  −6·75  −5  −2·75
―――――――――――――――――――――――――――――――――――――――――――――――――――――――――――――――――――――――
u | 0   0       0   0       0   0       0   0       0   0       0   0
v |     5·5    10   13·5   16   17·5   18   17·5   16   13·5   10   5·5
```

Apparently, $u_i = p_i = q_i = k_i = l_i = 0$; thus, it suffices to find only the values of r and s:

```
v | 5·5   10    13·5   16    17·5   18
  | 5·5   10    13·5   16    17·5
――――――――――――――――――――――――――――――――――――
r | 11    20    27     32    35     18
s |  0     0     0      0     0
```

It follows that also $s_i = m_i = n_i = 0$. Using formulas (6.12), we obtain finally $a_0 = a_1 = a_2 = a_3 = a_4 = a_5 = a_6 = b_2 = b_4 = 0$ and

$$b_1 \cong 9{\cdot}2884 \, ; \quad b_3 \cong 0{\cdot}3434 \, ; \quad b_5 \cong 0{\cdot}0733 \, .$$

APPROXIMATE HARMONIC ANALYSIS   221

The Fourier coefficients (correct to four decimal places) determined by the known Fourier expansion of f are given by numbers

$$b_1 \cong 9\cdot 2884 \;;\quad b_3 \cong 0\cdot 3440 \;;\quad b_5 \cong 0\cdot 0743\;.$$

Hence, the results are quite accurate here.

Example 6.10. Let us find the Fourier coefficients b_1, b_3 and b_5 of the function f from Example 6.8 by using the 24 points-scheme. As in the preceding example, Runge's scheme has a very simple form here; thus, we get

$$b_1 \cong 4\cdot 8914 \;;\quad b_3 \cong -0\cdot 5690 \;;\quad b_5 \cong 0\cdot 2248\;,$$

while the Fourier coefficients, correct to four decimal places, are

$$b_1 \cong 4\cdot 8634 \;;\quad b_3 \cong -0\cdot 5404 \;;\quad b_5 \cong 0\cdot 1945\;.$$

Clearly, the results are less satisfactory in this case. This is due to the fact that the function f is less smooth than that of Example 6.7.

Let us note that the error of the method may be rather large even for some smooth functions.

As mentioned above, the schemes described are usually unsuitable for the calculation of Fourier coefficients a_m and b_m with m large. If such a coefficient is to be calculated, it is better to employ *Filon's method* (cf. [13]), which yields quite good results, provided the interval is subdivided sufficiently densely. Namely, for large n, the functions $\cos nx$ and $\sin nx$ oscillate very rapidly; consequently, the subdivision points must be chosen so as to grasp the behaviour of the integrated functions sufficiently accurately.

Filon's method is a certain modification of Simpson's rule for the calculation of integrals of type

$$\int_a^b f(x) \cos px \, dx \;,\quad \int_a^b f(x) \sin px \, dx \;.$$

We proceed as follows. Let us subdivide the interval $[a, b]$ by the points $x_i = a + i(b - a)/2n$ $(i = 0, 1, 2, ..., 2n)$ into

2n equal intervals of length $h = (b - a)/2n$. Then introduce functions α, β and γ defined by power series of the variable $t = ph = p(b - a)/2n$; instead of giving the explicit form of these functions we present here only a table of values, which is sufficient for practical purposes.

t	α	β	γ
0·0	0·000 00	1·333 33	0·666 67
0·025	0·000 00	1·333 25	0·666 75
0·05	0·000 01	1·333 00	0·667 00
0·10	0·000 04	1·332 00	0·668 00
0·15	0·000 15	1·330 34	0·669 65
0·20	0·000 35	1·328 01	0·671 94
0·25	0·000 69	1·325 02	0·674 85
0·30	0·001 18	1·321 37	0·678 36
0·40	0·002 78	1·312 12	0·687 04
0·50	0·005 36	1·300 30	0·697 67
0·75	0·017 30	1·259 82	0·730 22
1·00	0·038 50	1·204 67	0·765 26
1·50	0·108 40	1·056 46	0·809 71

If t lies between two tabulated values, the corresponding values of α, β and γ are obtained by a linear interpolation.

Filon's method is based on the following formulas (α, β and γ are functions of n),

$$\int_a^b f(x) \cos px \, dx$$
$$= \frac{b-a}{2n} \left\{ \alpha [f(b) \sin pb - f(a) \sin pa] + \beta C_1 + \gamma C_2 \right\},$$

where

$$C_1 = \sum_{i=1}^{n} f(x_{2i-1}) \cos px_{2i-1},$$
$$C_2 = \tfrac{1}{2} f(a) \cos pa + \sum_{i=1}^{n-1} f(x_{2i}) \cos px_{2i} + \tfrac{1}{2} f(b) \cos pb;$$

and

$$\int_a^b f(x) \sin px \, dx$$
$$= \frac{b-a}{2n} \{-\alpha[f(b) \cos pb - f(a) \cos pa] + \beta S_1 + \gamma S_2\},$$

where

$$S_1 = \sum_{i=1}^n f(x_{2i-1}) \sin px_{2i-1},$$

$$S_2 = \tfrac{1}{2}f(a) \sin pa + \sum_{i=1}^{n-1} f(x_{2i}) \sin px_{2i} + \tfrac{1}{2}f(b) \sin pb.$$

This, of course, is the general Filon's method; if calculating Fourier coefficients we set $b - a = 2\pi$; then the last formulas can be simplified (cf. Problem (6.7)).

Problems

6.1. Find the Fourier coefficients of the function $f(x) = \tan^{-1} |x|$ for $x \in (-\pi, \pi)$ with accuracy of 10^{-2}.

6.2. Can the function $f(x) = \sqrt{x}$ in $(0, 2\pi)$ be written as

$$f(x) = \alpha \, \sigma_0(x) + \beta \, \sigma_1(x) + \sum_{n=-\infty}^{\infty} \gamma_n e^{inx},$$

where $|\gamma_n| \leq C|n|^{-3}$?

6.3. Increase the convergence of the series

$$f(x) = \sum_{n=1}^{\infty} (-1)^n \frac{n^5 - 3n + 6}{n(n^5 + 1)} \cos nx.$$

Where and of what kind are the discontinuities of f?

6.4. Increase the convergence of the series

$$\sum_{n=1}^{\infty} \sin \frac{1}{n} \sin nx .$$

What discontinuities does this function have?

6.5. Show that

$$S_q(x) = 2 \sum_{n=1}^{\infty} \frac{\cos nx}{n^q}$$

for q even and that

$$S_q(x) = 2i \sum_{n=1}^{\infty} \frac{\sin nx}{n^q}$$

for q odd. Express the functions $S_q(x)$ in terms of $\sigma_q(x)$.

6.6. Carry out in detail the calculations in Examples 6.7 to 6.10.

6.7. Modify Filon's method for the calculation of the Fourier coefficients in the interval $(-\pi, \pi)$.

6.8. Using Filon's method, find the coefficients b_1 and b_3 for Example 6.8.

CHAPTER 7 SOME SPECIAL CRITERIA FOR CONVERGENCE

7.1 GIBBS PHENOMENON

In preceding chapters we considered the Fourier series of various types of functions. We have established that the Fourier series of a function $f \in L_2(0, 2\pi)$ converges to f in the norm of the space $L_2(0, 2\pi)$; furthermore, if the function f satisfies the assumptions of Theorem 2.11, then the corresponding Fourier series converges uniformly.

An example of a function not admitting the application of Theorem 2.11 is furnished by the function

$$\sigma_0(x) = \frac{\pi - x}{2} \quad \text{for} \quad x \in (0, 2\pi),$$

which is periodically extended onto the entire axis. From the preceding chapter we know that

$$\sigma_0(x) = \sum_{k=1}^{\infty} \frac{\sin kx}{k}, \tag{7.1}$$

where the equality holds either in the sense of convergence in $L_2(a, a + 2\pi)$ with an arbitrary real number a, or in the sense of uniform convergence on any interval $(2k\pi + \varepsilon, 2(k + 1)\pi - \varepsilon)$ with $0 < \varepsilon < \pi$ (see Theorem 2.4). Thus, we do not know what is happening in a neighbourhood of a point of discontinuity of the function $\sigma_0(x)$, for example in a neighbourhood of zero. At the point $x = 0$ the sum of the series (7.1) is equal to zero, i.e. to the arithmetic mean of limits from the right and from the left,

$$\tfrac{1}{2}[\sigma_0(0 + 0) - \sigma_0(0 - 0)] = 0.$$

Let us investigate the error of a partial sum of the series (7.1). We have

$$R_n(x) = s_n(x) - \sigma_0(x) = \sum_{k=1}^{n} \frac{\sin kx}{k} - \frac{\pi - x}{2}$$

for $x \in (0, 2\pi)$. Clearly,

$$R'_n(x) = \tfrac{1}{2} + \sum_{k=1}^{n} \cos kx = \frac{\sin\left(n + \tfrac{1}{2}\right)x}{2 \sin \tfrac{1}{2} x}$$

and

$$R_n(0) = -\frac{\pi}{2},$$

or, more precisely, $R_n(0 + 0) = -\pi/2$, because the value of $\sigma_0(x)$ at $x = 0$ has not been defined yet. Hence,

$$R_n(x) = -\frac{\pi}{2} + \int_0^x \frac{\sin\left(n + \tfrac{1}{2}\right) t}{2 \sin \tfrac{1}{2} t}\, \mathrm{d}t.$$

The function $s_n(x)$ is increasing in a neighbourhood of zero. Next, let us find the point $x_n > 0$ closest to zero such that the function $R_n(x)$ attains a local extremum at x_n. From the equation $R'_n(x) = 0$ it follows that $\sin\left(n + \tfrac{1}{2}\right) x = 0$, and consequently, $x_n = \pi(n + \tfrac{1}{2})^{-1}$. Denoting $n + \tfrac{1}{2} = p$ and using the substitution $pt = s$ we obtain

$$R_n(x_n) = \int_0^{\pi/p} \frac{\sin pt}{2 \sin \tfrac{1}{2} t}\, \mathrm{d}t - \frac{\pi}{2} = \int_0^{\pi} \frac{\sin s}{2p \sin \dfrac{s}{2p}}\, \mathrm{d}s - \frac{\pi}{2}.$$

For p sufficiently large, i.e. for n sufficiently large, we have $2p \sin(s/2p) \geqq 0$ for $s \in (0, \pi)$ and $\lim_{p \to \infty} 2p \sin(s/2p) = s$. Thus, (find the integrable majorant!)

$$\lim_{n \to \infty} R_n(x_n) = \int_0^{\pi} \frac{\sin s}{s}\, \mathrm{d}s - \frac{\pi}{2} \cong 0{\cdot}18\, \frac{\pi}{2} \qquad (7.2)$$

and

$$s_n(x_n) \cong 1.18 \frac{\pi}{2}$$

for large n.

Summarizing these results we see that each partial sum $s_n(x)$ has a maximum exceeding the maximum of the function $\sigma_0(x)$ by approximately 18 percent. (See Fig. 7.1.) This fact is called

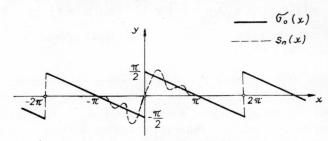

Fig. 7.1

the *Gibbs phenomenon*. For n increasing the maximum of $s_n(x)$ differs substantially from the maximum of the function $\sigma_0(x)$, and the point x_n, at which the maximum is attained, approaches zero. Therefore, none of the partial sums of the series (7.1) can approximate the function $\sigma_0(x)$ *uniformly* in a neighbourhood of zero.

In Chapter 6 we have shown that any function f having a finite number of discontinuities of the first kind and being sufficiently smooth elsewhere can be represented as a sum of two functions,

$$f(x) = g(x) + h(x),$$

where $g(x)$ is a function satisfying the assumptions of Theorem 2.11 whose Fourier series thus converges uniformly, and where

$$h(x) = \frac{1}{\pi} \sum_{i=1}^{m} c_i \, \sigma_0(x - x_i)$$

is a function compensating the jumps of the function $f(x)$, because x_i are the points of discontinuity and c_i the corresponding jumps of $f(x)$. Thus, from the above discussion concerning the function

$\sigma_0(x)$ it follows that the Fourier series of the function f converges at any point x to the value $\frac{1}{2}[f(x + 0) + f(x - 0)]$ and consequently, to the value $f(x)$ at any point x, where f is continuous. Furthermore, the Gibbs phenomenon appears in a neighbourhood of any discontinuity of the function f, i.e. in a neighbourhood of a discontinuity the partial sums of the Fourier series of f attain, up to a negligible error, the values

$$\tfrac{1}{2}[f(x_i + 0) + f(x_i - 0)] \pm \tfrac{1}{2} \cdot 1\cdot 18 |f(x_i + 0) - f(x_i - 0)| ;$$

hence, the partial sums do not converge uniformly in any neighbourhood of a discontinuity.

The function $f(x)$ just considered was reasonably well-behaved; in theoretical considerations and in practice, however, we encounter functions whose behaviour is much worse. Since we do not want to be content only with the fact that the convergence of the Fourier series of these functions is guaranteed in the space $L_2(a, a + 2\pi)$, we are going to present certain deeper assertions about the convergence of Fourier series in this Chapter. We should note that the following sections are intended primarily for those readers who were attracted enough by the theory of Fourier series and who will accept with understanding a slightly more brief presentation. This has been done mainly in order not to enlarge the scope of the book too much.

7.2 DIRICHLET'S KERNEL

As known the sum of a Fourier series is defined as the limit of partial sums. Now, we are going to derive a formula for the partial sums which does not involve the Fourier coefficients. Let us remind the reader of the fact that, as the Euler's formulas (1.19) show, a partial sum of a Fourier series is independent of whether we consider the real or the complex form of the Fourier series. If not stated otherwise, in what follows we shall always assume that the function f is defined on $(-\infty, \infty)$ and is periodic with period 2π.

By definition,
$$s_n(x) = \sum_{k=-n}^{n} c_k e^{ikx}, \tag{7.3}$$

where
$$c_k = \frac{1}{2\pi} \int_{-\pi}^{\pi} f(\xi) e^{-ik\xi} \, d\xi, \tag{7.4}$$

i.e.
$$s_n(x) = \frac{1}{2\pi} \int_{-\pi}^{\pi} f(\xi) \sum_{k=-n}^{n} e^{ik(x-\xi)} \, d\xi.$$

Using the substitution $\xi - x = t$ and the periodicity of the integrated function (see (1.17)) we obtain

$$\begin{aligned} s_n(x) &= \frac{1}{2\pi} \int_{-\pi}^{\pi} f(x+t) \sum_{k=-n}^{n} e^{ikt} \, dt \\ &= \frac{1}{2\pi} \int_{-\pi}^{\pi} f(x+t) \frac{\sin(n+\tfrac{1}{2})t}{\sin \tfrac{1}{2} t} \, dt. \end{aligned} \tag{7.5}$$

(Here we have used formula (1.26) from Chapter 1.) Let us denote

$$D_n(t) = \frac{1}{2\pi} \frac{\sin(n+\tfrac{1}{2})t}{\sin \tfrac{1}{2} t}$$

and call the function $D_n(t)$ the *Dirichlet's kernel*.

If $f(x) \equiv 1$, we have also $s_n(x) \equiv 1$ for all $n = 0, 1, 2, \ldots$, and consequently,

$$1 = \int_{-\pi}^{\pi} D_n(t) \, dt \quad \left(= 2 \int_{0}^{\pi} D_n(t) \, dt \right). \tag{7.6}$$

Taking a trigonometric polynomial $T_n(x)$ of degree n for the function $f(x)$, then the corresponding partial sum $s_m(x)$ will equal

$T_n(x)$ for $m \geqq n$. Thus, we have

$$T_n(x) = \int_{-\pi}^{\pi} T_n(x+t) D_m(t) \, dt \quad \text{for} \quad m \geqq n. \tag{7.7}$$

Moreover, by (7.6),

$$f(x) = \int_{-\pi}^{\pi} f(x) D_m(t) \, dt \quad (m = 0, 1, 2, \ldots),$$

because the function $f(x)$ can be carried in front of the integral; hence,

$$s_n(x) - f(x) = \int_{-\pi}^{\pi} [f(x+t) - f(x)] D_n(t) \, dt$$

$$= \frac{1}{2\pi} \int_{-\pi}^{\pi} [f(x+t) - f(x)] \frac{\sin(n + \tfrac{1}{2})t}{\sin \tfrac{1}{2} t} \, dt. \tag{7.8}$$

The following representations are also useful,

$$s_n(x) = \int_0^{\pi} [f(x+t) + f(x-t)] D_n(t) \, dt; \tag{7.9}$$

$$s_n(x) - K = \int_0^{\pi} [f(x+t) + f(x-t) - 2K] D_n(t) \, dt, \tag{7.10}$$

where K is a constant.

The next proposition follows easily from formula (7.9).

Theorem 7.1. *Let f be a bounded, measurable function on the interval $[-\pi, \pi]$. Then positive constants A and B exist such that*

$$|s_n(x)| \leq (A \log n + B) \sup_{x \in [-\pi, \pi]} |f(x)|.$$

Proof: By (7.9) we have

$$|s_n(x)| \leq \frac{2 \sup |f(x)|}{2\pi} \int_0^{\pi} \left| \frac{\sin(n + \tfrac{1}{2})t}{\sin \tfrac{1}{2} t} \right| dt.$$

CONVERGENCE OF FOURIER SERIES

Using the substitution $(n + \tfrac{1}{2})\, t = y$ and the fact that $\sin \tfrac{1}{2}t \geqq t/\pi$ on the interval $(0, \pi)$, we obtain the estimate

$$\int_0^\pi \left| \frac{\sin (n + \tfrac{1}{2})\, t}{\sin \tfrac{1}{2}t} \right| dt \leqq \pi \int_0^{(n+\frac{1}{2})\pi} \frac{|\sin y|}{y} \, dy$$

$$\leqq \pi \left(\int_0^1 dy + \int_1^{(n+\frac{1}{2})\pi} \frac{dy}{y} \right);$$

from this the assertion of the theorem follows immediately.

Earlier we have proved (cf. Theorem 3.18) that for $f \in L_1(a, b)$ and λ real we have

$$I(\lambda; f; a, b) = \int_a^b f(x) \, e^{i\lambda x} \, dx \to 0 \quad \text{as} \quad |\lambda| \to \infty. \qquad (7.11)$$

Theorem 7.1 shows that the partial sums of the Fourier series of a bounded function can not increase more rapidly than the sequence $\{\log n\}$. On the other hand, formula (7.8) together with (7.11) permit us to state an assertion concerning the convergence of the partial sums, i.e. we can prove the so-called *Dirichlet's criterion for convergence*.

Theorem 7.2. *Let $f \in L_1(-\pi, \pi)$ and let for x_0 a number $\varepsilon > 0$ exist such that the integral*

$$\int_{-\varepsilon}^\varepsilon \frac{|f(x_0 + t) - f(x_0)|}{|t|} \, dt \qquad (7.12)$$

converges. Then $\lim_{n \to \infty} s_n(x_0) = f(x_0)$.

Remark 7.1. The integral (7.12) is meaningful for $x_0 = \pm\pi$, too, because the function f is defined and periodic on $(-\infty, \infty)$.

Proof of Theorem 7.2: By (7.8),

$$s_n(x_0) - f(x_0) = \frac{1}{2\pi} \int_{-\pi}^\pi \varphi(t) \sin (n + \tfrac{1}{2})\, t \, dt,$$

where

$$\varphi(t) = \frac{f(x_0 + t) - f(x_0)}{t} \cdot \frac{t}{\sin \frac{1}{2}t} \in L_1(-\pi, \pi),$$

because by (7.12) $t^{-1}[f(x_0 + t) - f(x_0)] \in L_1(-\pi, \pi)$ and the function $t/\sin \frac{1}{2}t$ is bounded. Now, it suffices to use the relation (7.11).

Example 7.1. Let $f \in L_1(-\pi, \pi)$ and let

$$|f(x_0 + t) - f(x_0)| \leq C|t|^\alpha \quad (\alpha > 0). \tag{7.13}$$

Then the assumptions of Theorem 7.2 are satisfied and thus $s_n(x_0) \to f(x_0)$ as $n \to \infty$.

A function f satisfying the inequality (7.13) for all $x_0 \in (-\infty, \infty)$ with a fixed constant C independent of x_0 is called *Hölder-continuous with exponent* α, or briefly α-*Hölder-continuous*. Hence, the partial sums of an α-Hölder-continuous function f converge to the function f everywhere.

In Problem 3.17 we introduced the modulus of continuity of a function f by

$$\omega(\delta, f) = \max_{|x-y|<\delta} |f(x) - f(y)|;$$

it is clear that for an α-Hölder-continuous function f we have $\omega(\delta, f) \leq C\delta^\alpha$. The relation of the convergence of a Fourier series to the modulus of continuity of a function as well as to the behaviour of derivatives of a continuous function is treated in the next section.

7.3 CONTINUOUS PERIODIC FUNCTIONS

The following proposition, appearing as a certain analogue of Theorem 7.1 and concerning a differentiable function, is given without proof.

Theorem 7.3. *Let a periodic function f with period 2π possess a derivative $f^{(r)}$ of order r (in the sense of an absolutely continuous function) on the interval $(-\infty, \infty)$ and let $f^{(r)}$ be bounded by a constant M_r. Then positive constants A and B exist such that*

$$|f(x) - s_n(x)| \leq \frac{A \log n + B}{n^r} M_r. \tag{7.14}$$

Remark 7.2. From the assumptions of the theorem it follows that $f(\pi) = f(-\pi)$, $f'(\pi) = f'(-\pi), ..., f^{(r-1)}(\pi) = f^{(r-1)}(-\pi)$. If we omit the assumption on the periodicity of the function f, the estimate (7.14) need not hold. (See also Problem 7.3.)

Remark 7.3. Theorem 2.11 shows that, under some additional assumptions concerning the behaviour of the $(r + 1)$-th derivative of the function f, we may put $A = 0$ in (7.14). Actually, for a function satisfying the assumptions of Theorem 2.11 and for $r = 1$, we have

$$|f(x) - s_n(x)| \leq \sum_{k=n+1}^{\infty} (|a_k| + |b_k|) \leq C_1 \sum_{k=n+1}^{\infty} \frac{1}{k^2} \leq C_2 \frac{1}{n}.$$

The following assertion estimates the dependence of the speed of convergence on the modulus of continuity.

Theorem 7.4. *Let f be a continuous periodic function with period 2π. Then positive constants A and B exist such that*

$$|f(x) - s_n(x)| \leq (A \log n + B) \omega\left(\frac{\pi}{n}, f\right). \tag{7.15}$$

The proof of this theorem is based on an approximation of the function f by a polygon (see Problem 3.17) and on Theorem 7.3 for $r = 1$. The proofs of Theorems 7.3 and 7.4 may be found in [3].

Remark 7.4. Theorem 7.4 yields the following, so-called *Dini-Lipschitz criterion* for uniform convergence:

Let f be a continuous and periodic function, and let

$$\lim_{\delta \to 0+} \omega(\delta, f) \log \frac{1}{\delta} = 0. \tag{7.16}$$

Then the partial sums of the corresponding Fourier series converge to f uniformly.

Remark 7.5. From Remark 7.4 it follows that for an α-Hölder-continuous function f we have $f(x) = \sum_{k=-\infty}^{\infty} c_k e^{-ikx}$ in the sense of uniform convergence.

The following *Bernstein theorem* concerns absolute and uniform convergence. (See [3].)

Theorem 7.5. *If a function f is periodic with period 2π and α-Hölder-continuous with $\alpha > \frac{1}{2}$, then the series*

$$\sum_{n=1}^{\infty} (|a_n| + |b_n|) n^\beta$$

converges for $0 \leqq \beta < \alpha - \frac{1}{2}$. Thus, the Fourier series of f converges absolutely and uniformly.

7.4 PRINCIPLE OF LOCALIZATION

The criteria considered in Section 7.3 are inconvenient from a certain point of view, because the continuity of the function has to be assumed on the entire axis. On the other hand, the behaviour of the Fourier series of a function, which is well-behaved only on a part of the interval $(-\pi, \pi)$, can be tested with the aid of the following proposition.

Theorem 7.6. (Principle of localization.) *Let $f \in L_1(-\pi, \pi)$ and let $f(x) = 0$ for every $x \in (a, b)$, where $[a, b] \subset (-\pi, \pi)$. Further-*

CONVERGENCE OF FOURIER SERIES

more, let $a < a_1 < b_1 < b$. Then the Fourier series of the function f converges uniformly to zero on the interval (a_1, b_1).

Proof: Denote $a_2 = \frac{1}{2}(a + a_1)$, $b_2 = \frac{1}{2}(b + b_1)$ and let $\varepsilon > 0$. According to Problem 3.24, a trigonometric polynomial T_n exists such that $|T_n(x)| < \varepsilon$ for $x \in (a_2, b_2)$ and $\|f - T_n\|_{L_1(-\pi,\pi)} < \varepsilon$. From the formulas (7.5) and (7.7) it follows that, for $k \geq n$,

$$s_k(x) - T_n(x) = \frac{1}{2\pi} \int_{-\pi}^{\pi} [f(x + t) - T_n(x + t)] \frac{\sin(k + \frac{1}{2})t}{\sin \frac{1}{2}t} \, dt \,.$$

Denote $\delta = \frac{1}{2} \min(a_1 - a, b - b_1)$. If $x \in (a_1, b_1)$, then $f(x + t) = 0$ for $|t| < \delta$. Hence,

$$|s_k(x) - T_n(x)| \leq \frac{1}{2\pi} \int_{\pi > |t| > \delta} |f(x + t) - T_n(x + t)| \sin^{-1} \tfrac{1}{2}\delta \, dt$$

$$+ \left| \frac{1}{2\pi} \int_{\pi > |t| > \delta} T_n(x + t) \frac{\sin(k + \frac{1}{2})t}{\sin \frac{1}{2}t} \, dt \right|$$

$$+ \left| \frac{1}{2\pi} \int_{-\pi}^{\pi} T_n(x + t) \frac{\sin(k + \frac{1}{2})t}{\sin \frac{1}{2}t} \, dt \right|.$$

(We used the fact that

$$\int_{|t|<\delta} T_n(x + t) D_k(t) \, dt = \int_{-\pi}^{\pi} - \int_{\pi > |t| > \delta} \,.)$$

The first integral on the right hand side of the inequality is less than $\varepsilon/(2\pi \sin \frac{1}{2}\delta)$, because $\|f - T_n\|_{L_1(-\pi,\pi)} < \varepsilon$; the second integral is less than K/k by Theorem 2.10. (Observe that the latter integral is equal to the sum of the Fourier coefficient of $\cos kt$ for the function vanishing for $|t| < \delta$ and equalling $T_n(x + t)$ for $|t| > \delta$, and of the Fourier coefficient of $\sin kt$ for the function vanishing for $|t| < \delta$ and equalling $T_n(x + t) \sin^{-1} \frac{1}{2}t \cos \frac{1}{2}t$ for $|t| > \delta$; verify that for these functions the number K appearing in the proof of Theorem 2.10 does not depend on x.) Finally, the third integral is equal to $T_n(x)$ by (7.7). Since $|T_n(x)| < \varepsilon$ for

$x \in (a_1, b_1)$, the difference $s_k(x) - T_n(x)$, and consequently, the partial sum $s_k(x)$, too, is small for k sufficiently large uniformly with respect to $x \in (a_1, b_1)$.

Thus, the principle of localization states that the behaviour of the Fourier series of a function $f(x)$ on an interval (a_1, b_1) depends only on the behaviour of $f(x)$ on an imperceptibly larger interval (a, b).

The following proposition furnishes a simple example of the application of the localization theorem.

Theorem 7.7. *Let f be periodic with period 2π and let $f \in L_1(-\pi, \pi)$. Furthermore, let f be continuous on an interval (a, b) such that $[a, b] \subset (-\pi, \pi)$, and let the modulus of continuity $\omega(\delta, f)$ of the function f satisfy the condition*

$$\lim_{\delta \to 0+} \omega(\delta, f) \log \frac{1}{\delta} = 0$$

on the interval (a, b). Then the Fourier series of the function f converges to f uniformly on any interval (a_1, b_1) with $a < a_1 < b_1 < b$.

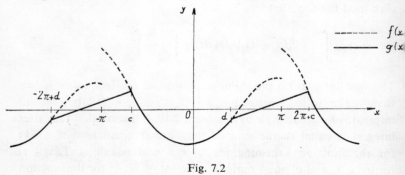

Fig. 7.2

Proof: Denote $c = \frac{1}{2}(a + a_1)$, $d = \frac{1}{2}(b + b_1)$ and define the function g as follows: let $g(x) = f(x)$ for $x \in (c, d)$, and in the intervals $[-\pi, c]$, $[d, \pi]$ let g be a linear function such that

$g(c) = f(c)$, $g(d) = f(d)$ and $g(-\pi) = g(\pi)$. (See Fig. 7.2.) The function g defined in this way satisfies the assumptions stated in Remark 7.4; thus, the partial sums of the Fourier series of g converge uniformly to g, i.e. to the function f for $x \in (c, d)$. Moreover, by Theorem 7.6, the Fourier series of the function $f - g$ converges uniformly to zero on the interval (a_1, b_1). From this it follows that the Fourier series of $f = (f - g) + g$ converges uniformly to the function f on the interval (a_1, b_1); hence the theorem is proved.

Remark 7.6. As shown above, we can often derive propositions concerning the local convergence of the series from the principle of localization. However, in this way we are unable to obtain local estimates of the speed of convergence, i.e. estimates analogous to (7.14) which hold locally. This fact is demonstrated by the following example.

Example 7.2. The function $f(x) = \operatorname{sgn} \sin x$ equals a constant on the interval $(0, \pi)$, and consequently, is perfectly smooth there. In spite of that the partial sums of the series (2.15) from Example 2.1 converge to $f(x)$ as fast as the sequence $\{1/n\}$. In order to see it, it suffices to put $x = \pi/2$ and estimate the difference $s_n(\pi/2) - f(\pi/2)$.

7.5 ABSOLUTELY CONTINUOUS FUNCTIONS

To begin with let us recall the definition of an absolutely continuous function. A function f is called *absolutely continuous* on an interval $[a, b]$ of finite length, if a function $g \in L_1(a, b)$ exists such that

$$f(x) - f(y) = \int_y^x g(t)\, dt \tag{7.17}$$

for any $x, y \in [a, b]$. The function $g(t)$ is called a *derivative of f in the sense of an absolutely continuous function* on the interval (a, b), and we write $g(t) = f'(t)$.

A function has a derivative in the sense of an absolutely continuous function on an interval of infinite length, if it is absolutely continuous on every bounded part of this interval.

The following proposition on approximation is true for absolutely continuous functions.

Theorem 7.8. *If a function f is absolutely continuous on the interval $[-\pi, \pi]$ and periodic with period 2π, then for every $\varepsilon > 0$ a trigonometric polynomial T_n exists such that*

$$|T_n(x) - f(x)| < \varepsilon \quad \text{for} \quad x \in [-\pi, \pi]$$

and

$$\|T_n' - f'\|_{L_1(-\pi,\pi)} < \varepsilon.$$

Proof: By the assumptions of the Theorem, $f' \in L_1(-\pi, \pi)$ and

$$\int_{-\pi}^{\pi} f'(t) \, dt = f(\pi) - f(-\pi) = 0.$$

From the considerations in the Appendix to Chapter 3 it follows in an analogous way as in the proof of Theorem 3.15 that a sequence of trigonometric polynomials T_k^* exists such that $T_k^* \to f'$ in $L_1(-\pi, \pi)$ as $k \to \infty$. Since

$$\left| \int_{-\pi}^{\pi} T_k^*(t) \, dt \right| = \left| \int_{-\pi}^{\pi} [T_k^*(t) - f'(t)] \, dt \right| \leq \|T_k^* - f'\|_{L_1(-\pi,\pi)},$$

the first integral also converges to zero, and consequently,

$$T_k^{**}(x) = T_k^*(x) - \frac{1}{2\pi} \int_{-\pi}^{\pi} T_k^*(t) \, dt \to f'(x) \quad \text{in} \quad L_1(-\pi, \pi)$$

as $k \to \infty$. Furthermore,

$$\int_{-\pi}^{\pi} T_k^{**}(t) \, dt = 0,$$

CONVERGENCE OF FOURIER SERIES

i.e. the constant term of the trigonometric polynomial $T_k^{**}(x)$ vanishes. Then, of course, the function

$$T_k(x) = \int_0^x T_k^{**}(t) \, dt + f(0)$$

is also a trigonometric polynomial and we have

$$|T_n(x) - f(x)| = \left| \int_0^x [T_n^{**}(t) - f'(t)] \, dt \right| \leq \|T_n^{**} - f'\|_{L_1(-\pi,\pi)}$$

for $x \in [-\pi, \pi]$. Since $T_n' = T_n^{**}$ and $T_n^{**} \to f'$ in $L_1(-\pi, \pi)$ as $n \to \infty$, the proof is completed.

The main theorem in this section reads as follows.

Theorem 7.9. *If f is a function absolutely continuous on the interval $[-\pi, \pi]$ and periodic with period 2π, then its Fourier series converges uniformly to f.*

In order to prove this theorem we have to state two auxiliary propositions first.

Theorem 7.10. *Let $\tau \in (0, \pi]$; then*

$$\left| \int_\tau^\pi D_n(t) \, dt \right| = \left| \int_{-\pi}^{-\tau} D_n(t) \, dt \right| < M,$$

where M is a constant independent of τ and n.

Proof: Since the function $D_n(t)$ is even the integrals are actually equal; thus, let us estimate only the first one. From the formula (7.6) it follows that

$$\int_\tau^\pi D_n(t) \, dt = \int_0^\pi D_n(t) \, dt - \int_0^\tau D_n(t) \, dt = \frac{1}{2} - \int_0^\tau D_n(t) \, dt \, .$$

For the last integral we obtain

$$\int_0^\tau D_n(t)\,dt = \frac{1}{\pi}\int_0^\tau \left(\tfrac{1}{2}\sin^{-1}\tfrac{1}{2}t - \frac{1}{t}\right)\sin\left(n+\tfrac{1}{2}\right)t\,dt$$

$$+ \frac{1}{\pi}\int_0^\tau \frac{1}{t}\cdot\sin\left(n+\tfrac{1}{2}\right)t\,dt.$$

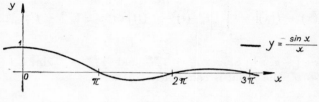

Fig. 7.3

The function $\tfrac{1}{2}\sin^{-1}\tfrac{1}{2}t - 1/t$ is clearly bounded; thus, the first integral is bounded independently of n and τ. Introducing the substitution $(n+\tfrac{1}{2})t = s$ into the second integral, we obtain the integral

$$\frac{1}{\pi}\int_0^{(n+1/2)\tau} \frac{\sin s}{s}\,ds\,;$$

this integral is a bounded function of variables n and τ, because

$$\left|\int_0^x \frac{\sin t}{t}\,dt\right| \leq \int_0^\pi \frac{\sin t}{t}\,dt.$$

(see Fig. 7.3). Hence, the theorem is proved.

Theorem 7.11. *If f is a function absolutely continuous on the interval $[-\pi, \pi]$ and periodic with period 2π, then*

$$|s_n(0) - f(0)| \leq M\|f'\|_{L_1(-\pi,\pi)}. \tag{7.18}$$

CONVERGENCE OF FOURIER SERIES

Proof: In view of (7.8),

$$s_n(0) - f(0) = \int_{-\pi}^{\pi} [f(t) - f(0)] D_n(t)\, dt$$

$$= \int_{-\pi}^{\pi} \left(\int_0^t f'(\tau)\, d\tau \right) D_n(t)\, dt$$

$$= \int_0^{\pi} \left(\int_0^t f'(\tau)\, d\tau \right) D_n(t)\, dt$$

$$+ \int_{-\pi}^0 \left(\int_0^t f'(\tau)\, d\tau \right) D_n(t)\, dt.$$

Applying the Fubini theorem 1.11 to the last two integrals, we obtain

$$|s_n(0) - f(0)| = \left| \int_0^{\pi} \left(f'(\tau) \int_{\tau}^{\pi} D_n(t)\, dt \right) d\tau \right.$$

$$\left. - \int_{-\pi}^0 \left(f'(\tau) \int_{-\pi}^{\tau} D_n(t)\, dt \right) d\tau \right|;$$

thus, by Theorem 7.10,

$$|s_n(0) - f(0)| \leq M \|f'\|_{L_1(-\pi,\pi)}.$$

Remark 7.7. A linear substitution in (7.18) yields immediately the formula

$$|s_n(x) - f(x)| \leq M \|f'\|_{L_1(-\pi,\pi)}. \qquad (7.19)$$

Proof of Theorem 7.9: Let T_n be the trigonometric polynomial defined in Theorem 7.8. Then the function T_n is a partial sum S_k of its Fourier series for any $k > k_0\ (=n)$. Applying the estimate (7.19) to the function $f - T_n$, we obtain for $k > k_0$,

$$|s_k(x) - f(x)| = |s_k(x) - S_k(x) - (f(x) - T_n(x))|$$

$$\leq M \|T_n' - f'\|_{L_1(-\pi,\pi)} < \varepsilon;$$

this finishes the proof.

The following proposition is a local analogue of Theorem 7.9.

Theorem 7.12. *Let f be a function periodic with period 2π and let $f \in L_1(-\pi, \pi)$. Furthermore, let f be absolutely continuous on the interval (a, b), where $[a, b] \subset (-\pi, \pi)$. Then the Fourier series of the function f converges to f uniformly on any interval (a_1, b_1) with $a < a_1 < b_1 < b$.*

The proof, which is analogous to that of Theorem 7.7, is left to the reader.

7.6 FUNCTIONS OF BOUNDED VARIATION

Hitherto we have considered only continuous periodic functions. In the preceding chapters however, we have seen that the expansion of a non-periodic function leads us always to a function which has points of discontinuity. The so-called *functions of bounded variation* constitute a rather general class of such functions.

A function f is said to have a *bounded variation* on the interval $[a, b]$, if two non-decreasing functions f_1 and f_2 defined on $[a, b]$ exist such that $f = f_1 - f_2$, i.e. f can be represented as a difference of two non-decreasing functions.

Recalling Problem 7.9 we see that we can always assume that the functions f_1 and f_2 are non-negative and bounded.

A quite frequent type of function of bounded variation is furnished by any function satisfying the assumption of Theorem 2.10 (see Problem 7.7). For such functions we have shown in Theorem 2.10 that their Fourier coefficients satisfy the inequality $|a_n| + |b_n| \leq K/n$. The same estimate is true for functions of bounded variation, i.e. we have:

Theorem 7.13. *Let f be a non-negative non-decreasing function bounded by a constant M on the interval $[-\pi, \pi]$. Then the*

Fourier coefficients a_n and b_n of f satisfy the inequality

$$|a_n| + |b_n| \leq \frac{4M}{n}. \tag{7.20}$$

Proof: The substitution $nx = y$ yields

$$\pi b_n = \frac{1}{n}\int_{-n\pi}^{n\pi} f\left(\frac{y}{n}\right)\sin y \, dy$$

$$= \frac{1}{n}\sum_{k=-n}^{n-1}(-1)^k \int_{k\pi}^{(k+1)\pi} f\left(\frac{y}{n}\right)|\sin y|\, dy.$$

Introducing the substitution $t = y - k\pi$ into the respective integrals we obtain

$$\pi b_n = \frac{1}{n}\int_0^\pi \sum_{k=-n}^{n-1}(-1)^k f\left(\frac{t}{n} + \frac{k\pi}{n}\right)\sin t \, dt.$$

However, because the function f is non-decreasing and $t/n + k\pi/n < t/n + (k+1)\pi/n$ for $k = -n, -(n-1), \ldots, -1, 0, 1, 2, \ldots$ $\ldots, n-1$, it follows that

$$\left|\sum_{k=-n}^{n-1}(-1)^k f\left(\frac{t}{n} + \frac{k\pi}{n}\right)\right| \leq 2M$$

and consequently, $|\pi b_n| \leq (1/n) 2M\pi$. The coefficients a_n can be estimated in a similar way. Hence, we obtain (7.20) which concludes the proof.

From Theorem 7.13 and from the definition of a function of bounded variation we obtain immediately the following proposition.

Theorem 7.14. *If f is a function of bounded variation, then the Fourier coefficients a_n and b_n of f satisfy the inequality*

$$|a_n| + |b_n| \leq \frac{C}{n}.$$

In what follows the following generalization of Theorem 7.13 will be useful.

Theorem 7.15. *Let f be a function which is non-negative, non-decreasing and bounded by a constant M on an interval $[a, b] \subset [-\pi, \pi]$. Then a constant C exists such that*

$$\left| \int_a^b f(t) \sin nt \, dt \right| \leq \frac{CM}{n}$$

for every integer $n > 0$.

Proof: Define the function g by

$$g(t) = f(t) \quad \text{for} \quad t \in [a, b] \quad \text{and} \quad g(t) = 0 \quad \text{elsewhere}.$$

The function g has a bounded variation, because it can be written as $g = g_1 - g_2$, where the functions g_1 and g_2 are non-decreasing and defined as follows:

$$g_1(t) = \begin{cases} 0 & \text{for } t < a, \\ f(t) & \text{for } t \in [a, b], \\ M & \text{for } t > b, \end{cases} \quad g_2(t) = \begin{cases} 0 & \text{for } t \leq b, \\ M & \text{for } t > b. \end{cases}$$

Now, Theorem 7.15 follows from Theorem 7.14, because we in fact estimate the Fourier coefficient b_n of the function g having a bounded variation on $[-\pi, \pi]$.

Remark 7.8. Theorems 7.13 and 7.15 remain true if the interval $[-\pi, \pi]$ is replaced by the interval $[-2\pi, 2\pi]$, or, which is the same, if the variable x is replaced by the variable $t = x/2$.

Theorem 7.16. *Let $f(x)$ be a non-negative, non-decreasing and bounded function on $(0, \pi]$ such that $\lim_{x \to 0+} f(x) = 0$. Then*

$$\lim_{n \to \infty} \int_0^\pi f(t) D_n(t) \, dt = 0.$$

CONVERGENCE OF FOURIER SERIES

Proof: Choose $\varepsilon > 0$. Then for $0 < \delta < \pi$ we have

$$\int_0^\pi f(t)\, D_n(t)\, dt = \int_0^\delta f(t)\, D_n(t)\, dt + \int_\delta^\pi f(t)\, D_n(t)\, dt.$$

The function $f(t) \sin^{-1} \tfrac{1}{2} t$ is bounded on the interval $(\delta, \pi]$; consequently, by (7.11), the second integral on the right hand side is less than $\varepsilon/2$ for $n > n(\delta)$.

Let $p = n + \tfrac{1}{2} > 2\pi/\delta$. The function $(1/2\pi)\, t \sin^{-1} \tfrac{1}{2} t$ is increasing and bounded on the interval $(0, \pi)$; thus, the function

$$g(t) = \frac{1}{2\pi} f(t)\, t \sin^{-1} \tfrac{1}{2} t$$

is non-decreasing and bounded by the constant $\tfrac{1}{2} f(\delta)$ on the interval $(0, \delta)$. Furthermore, we have

$$\int_0^\delta f(t)\, D_n(t)\, dt = \int_0^{2\pi/p} g(t) \frac{\sin pt}{t}\, dt$$

$$+ \int_{2\pi/p}^\delta g(t) \frac{p}{2\pi} \sin pt\, dt$$

$$+ \int_{2\pi/p}^\delta g(t) \left(\frac{1}{t} - \frac{p}{2\pi}\right) \sin pt\, dt.$$

The functions $(p/2\pi)\, g(t)$ and $(p/2\pi - 1/t)\, g(t)$ are non-negative, non-decreasing and bounded by the constant $(p/4\pi) f(\delta)$ on the interval $[2\pi/p, \delta]$; thus, in view of Remark 7.8, the last two integrals are bounded by the constant $(C/4\pi) f(\delta)$.

Moreover

$$\left| \int_0^{2\pi/p} g(t) \frac{\sin pt}{t}\, dt \right| \leq f(\delta)\, \pi,$$

because $|g(t)| \leq \tfrac{1}{2} f(\delta)$ and $|(\sin pt)/pt| \leq 1$. Hence, we have finally,

$$\left| \int_0^\delta f(t)\, D_n(t)\, dt \right| \leq f(\delta) \left(\pi + \frac{C}{4\pi}\right).$$

Choosing δ_0 such that $f(\delta_0)(\pi + C/4\pi) < \varepsilon/2$, we obtain for $n > n(\delta_0)$,

$$\left| \int_0^\pi f(t) D_n(t) \, dt \right| \leq \varepsilon.$$

This concludes the proof.

For a function non-decreasing on an interval $[a, b]$ the limit from the right and from the left exists at any point from $[a, b]$. (At the endpoints a and b, of course, only one unilateral limit is defined.) Consequently, the same statement is true for any function of bounded variation. The relation between these limits and the Fourier series is given by the following, important proposition.

Theorem 7.17. *If a function f is periodic with period 2π and has a bounded variation on the interval $[-\pi, \pi]$, then the partial sums $s_n(x)$ of the Fourier series of f converge at every point x and we have*

$$\lim_{n \to \infty} s_n(x) = \tfrac{1}{2}[f(x + 0) + f(x - 0)]. \tag{7.21}$$

Proof: Since the function f has a bounded variation, we can write $f = f_1 - f_2$, where f_1 and f_2 are non-decreasing on $(-\pi, \pi)$. Thus, without loss of generality we can assume that the function f itself is non-decreasing. From formula (7.10) it follows that

$$s_n(x) - \tfrac{1}{2}[f(x + 0) + f(x - 0)]$$
$$= \int_0^\pi [f(x + t) + f(x - t) - f(x + 0) - f(x - 0)] D_n(t) \, dt.$$

The assertion of the theorem follows now by applying Theorem 7.16 to the functions

$$f(x + t) - f(x + 0) \quad \text{and} \quad -f(x - t) + f(x - 0).$$

Example 7.3. The function $f(x) = x^2$ extended periodically from the interval $(0, 2\pi)$ onto the entire axis has for $x = 0$ the limit from the right and from the left equal to zero and $4\pi^2$ respectively. Hence, at the point $x = 0$ and at all points $x = 2k\pi$ with k being an integer the sum of the Fourier series of this function on $(0, 2\pi)$ assumes the value $2\pi^2$. The reader is advised to realize which actual series is under consideration in this situation (see Example 4.3 for $l = \pi$).

In Chapter 5 we have defined the value of a function at a point of discontinuity of the first kind as the arithmetic mean of the limits from the right and from the left; Theorem 7.17 indicates that this definition is reasonable and appropriate.

Remark 7.9. The relation (7.21) remains true if a function $f \in L_1(-\pi, \pi)$ has a bounded variation only on an interval $[a, b] \subset (-\pi, \pi)$; its validity, however, is then confined to $x \in (a, b)$.

7.7 FOURIER COEFFICIENTS AND THE PROPERTIES OF THE SUM

Let us perform the substitution $x = t + \pi/n$ in the formulas (2.13) and (2.19) defining the Fourier coefficients a_n, b_n and c_n, respectively. Using the properties of functions $\cos nx$, $\sin nx$, e^{inx} and the relation (1.17), we obtain another expression for the Fourier coefficient a_n of a periodic function f, i.e.

$$a_n = -\frac{1}{\pi} \int_{-\pi}^{\pi} f\left(t + \frac{\pi}{n}\right) \cos nt \, dt \, ; \tag{7.22}$$

analogous formulas are true for b_n and c_n. Comparing the formulas (7.22) and (2.13) we see that

$$a_n = \frac{1}{2\pi} \int_{-\pi}^{\pi} \left[f(t) - f\left(t + \frac{\pi}{n}\right)\right] \cos nt \, dt \, . \tag{7.23}$$

Estimating (7.23) by the integral of the magnitude of the integrated function, we get finally

$$|a_n| \leq \omega\left(\frac{\pi}{n}, f\right). \tag{7.24}$$

Entirely analogous estimates may be obtained for b_n and c_n. (In the estimate for c_n, of course, n is replaced by $|n|$.)

Assume that f is a periodic function having a continuous derivative of the k-th order with the modulus of continuity $\omega(\delta, f^{(k)})$; applying Theorem 2.9 and the relation (7.24) to the function $f^{(k)}$, we immediately get the following proposition.

Theorem 7.18. *Let f be a function periodic with period 2π which has a continuous derivative of the k-th order. Then,*

$$|a_n| + |b_n| \leq \frac{C}{n^k} \omega\left(\frac{\pi}{n}, f^{(k)}\right).$$

Remark 7.10. If in addition the k-th derivative of a function f is α-Hölder-continuous, then we have in view of Example 7.1,

$$|a_n| + |b_n| \leq \frac{C}{n^{k+\alpha}}.$$

Now we are going to prove a theorem which permits us to establish the differential properties of a function from the behaviour of the corresponding Fourier coefficients a_n and b_n.

Theorem 7.19. *Let $k > 0$ and let the series $\sum_{n=1}^{\infty} n^k(|a_n| + |b_n|)$ converge. Then the sum f of the corresponding Fourier series has a continuous derivative of $[k]$-th order and $f^{([k])}$ is Hölder-continuous with exponent $k - [k]$.**

---

*   The symbol $[k]$ signifies the integral part of the number k, i.e. the largest integer not exceeding k.

Proof: According to Theorems 2.9 and 1.2 we can confine ourselves to the case for which $0 < k < 1$. We have

$$\frac{|\cos nx - \cos ny|}{|x - y|^k} = n^k \frac{|\cos nx - \cos ny|}{|nx - ny|^k} \leq n^k C,$$

because

$$\frac{|\cos \tilde{x} - \cos \tilde{y}|}{|\tilde{x} - \tilde{y}|^k} \leq C$$

for any $\tilde{x}$, $\tilde{y}$ and $0 < k < 1$, and a similar estimate holds for the sine, too.

From this it follows that

$$|s_m(x) - s_m(y)| \leq \sum_{n=1}^{m} (|a_n| + |b_n|) C n^k |x - y|^k$$

$$= |x - y|^k C \sum_{n=1}^{m} n^k (|a_n| + |b_n|) \leq C_1 |x - y|^k.$$

Passing to the limit as $m \to \infty$, we obtain

$$|f(x) - f(y)| \leq C_1 |x - y|^k,$$

which is what we wished to show.

The theorems proved in this section do not characterize the behaviour of a function f completely. Actually, if, for example, the assumptions of Theorem 7.19 are satisfied, then the function f has a $[k]$-th derivative which is Hölder-continuous with exponent $k - [k]$. On the other hand, for such a function we have $(|a_n| + |b_n|) n^k \leq C$ by Remark 7.10; this fact, however, does not imply anything about the convergence of the series $\sum_{n=1}^{\infty} n^k (|a_n| + |b_n|)$.

This example shows that in general it is impossible to establish a one-to-one correspondence relating the properties of a function to properties of the corresponding Fourier coefficients. In certain specific cases, however, such a correspondence exists. For example, a function is square-integrable exactly if the series $\sum_{n=1}^{\infty}(|a_n|^2 + |b_n|^2)$ converges. An analogous situation occurs for functions which are Hölder-continuous in a certain integral sense; these functions will be dealt with in the next section.

7.8 FUNCTIONS IN $W_2^{(k)}$

Consider a function f periodic with period 2π which has a derivative $f^{(k)}$ of k-th order (k being a positive integer) in the sense of an absolutely continuous function and assume that $f^{(k)}$ is square-integrable. Then, by Theorem 5.1, the Fourier series of the function $f^{(k)}$ is the k-times formally differentiated Fourier series of f. Thus, we have $f^{(k)} \in L_2(-\pi, \pi)$, if and only if the series

$$\sum_{n=1}^{\infty}(|a_n|^2 + |b_n|^2) n^{2k} \tag{7.25}$$

converges.

Our next objective will be the study of those properties of the function f which are equivalent to the convergence of the series (7.25) for a non-integral k in general. First, let us present two remarks.

Remark 7.11. Let k be an arbitrary non-negative number; the space of functions such that the series (7.25) converges, will be denoted by $W_2^{(k)}$ and called *Sobolev space*. Thus, for k integral, $W_2^{(k)}$ is the space of all functions whose derivatives of k-th order are square-integrable. This characterization is also possible for non-integral k; to do it, however, would require defining the concept of a derivative of non-integral order and thus lead us far beyond the scope of this book. (The interested reader is referred to [3].)

Remark 7.12. It can be shown that the space $W_2^{(k)}$ with an appropriately defined scalar product becomes a complete Hilbert space. The reader is recommended to try to define a scalar product

CONVERGENCE OF FOURIER SERIES

via a norm defined as the square-root of the series (7.25) (see (3.9)) and prove the completeness. For $k = 0$ the space $W_2^{(k)}$ is clearly identical with the Hilbert space $L_2(-\pi, \pi)$.

Theorem 7.20. *Let k be a non-integer. Then the series (7.25) converges exactly if the function f has an integrable derivative of $[k]$-th order and the double integral*

$$I(f) = \int_{-\pi}^{\pi} \int_{-\pi}^{\pi} \frac{|f^{([k])}(x) - f^{([k])}(y)|^2}{\left|\sin \dfrac{x-y}{2}\right|^{1+2(k-[k])}} \, dx \, dy \qquad (7.26)$$

converges.

Proof: Here again it suffices to consider the case for which $0 < k < 1$. Putting $y = x + t$ in (7.26) and using the periodicity of the integrated function and Fubini's Theorem 1.11, we obtain

$$I(f) = \int_{-\pi}^{\pi} \int_{-\pi}^{\pi} \frac{|f(x+t) - f(x)|^2}{|\sin \tfrac{1}{2}t|^{1+2k}} \, dx \, dt$$

$$= \int_{-\pi}^{\pi} |g(t)|^2 |\sin \tfrac{1}{2}t|^{-1-2k} \, dt ,$$

where we have set $g(t) = \|f(x+t) - f(x)\|_{L_2(-\pi,\pi)}$ for a fixed t. If c_n signifies the Fourier coefficient in the expansion of the function f in the system $\{e^{inx}\}_{-\infty}^{\infty}$, then $c_n(1 - e^{int})$ is the corresponding Fourier coefficient for the function $f(x) - f(x + t)$. (Prove it!) By Parseval's equality,

$$\|f(x) - f(x+t)\|_{L_2(-\pi,\pi)}^2 = 8\pi \sum_{n=-\infty}^{\infty} |c_n|^2 \left(\sin \frac{nt}{2}\right)^2 .$$

Hence,

$$I(f) = C \sum_{n=-\infty}^{\infty} |c_n|^2 \int_{-\pi}^{\pi} \frac{\left(\sin \dfrac{nt}{2}\right)^2}{|\sin \tfrac{1}{2}t|^{1+2k}} \, dt$$

$$= 2C \sum_{n=-\infty}^{\infty} |c_n|^2 \int_{0}^{\pi} \frac{\left(\sin \dfrac{nt}{2}\right)^2}{(\sin \tfrac{1}{2}t)^{1+2k}} \, dt , \qquad (7.27)$$

where C is a certain constant. Since

$$\frac{t}{\pi} \leq \sin \tfrac{1}{2}t \leq \tfrac{1}{2}t \qquad (*)$$

for $t \in (0, \pi)$, we have

$$2^{1+2k} \int_0^\pi \frac{\left(\sin \dfrac{nt}{2}\right)^2}{t^{1+2k}}\, dt \leq \int_0^\pi \frac{\left(\sin \dfrac{nt}{2}\right)^2}{(\sin \tfrac{1}{2}t)^{1+2k}}\, dt$$

$$\leq \pi^{1+2k} \int_0^\pi \frac{\left(\sin \dfrac{nt}{2}\right)^2}{t^{1+2k}}\, dt. \qquad (7.28)$$

The substitution $nt/2 = s$ yields

$$\int_0^\pi \frac{\left(\sin \dfrac{nt}{2}\right)^2}{t^{1+2k}}\, dt = \left(\frac{n}{2}\right)^{2k} \int_0^{n\pi/2} \frac{\sin^2 s}{s^{1+2k}}\, ds. \qquad (7.29)$$

Since the last integral has a non-zero limit as $n \to \infty$, the convergence of the integral (7.26) is equivalent to the convergence of the series (7.25) in view of (7.28), (7.29) and (7.27). This finishes the proof.*

Remark 7.13. If a function f is α-Hölder-continuous, then, by definition, the ratio

$$\frac{|f(x) - f(y)|}{|x - y|^\alpha}$$

is bounded by a constant. Since $|\sin \tfrac{1}{2}(x - y)|$ is equivalent to $|x - y|$ on $(-\pi, \pi)$, (see (*)), Theorem 7.20 in fact requires such

---

* Because the complex form of the Fourier series was used in the proof, the series $\sum_{n=-\infty}^{\infty} |c_n|^2 |n|^{2k}$ instead of (7.25) was under consideration. However, the difference between these two series is merely formal.

a ratio to be square-integrable with respect to x and y. This, however, is the α-Hölder-continuity in an "integral sense" mentioned above.

Sobolev spaces $W_2^{(k)}$ play a significant role in the theory of partial differential equations, for example, and are thoroughly studied. The following proposition is a particular example of Sobolev's *embedding theorems* (see [11]).

Theorem 7.21. *Let $f \in W_2^{(k)}$ with $\frac{1}{2} < k < \frac{3}{2}$. Then the function f is α-Hölder-continuous with $0 \leq \alpha < k - \frac{1}{2}$.*

Proof: Since $|ab| \leq \frac{1}{2}(|a|^2 + |b|^2)$, (cf. Problem 1.11), we have

$$|c_n| n^\alpha = |c_n| n^{\alpha + (1+\varepsilon)/2} n^{-(1+\varepsilon)/2} \leq \tfrac{1}{2}(|c_n|^2 n^{2\alpha + 1 + \varepsilon} + n^{-1-\varepsilon}).$$

Choosing ε so that $2\alpha + 1 + \varepsilon = 2k$, the number ε will be positive due to conditions imposed on α; then the series $\sum_{n=1}^{\infty} |c_n|^2 n^{2\alpha + 1 + \varepsilon}$ and $\sum_{n=1}^{\infty} n^{-1-\varepsilon}$ converge, and consequently, $\sum_{n=1}^{\infty} |c_n| n^\alpha$ converges too. For concluding the proof it suffices to use Theorem 7.19.

7.9 CONVERGENCE OF ARITHMETIC MEANS

As shown earlier, the partial sums s_n of the Fourier series of a function $f \in L_2(-\pi, \pi)$ converge to f in the norm of the space $L_2(-\pi, \pi)$. However, an entirely different situation occurs in the space $L_1(-\pi, \pi)$. Namely, a function $f \in L_1(-\pi, \pi)$ exists such that the corresponding partial sums s_n do not converge to f in the norm of the space $L_1(-\pi, \pi)$. Thus, the important question arises of how to recover the function f from the sequence s_n, i.e. how to "sum" the Fourier series of a function f from $L_1(-\pi, \pi)$. This question is answered by Theorem 7.22 below.

First, let us derive several auxiliary formulas. Denote

$$S_n(x) = \frac{1}{n+1} [s_0(x) + s_1(x) + \ldots + s_n(x)].$$

Starting from formula (7.5) and Problem 1.10, and using a similar procedure to that which led us to the concept of Dirichlet's kernel in Section 7.2, we obtain the following expression for S_n,

$$S_n(x) = \int_{-\pi}^{\pi} f(x+t) F_n(t) \, dt, \qquad (7.30)$$

where

$$F_n(t) = \frac{1}{2\pi(n+1)} \frac{\sin^2 \tfrac{1}{2}(n+1)t}{\sin^2 \tfrac{1}{2} t} \qquad (7.31)$$

is the so-called *Fejér's kernel*. Observe that $F_n(t) \geqq 0$ for all t; this fact appears to be the most powerful property of the kernel $F_n(t)$.

We have again

$$\int_{-\pi}^{\pi} F_n(t) \, dt = 1, \qquad (7.32)$$

and for a trigonometric polynomial T_n of n-th degree,

$$T_n(x) = \int_{-\pi}^{\pi} T_n(x+t) F_m(t) \, dt \quad \text{for} \quad m \geqq n. \qquad (7.33)$$

Furthermore,

$$|S_n(x)| \leqq \int_{-\pi}^{\pi} |f(x+t)| F_n(t) \, dt \qquad (7.34)$$

and consequently,

$$\int_{-\pi}^{\pi} |S_n(x)| \, dx \leqq \int_{-\pi}^{\pi} F_n(t) \left(\int_{-\pi}^{\pi} |f(x+t)| \, dx \right) dt = \int_{-\pi}^{\pi} |f(x)| \, dx. \qquad (7.35)$$

(Since the function f is periodic, we have taken advantage of formula (1.17) and (7.32) here.)

Theorem 7.22. *Let $f \in L_1(-\pi, \pi)$. Then the functions S_n converge to the function f in the space $L_1(-\pi, \pi)$ as $n \to \infty$.*

Proof: Using the considerations carried out in the Appendix to Chapter 3 we can show in a similar manner to the proof of Theorem 7.8 that for any $\varepsilon > 0$ a trigonometric polynomial T_n exists such that $\|f - T_n\|_{L_1(-\pi,\pi)} < \varepsilon/2$. In view of (7.33) with $m \geq n$ we have by the inequality (7.35) applied to the function $f - T_n$,

$$\|S_m - T_n\|_{L_1(-\pi,\pi)} = \int_{-\pi}^{\pi} |S_m(x) - T_n(x)|\, dx$$

$$\leq \int_{-\pi}^{\pi} |f(x) - T_n(x)|\, dx < \frac{\varepsilon}{2},$$

and consequently,

$$\|f - S_m\|_{L_1(-\pi,\pi)} \leq \|S_m - T_n\|_{L_1(-\pi,\pi)} + \|f - T_n\|_{L_1(-\pi,\pi)} < \varepsilon$$

for $m \geq n$; this finishes the proof.

Thus, it is clear that the arithmetic means S_n approximate the function $f \in L_1(-\pi, \pi)$ better than the partial sums s_n themselves. The same is true in the case of a continuous function as the following proposition shows.

Theorem 7.23. *Let f be a function continuous and periodic with period 2π. Then the arithmetic means S_n converge to f uniformly on the interval $[-\pi, \pi]$.*

Proof: By Problem 3.19, for any $\varepsilon > 0$ a trigonometric polynomial T_n exists such that $|f(x) - T_n(x)| < \varepsilon/2$ for every $x \in [-\pi, \pi]$. Due to (7.33), (7.30) and (7.32) we have for $m \geq n$,

$$|S_m(x) - T_n(x)| \leq \int_{-\pi}^{\pi} |f(x+t) - T_n(x+t)| F_m(t)\, dt < \frac{\varepsilon}{2}.$$

Hence
$$|S_m(x) - f(x)| \leq |S_m(x) - T_n(x)| + |f(x) - T_n(x)| < \varepsilon$$
for $m \geq n$ and $x \in [-\pi, \pi]$; the theorem is proved.

Remark 7.14. (Principle of localization.) *If the assumptions of Theorem 7.6 are satisfied, then the arithmetic means S_n converge to zero uniformly on the interval (a_1, b_1).* This proposition follows immediately from Theorem 7.6 and from the equality

$$S_n(x) = \frac{s_0(x) + \ldots + s_m(x)}{n + 1} + \frac{n - m}{n + 1} \frac{s_{m+1}(x) + \ldots + s_n(x)}{n - m}.$$

Actually, if $|s_k(x)| < \varepsilon/2$ for $k > m$, then the second term is less than $\varepsilon/2$. On the other hand, for a fixed m, the first term is also less than $\varepsilon/2$ in magnitude if n is sufficiently large; hence, $|S_n(x)| < \varepsilon$ for large n.

From the principle of localization we obtain immediately the following statement.

Theorem 7.24. *If a function $f \in L_1(-\pi, \pi)$ is periodic with period 2π and continuous on an interval $(a, b) \subset (-\pi, \pi)$, then the arithmetic means S_n converge to f uniformly on every interval (a_1, b_1) with $a < a_1 < b_1 < b$.*

Even a proposition analogous to Theorem 7.17 is true.

Theorem 7.25. *Let f be a function periodic with period 2π and let $f \in L_1(-\pi, \pi)$. Furthermore, at a point x_0, let the limits $f(x_0 + 0)$ and $f(x_0 - 0)$ exist. Then*

$$\lim_{n \to \infty} S_n(x_0) = \tfrac{1}{2}[f(x_0 + 0) + f(x_0 - 0)].$$

Proof: Let $g(t) = \tfrac{1}{2}[f(x_0 + t) + f(x_0 - t)]$ for $t \neq 0$ and $g(0) = \tfrac{1}{2}[f(x_0 + 0) + f(x_0 - 0)]$. The function $g(t)$ is continuous at the point $t = 0$, and consequently, a $\delta > 0$ exists such that $|g(t) - g(0)| < \varepsilon/2$ for $|t| < \delta$. Put

$$g(t) = g_1(t) + g_2(t) + g(0),$$

where $g_1(t) = 0$ for $|t| \geq \delta$ and $g_2(t) = 0$ for $|t| < \delta$. Next, we can easily verify that

$$S_n(x_0) - g(0) = \int_{-\pi}^{\pi} [g(t) - g(0)] F_n(t) \, dt$$

$$= \int_{-\pi}^{\pi} g_1(t) F_n(t) \, dt + \int_{-\pi}^{\pi} g_2(t) F_n(t) \, dt.$$

Considering the last two integrals, the first one is less than $\varepsilon/2$ in magnitude, because $|g_1(t)| < \varepsilon/2$; by (7.30), the second one is the arithmetic mean of the partial sums of the Fourier series for the function $g_2(t)$ at the point zero, so that it converges to zero as $n \to \infty$ in view of Remark 7.14. Thus, we have $|S_n(x_0) - g(0)| < \varepsilon$ for $n > n_0$ and the theorem is proved.

Often it is useful to be familiar with the fact that if all the Fourier coefficients of a function $f \in L_1(-\pi, \pi)$ vanish, then f vanishes, too. This proposition may be used, for example, for identifying two functions whose Fourier coefficients coincide; here we in fact use the following statement.

Theorem 7.26. *Let $f \in L_1(-\pi, \pi)$; if all Fourier coefficients of f vanish, then $f(x) = 0$ for almost all x.*

Proof: Clearly, $S_n(x) = 0$ for every x and n. By Theorem 7.22, however, the arithmetic means $\overline{S}_n$ converge to the function f in the space $L_1(-\pi, \pi)$; hence, $f(x) = 0$ almost everywhere.

Remark 7.15. In all considerations prior to those in this section, by the sum of the series $\sum_{k=1}^{\infty} f_k(x)$ we understood the limit of the partial sums $s_n(x) = \sum_{k=1}^{n} f_k(x)$ as $n \to \infty$; the limit, of course, has been interpreted in various ways, i.e., as a uniform limit, or as a limit in the sense of convergence in $L_1(-\pi, \pi)$, etc. In this section, however, by the "sum" of a series we understood the limit of functions $S_n(x)$ defined as the arithmetic means of the partial sums $s_n(x)$. This method of summation was introduced for the first time by E. CESÀRO, and is, therefore, called *Cesàro's summation*.

The theory of Fourier series is a rather old mathematical discipline which is rich in magnificent and deep results. From the entire abundance of results only a small fraction was presented in this chapter; we selected those propositions which seemed to us to be most important. The reader who has become attracted more deeply by the theory of Fourier series after having studied this book, may widen and complete his knowledge by studying further literature which is very extensive in this field. At the end of this book a brief list of recommended supplementary reading may be found.

Problems

7.1. Prove that the number $I(\lambda; f; a, b)$ defined by (7.11) is a continuous function of the variable λ.

7.2. Show that the inequality (7.13) is true for the function $f(x) = |x|^\varepsilon$ for $x \in (-\pi, \pi)$, $\varepsilon > 0$, which is extended periodically onto $(-\infty, \infty)$.

7.3. Show that the relation (7.14) fails to hold for the function $f(x) = x$ on $(-\pi, \pi)$.

7.4. Show that the Fourier series of the function $f(x) = \log^{-1}|x|$ on $(-\pi, \pi)$ converges uniformly.

7.5. Show that the Fourier series of the function $f(x) = \sqrt{|x|^3}$ on $(-\pi, \pi)$ converges absolutely and uniformly.

CONVERGENCE OF FOURIER SERIES

7.6. Prove that any function f satisfying the assumptions of Theorem 2.11 is absolutely continuous.

7.7. Prove that a function f satisfying the assumptions of Theorem 2.10 has a bounded variation.

(*Hint*: Put $f^+ = \max(f', 0)$ and $f^- = \min(f', 0)$. Then $f(x) = g(x) + \int_{-\pi}^{x} f^+(t)\, dt + \int_{-\pi}^{x} f^-(t)\, dt$, where $g(x)$ is a piecewise constant function.)

7.8. Show that the function $f(x) = 0$ for $x \neq 0$ and $f(0) = 1$ has a bounded variation.

7.9. Prove that a function of bounded variation on an interval $[a, b]$ is the difference of two non-negative, non-decreasing and bounded functions.

7.10. Using Problem 7.9 prove that the sum and the product of functions of bounded variation are again functions of bounded variation.

7.11. Prove the proposition given in Remark 7.8.

7.12. Prove Remark 7.14 in detail.

7.13. Find the sum of each series given in Chapter 5 at a point of discontinuity of the function whose Fourier series is the considered series. Which series of numbers may be summed in this way?

CHAPTER 8 FOURIER TRANSFORMS

8.1 FOURIER TRANSFORMS AS A LIMIT CASE OF FOURIER SERIES

In Chapter 4, Section 4.3 and in Chapter 5, Section 5.3 we have discussed the expansion of a *non-periodic* function $f(x)$ in $(-\infty, \infty)$ in a Fourier series on the interval $(-l, l)$. These results will now be used with a slight formal change, i.e. we are going to write πl instead of l. Under certain assumptions on f, we have for $x \in (-\pi l, \pi l)$,

$$f_l(x) = \sum_{n=-\infty}^{\infty} c_n^{(l)} e^{in(x/l)}, \qquad (8.1)$$

where

$$c_n^{(l)} = \frac{1}{2\pi l} \int_{-\pi l}^{\pi l} f(\xi) e^{-in(\xi/l)} d\xi, \qquad (8.2)$$

and

$$f_l(x) = f(x) \quad \text{for} \quad x \in (-\pi l, \pi l).$$

However, the sum of the series (8.1) does not coincide with f outside the interval $(-\pi l, \pi l)$ unless f is periodic with period $2\pi l$. The larger the number l is, the larger the interval is, where the series (8.1) describes the function f.

Consider now an infinitely differentiable function $f(x)$ defined in $(-\infty, \infty)$, which vanishes for $|x|$ sufficiently large; i.e., there exists a positive number K such that $f(x) = 0$, whenever $|x| > K$. Such a function is called a *smooth finitary function*

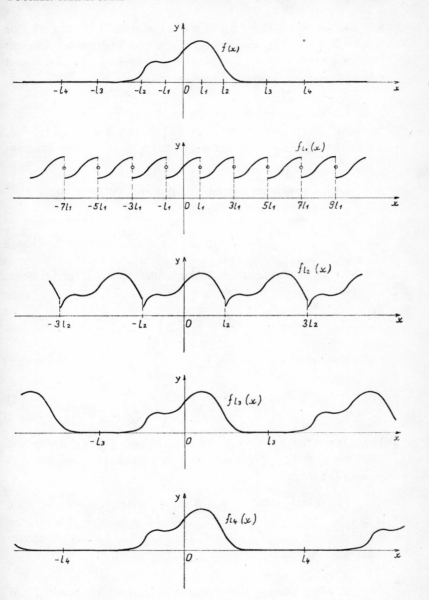

Fig. 8.1

and the set of all these functions is denoted by $\mathscr{D}(-\infty, \infty)$ or simply $\mathscr{D}$. (Note that the number K may be different for different functions in $\mathscr{D}$.) In Fig. 8.1, a smooth finitary function f and related functions f_l for various l are plotted.

Let x be a fixed real number and let f be a smooth finitary function. Then we have $f_l(x) = f(x)$ for $l > |x|/\pi$. Hence, $\lim_{l \to \infty} f_l(x) = f(x)$ for any fixed $x \in (-\infty, \infty)$ and for any function $f \in \mathscr{D}$.

Let us now write formula (8.1) in another form. Denote

$$F(\xi) = \frac{1}{2\pi} \int_{-\infty}^{\infty} f(x) \, e^{-i\xi x} \, dx \, . \tag{8.3}$$

If f belongs to $\mathscr{D}$ and if l is so large that $f(x) = 0$ for $|x| > \pi l$, we have

$$c_n^{(l)} = \frac{1}{l} F\left(\frac{n}{l}\right)$$

and consequently,

$$f_l(x) = \sum_{n=-\infty}^{\infty} \frac{1}{l} e^{i(n/l)x} F\left(\frac{n}{l}\right). \tag{8.4}$$

Thus, from the function F we can, by using (8.4), derive the functions f_l for large ls and thus reconstruct the function $f(x) = \lim_{l \to \infty} f_l(x)$.

The function $F(\xi)$ is called the *Fourier transform* of $f(x)$. It can be easily seen from (8.3) that the definition of a Fourier transform is meaningful not only for $f \in \mathscr{D}$, but for any absolutely (Lebesgue) integrable function $f \in L_1(-\infty, \infty)$.

In what follows, these properties of $F(\xi)$ are used: if f is smooth and finitary, then both $F(\xi)$ and its derivative $F'(\xi)$ are continuous and absolutely integrable. As for continuity, cf.

Problem 7.1; as for integrability, we can prove it as follows. Integrating twice by parts, we get

$$\left|(1 + \xi^2) F(\xi)\right| = \frac{1}{2\pi} \left|\int_{-\infty}^{\infty} [f(x) + f''(x)] e^{-i\xi x} \, dx\right|$$

$$\leq \frac{1}{2\pi} \left(\int_{-\infty}^{\infty} |f(x)| \, dx + \int_{-\infty}^{\infty} |f''(x)| \, dx\right) = C.$$

Consequently, $|F(\xi)| \leq C(1 + \xi^2)^{-1}$, i.e. $F \in L_1(-\infty, \infty)$. Similarly for $F'(\xi)$. Moreover, as it will be shown later, Fourier transforms of functions more general than those from $\mathscr{D}$ also have these properties.

Since the functions $F(\xi)$ and $F'(\xi)$ are continuous and absolutely integrable, the same is true for the function $F^x(\xi) = e^{i\xi x} F(\xi)$ and its derivative $dF^x(\xi)/d\xi$ (x fixed). The formula (8.4) can be rewritten as follows

$$f_l(x) = \sum_{n=-\infty}^{\infty} e^{i(n/l)x} F\left(\frac{n}{l}\right)\left(\frac{n}{l} - \frac{n-1}{l}\right)$$

$$= \sum_{n=-\infty}^{\infty} \int_{(n-1)/l}^{n/l} F^x\left(\frac{n}{l}\right) d\tau.$$

Then

$$\left|f_l(x) - \int_{-\infty}^{\infty} F^x(\tau) \, d\tau\right| = \left|\sum_{n=-\infty}^{\infty} \int_{(n-1)/l}^{n/l} \left[F^x\left(\frac{n}{l}\right) - F^x(\tau)\right] d\tau\right|$$

$$= \left|\sum_{n=-\infty}^{\infty} \int_{(n-1)/l}^{n/l} \left(\int_{\tau}^{n/l} \frac{d}{d\sigma} F^x(\sigma) \, d\sigma\right) d\tau\right|$$

$$\leq \sum_{n=-\infty}^{\infty} \int_{(n-1)/l}^{n/l} \left(\int_{\tau}^{n/l} \left|\frac{d}{d\sigma} F^x(\sigma)\right| d\sigma\right) d\tau$$

$$\leq \sum_{n=-\infty}^{\infty} \frac{1}{l} \int_{(n-1)/l}^{n/l} \left|\frac{d}{d\sigma} F^x(\sigma)\right| d\sigma$$

$$= \frac{1}{l} \int_{-\infty}^{\infty} \left|\frac{d}{d\sigma} F^x(\sigma)\right| d\sigma.$$

Thus, we have

$$\lim_{l \to \infty} \left| f_l(x) - \int_{-\infty}^{\infty} e^{i\tau x} F(\tau) \, d\tau \right| = \lim_{l \to \infty} \frac{1}{l} \int_{-\infty}^{\infty} \left| \frac{d}{d\sigma} F^x(\sigma) \right| d\sigma = 0 \tag{8.5}$$

and the convergence is then uniform in every finite interval (cf. Problem 8.12).

Since $\lim_{l \to \infty} f_l(x) = f(x)$, for $f \in \mathcal{D}$ we get immediately from (8.5),

$$f(x) = \int_{-\infty}^{\infty} e^{i\xi x} F(\xi) \, d\xi \, . \tag{8.6}$$

Remark 8.1. This formula, up to the factor $1/2\pi$ and the sign in the exponential function, is identical with the formula (8.3). Thus, all theorems on Fourier transforms remain true if in (8.3) x is replaced by $-x$ in the argument of f and in the exponential function, and if $f/2\pi$ is replaced by F and F by f (cf. also Chapter 10, Part III, formula 1).

The formula (8.6) permits us to express a non-periodic function f as a superposition of simple periodic functions $e^{i\xi x} F(\xi)$ with magnitudes $|F(\xi)|$. We shall discuss some applications of this decomposition in Chapter 9.

So far, only functions in $\mathcal{D}$ have been considered. However, $F(\xi)$ is meaningful for any function integrable in $(-\infty, \infty)^*$. We then have the proposition

Theorem 8.1. *If f is integrable in $(-\infty, \infty)$ then the function $F(\xi)$ defined by (8.3) is continuous and*

$$\lim_{|\xi| \to \infty} F(\xi) = 0 \, .$$

---

* In the present chapter, by integrable functions we understand absolutely Lebesgue integrable functions $f \in L_1(-\infty, \infty)$.

Furthermore, we have the estimate,

$$|F(\xi)| \leq \frac{1}{2\pi} \int_{-\infty}^{\infty} |f(x)| \, dx . \tag{8.7}$$

Proof: The assertion follows from Problem 7.1 and Theorem 3.18. The estimate (8.7) is obtained from the inequality

$$\left| \frac{1}{2\pi} \int_{-\infty}^{\infty} f(x) e^{-i\xi x} \, dx \right| \leq \frac{1}{2\pi} \int_{-\infty}^{\infty} |f(x)| \, |e^{-i\xi x}| \, dx$$

$$= \frac{1}{2\pi} \int_{-\infty}^{\infty} |f(x)| \, dx .$$

The integrability of f alone does not imply the integrability of F, as indicated by the following example.

Example 8.1. Put $f(x) = 0$ for $|x| \geq 1$ and $f(x) = 1$ for $x \in (-1, 1)$. Then

$$F(\xi) = \frac{1}{2\pi} \int_{-1}^{1} e^{-i\xi x} \, dx = \frac{1}{2\pi} \frac{e^{i\xi} - e^{-i\xi}}{i\xi} = \frac{\sin \xi}{\pi \xi}.$$

This function is not integrable in $(-\infty, \infty)$, i.e., does not belong to $L_1(-\infty, \infty)$ (cf. Chapter 1, p. 10).

Thus, the integral in (8.6) need not be meaningful for every $f \in L_1(-\infty, \infty)$ and consequently, the formula (8.6) need not hold. This situation is the same as in the case of Fourier series: the sum of a Fourier series need not exist for $f \in L_1(-\pi, \pi)$.

Therefore, it is important to show how a function f can be obtained from the Fourier transform F.

Theorem 8.1 states that the Fourier transform of a function f defined by (8.3) is meaningful for each function $f \in L_1(-\infty, \infty)$. Note that for the Fourier transform of a function f also the notations $\hat{f}$ or $\mathscr{F}f$ are sometimes used (the former one will be used in Chapter 9).

8.2 INVERSION FORMULA

The formula (8.6) enables us to express a function f in terms of its Fourier transform F; thus, (8.6) is in fact an inversion of the formula (8.3). Hence, the formula (8.6) is sometimes called the *inversion formula for the Fourier transform* and written in the form

$$f = \mathscr{F}^{-1}F$$

whenever the notation $F = \mathscr{F}f$ is used. The difference between formulas (8.3) and (8.6) is more or less formal — cf. Remark 8.1.

The formula (8.6) is meaningful for $f \in \mathscr{D}$; however, for $f \in L_1(-\infty, \infty)$ difficulties arise. Thus, it is worth knowing what sense should be given to the inversion formula (8.6) for more general functions.

The reader can easily convince himself that the replacing of the series (8.4) by the integral (8.6) is also justified if the function $F(\xi)$ has a bounded variation. However, an additional assumption is needed, viz. that the series (8.4) converges to $f(x)$ for sufficiently large l's. Then we have (8.6) again and the integral on the right hand side of (8.6) is absolutely convergent (i.e. the integrand is absolutely integrable).

For the function $F(\xi)$ from Example 8.1 the integral (8.6) is not absolutely convergent; however, we can calculate the principal value of this integral, i.e. the limit $\lim_{N \to \infty} f_N(x)$, where

$$f_N(x) = \int_{-N}^{N} e^{i\xi x} F(\xi) \, d\xi \,. \tag{8.8}$$

For the function $F(\xi) = (\sin \xi)/\pi\xi$ from Example 8.1, this limit is actually equal to f everywhere except for the points $x = \pm 1$, where the principal value of the integral (8.6) is $\frac{1}{2}$, and consequently, equal to the mean of the limits from the left and from the right (cf. also Theorem 8.3).

The functions $f_N(x)$ from (8.8) are analogous to partial sums $s_n(x)$ of a Fourier series. This becomes more apparent, if $f_N(x)$ is expressed directly by $f(x)$, i.e.,

$$f_N(x) = \frac{1}{2\pi} \int_{-N}^{N} e^{i\xi x} \left(\int_{-\infty}^{\infty} e^{-i\xi s} f(s) \, ds \right) d\xi$$

$$= \frac{1}{2\pi} \int_{-\infty}^{\infty} \left(\int_{-N}^{N} e^{i\xi(x-s)} \, d\xi \right) f(s) \, ds$$

$$= \frac{1}{\pi} \int_{-\infty}^{\infty} \frac{\sin N(x-s)}{x-s} f(s) \, ds \,;$$

hence,

$$f_N(x) = \frac{1}{\pi} \int_{-\infty}^{\infty} f(x+t) \frac{\sin Nt}{t} \, dt$$

$$= \frac{1}{\pi} \int_{0}^{\infty} [f(x+t) + f(x-t)] \frac{\sin Nt}{t} \, dt \,. \qquad (8.9)$$

So the function $f_N(x)$ is expressed by f and the kernel $(\sin Nt)/t$ in a similar manner as the partial sum $s_n(x)$ of a Fourier series is by the function f and the Dirichlet kernel $D_n(t)$ (cf. Chapter 7),

$$s_n(x) = \frac{1}{2\pi} \int_{-\pi}^{\pi} f(x+t) \frac{\sin(n+\frac{1}{2})t}{\sin \frac{1}{2}t} \, dt \,.$$

The following theorem is further evidence for the discussed analogy.

Theorem 8.2. Let $f \in L_1(-\infty, \infty)$ and x be a real number. Then

$$\lim_{N \to \infty} [f_N(x) - s_{[N]}(x)] = 0 \,.^*$$

---

*   [N] denotes the integral part of the number N, cf. the footnote on p. 248.

In other words, the integral in (8.6) is finite (e.g. in the sense of a principal value) for all points x satisfying the condition: the periodic extension $\tilde{f}$ with period 2π of the function f restricted to the interval $(x - \pi, x + \pi)$ has a convergent Fourier series. The integral in (8.6) is then equal to the sum of this series.

The proof of Theorem 8.2 is omitted and may be found in BARI, N. K.: Trigonometric series, London 1964.

In a similar way as the convergence of partial sums of the Fourier series of a function f was investigated in Chapter 7, we could build up a theory concerning the convergence of f_N to f. The difference is only that the infinite interval is to be considered and trigonometric polynomials replaced by smooth finitary functions; then (8.6) implies that $f_N \to f$ uniformly in every finite interval, since $F(\xi)$ is absolutely integrable whenever $f \in \mathscr{D}$. Possibly, we could use Theorem 8.2.

For demonstrating these facts we state the following two theorems (cf. [1]):

Theorem 8.3. *If f is absolutely integrable in $(-\infty, \infty)$ and of bounded variation in (a, b) then*

$$\lim_{N \to \infty} f_N(x) = \tfrac{1}{2}[f(x + 0) + f(x - 0)]$$

for each $x \in (a, b)$.

Theorem 8.4. *If f is absolutely integrable in $(-\infty, \infty)$ and absolutely continuous in (a, b) then*

$$\lim_{N \to \infty} f_N(x) = f(x)$$

uniformly in each interval (a_1, b_1) with $a < a_1 < b_1 < b$.

The similarity with Theorems 7.17 and 7.12 is straightforward.

In order to show the difference between the proofs for Fourier transforms and the proofs for Fourier series, we are

FOURIER TRANSFORMS

going to derive *Dini's test for convergence*. We indicate only the main ideas of the proof instead of carrying it out in detail.

Theorem 8.5. *Let* $f \in L_1(-\infty, \infty)$ *and let for* $x \in (-\infty, \infty)$ *a* $\delta > 0$ *exist such that the integral*

$$\int_{-\delta}^{\delta} \frac{|f(x+t) - f(x)|}{|t|} \, dt$$

is finite. Then $f(x) = \lim_{N \to \infty} f_N(x)$.

Proof: We make use of the fact that

$$\lim_{T \to \infty} \int_{-T}^{T} \frac{\sin Nt}{t} \, dt = \pi$$

(cf. Problem 8.13, or Chapter 1, p. 10). Multiplying this relation by $f(x)$ and using the formula (8.9), we get

$$f_N(x) - f(x) = \frac{1}{\pi} \lim_{T \to \infty} \int_{-T}^{T} [f(x+t) - f(x)] \frac{\sin Nt}{t} \, dt. \quad (8.10)$$

For $T > T_0 > 0$ we have

$$\left| \int_{T_0}^{T} f(x+t) \frac{\sin Nt}{t} \, dt \right| \leq \frac{1}{T_0} \|f\|_{L_1(-\infty, \infty)}$$

and, in addition to it, for $N > 1$,

$$\left| \int_{T_0}^{T} f(x) \frac{\sin Nt}{t} \, dt \right| = |f(x)| \left| \frac{1}{T_0} \int_{T_0}^{\xi} \sin Nt \, dt \right|$$

$$\leq |f(x)| \frac{1}{T_0} \frac{2}{N} < \frac{1}{T_0} 2|f(x)|$$

(here we have used the mean-value theorem for integrals); analogous relations hold for integrals with limits $-T$ and $-T_0$. Thus, for a given $\varepsilon > 0$ a sufficiently large $T_0 > 0$ can be found such that the limit in (8.10) is approximated independently of N by an integral with limits $-T_0$ and T_0 with an error less than ε, i.e.

$$\left| f_N(x) - f(x) - \frac{1}{\pi} \int_{-T_0}^{T_0} \frac{f(x+t) - f(x)}{t} \sin Nt \, dt \right| < \varepsilon, \quad (8.11)$$

independently of $N > 1$. By assumption, the function $t^{-1}[f(x+t) - f(x)]$ is integrable in $(-\delta, \delta)$ and thus in the entire interval $(-T_0, T_0)$. By Theorem 3.18, the integral in (8.11) tends to zero as $N \to \infty$; thus, for N sufficiently large,

$$|f_N(x) - f(x)| < 2\varepsilon ;$$

which completes the proof.

Remark 8.2. If the function f satisfies the Hölder condition at x, i.e. if $|f(x+t) - f(x)| \leq Ct^\alpha$ for some $\alpha > 0$, then it satisfies the assumption of Theorem 8.5.

Remark 8.3. The arithmetic means of partial sums of a Fourier series are counterparted by the integral means

$$S_M(x) = \frac{1}{M} \int_0^M f_N(x) \, dN .$$

Similarly as in the case of arithmetic means, the integral means approximate the function f better than the functions f_N.

8.3 FOURIER TRANSFORMS OF FUNCTIONS IN L_2

Since not every function in $L_2(-\infty, \infty)$ is absolutely integrable in $(-\infty, \infty)$ (cf. Problem 8.11), it is senseless to define Fourier transforms of functions from L_2 in the way used in Section 8.1. However, we may use the following theorem for defining Fourier transforms of functions in $L_2(-\infty, \infty)$.

Theorem 8.6. (PLANCHEREL, 1910.) *Let $f \in L_2(-\infty, \infty)$. Then, for each $N > 0$, the function of the variable ξ*

$$F_N(\xi) = \frac{1}{2\pi} \int_{-N}^{N} f(x) \, e^{-i\xi x} \, dx \qquad (8.12)$$

belongs to $L_2(-\infty, \infty)$. The functions F_N converge to a function F as $N \to \infty$ in the norm of the space $L_2(-\infty, \infty)$, and F satisfies the Parseval equality

$$2\pi \int_{-\infty}^{\infty} |F(\xi)|^2 \, d\xi = \int_{-\infty}^{\infty} |f(x)|^2 \, dx. \qquad (8.13)$$

Remark 8.4. If, in addition, f is absolutely integrable in $(-\infty, \infty)$, then the limit $\lim_{N \to \infty} F_N(\xi)$ exists at each ξ and $F(\xi)$ is the Fourier transform of f. Hence, the function $F(\xi)$ defined in Theorem 8.6 is called the *Fourier transform of $f \in L_2$* even if f is not an element of $L_1(-\infty, \infty)$.

Proof of Theorem 8.6: Let φ_1, φ_2 be two functions from $\mathscr{D}$ and let Φ_1 and Φ_2 be the corresponding Fourier transforms. By (8.6), (8.3) and Fubini's theorem we have

$$\int_{-\infty}^{\infty} \varphi_1(x) \overline{\varphi_2(x)} \, dx = \int_{-\infty}^{\infty} \left\{ \int_{-\infty}^{\infty} \Phi_1(\xi) \, e^{i\xi x} \, d\xi \right\} \overline{\varphi_2(x)} \, dx$$

$$= \int_{-\infty}^{\infty} \Phi_1(\xi) \overline{\left\{ \int_{-\infty}^{\infty} e^{-i\xi x} \varphi_2(x) \, dx \right\}} \, d\xi = 2\pi \int_{-\infty}^{\infty} \Phi_1(\xi) \overline{\Phi_2(\xi)} \, d\xi.$$
$$(8.14)$$

By Problem 3.25, any function $f \in L_2(-\infty, \infty)$ can be approximated in the norm of L_2 by functions $\psi_n \in \mathscr{D}$. Denote Ψ_n the sequence of Fourier transforms of functions ψ_n; setting $\varphi_1 = \varphi_2 = \psi_n - \psi_m$ in (8.14), we have

$$2\pi \int_{-\infty}^{\infty} |\Psi_n(\xi) - \Psi_m(\xi)|^2 \, d\xi = \int_{-\infty}^{\infty} |\psi_n(x) - \psi_m(x)|^2 \, dx. \quad (8.15)$$

Since ψ_n is a Cauchy sequence in $L_2(-\infty, \infty)$, $\psi_n \to f$, so is also Ψ_n by (8.15). Because $L_2(-\infty, \infty)$ is complete, a limit F of the sequence Ψ_n exists in $L_2(-\infty, \infty)$. It can be easily verified that the limit F is independent of the choice of the sequence ψ_n.

Suppose now that f is absolutely integrable in $(-\infty, \infty)$. By Problem 3.26, the sequence ψ_n can be chosen such that $\psi_n \to f$ both in the norm of $L_2(-\infty, \infty)$ and in the norm of $L_1(-\infty, \infty)$. By (8.7), the functions Ψ_n converge uniformly to the Fourier transform of f and consequently, F is the Fourier transform of f.

Next, setting $\varphi_1 = \varphi_2 = \psi_n$ in (8.14) and calculating the limits as $n \to \infty$, we obtain for $f \in L_2(-\infty, \infty)$,

$$2\pi \int_{-\infty}^{\infty} |F(\xi)|^2 \, d\xi = \int_{-\infty}^{\infty} |f(x)|^2 \, dx . \qquad (8.16)$$

Define the function f_N as follows.

$$f_N(x) = f(x) \text{ for } |x| \leq N, \quad f_N(x) = 0 \text{ for } |x| > N$$

(this function is both in L_2 and L_1). Then the function F_N from (8.12) is the Fourier transform of f_N and, by (8.16),

$$2\pi \int_{-\infty}^{\infty} |F(\xi) - F_N(\xi)|^2 \, d\xi = \int_{-\infty}^{\infty} |f(x) - f_N(x)|^2 \, dx$$
$$= \int_{|x|>N} |f(x)|^2 \, dx .$$

The integral on the right hand side converges to zero as $N \to \infty$ and thus $F_N \to F$ in the norm of $L_2(-\infty, \infty)$. This finishes the proof.

Remark 8.5. If f_1 and f_2 are in $L_2(-\infty, \infty)$ and F_1 and F_2 are the corresponding Fourier transforms in the sense of Theorem 8.6, then a limiting process in (8.14) yields

$$2\pi \int_{-\infty}^{\infty} F_1(\xi) \overline{F_2(\xi)} \, d\xi = \int_{-\infty}^{\infty} f_1(x) \overline{f_2(x)} \, dx . \qquad (8.17)$$

Remark 8.6. Theorem 8.6 states that to each $f \in L_2(-\infty, \infty)$ there exists the Fourier transform $F = \lim_{N \to \infty} F_N$, where F_N are defined by (8.12). Conversely, it can be shown that each function $F \in L_2(-\infty, \infty)$ is the Fourier transform of a function $f \in L_2(-\infty, \infty)$, $f = \lim_{N \to \infty} f_N$ in the norm of $L_2(-\infty, \infty)$, where f_N are defined by

$$f_N(x) = \int_{-N}^{N} e^{i\xi x} F(\xi) \, d\xi \tag{8.18}$$

and that the equality (8.13) holds (cf. also Remark 8.1).

Let us outline the proof. Denote by g the limit of the sequence f_N defined by (8.18) and let G be the Fourier transform of g. Then, using (8.13) and Remark 8.1,

$$\int_{-\infty}^{\infty} |F(\xi)|^2 \, d\xi = \int_{-\infty}^{\infty} |G(\xi)|^2 \, d\xi \, .$$

Next, let ψ be a function from $\mathscr{D}$ and ψ_N the corresponding functions given by (8.18), where F is replaced by the Fourier transform Ψ of ψ. Then, by (8.6), $\lim_{N \to \infty} \psi_N = \psi$. Hence,

$$\int_{-\infty}^{\infty} |F(\xi) - \Psi(\xi)|^2 \, d\xi = \int_{-\infty}^{\infty} |G(\xi) - \Psi(\xi)|^2 \, d\xi$$
$$= \frac{1}{2\pi} \int_{-\infty}^{\infty} |g(x) - \psi(x)|^2 \, dx \, .$$

If $\{\psi_n\}$ is a sequence of functions from $\mathscr{D}$ converging to g in the metric of $L_2(-\infty, \infty)$ then $\Psi_n \to F$ and $\Psi_n \to G$. Hence, $F = G$.

8.4 PROPERTIES OF FOURIER TRANSFORMS

In the present section some formulas are derived which are useful for the calculation of Fourier transforms. The forthcoming theorems are of fundamental importance for practical applications.

A. From the definition of Fourier transforms it follows immediately that the Fourier transform of a sum of two functions is equal to the sum of the Fourier transforms of both summands and that the Fourier transform of a multiple of a function (by a complex number) is equal to the multiple of the Fourier transform. Using the notation introduced at the end of Section 8.1, the previous statements can be written as follows

$$\widehat{f_1 + f_2} = \hat{f}_1 + \hat{f}_2 \quad \text{and} \quad \widehat{\alpha f} = \alpha \hat{f}$$

or

$$\mathscr{F}(f_1 + f_2) = \mathscr{F}f_1 + \mathscr{F}f_2 \quad \text{and} \quad \mathscr{F}(\alpha f) = \alpha \mathscr{F}f.$$

B. The following theorem yields a means for finding (under certain assumptions) the Fourier transform of a derivative of a function f provided the Fourier transform of f is known.

Theorem 8.7. *Let $f(x)$ have a derivative $f'(x)$ (in the sense of absolutely continuous function) in every finite interval and let one of the following conditions be satisfied.*

(a) $f \in L_1(-\infty, \infty)$ *and* $f' \in L_1(-\infty, \infty)$;
(b) $f \in L_2(-\infty, \infty)$ *and* $f' \in L_2(-\infty, \infty)$.

Denote by F the Fourier transform of f and by G the Fourier transform of f'. Then

$$G(\xi) = i\xi \, F(\xi). \tag{8.19}$$

Proof: First, let us prove case (a). Since

$$f(x) = \int_0^x f'(t) \, dt + f(0),$$

an integration by parts yields

$$\frac{1}{2\pi} \int_{-N}^{N} f(x) \, e^{-i\xi x} \, dx$$
$$= \frac{1}{2\pi i \xi} \int_{-N}^{N} f'(x) \, e^{-i\xi x} \, dx - \frac{1}{2\pi i \xi} \left[f(x) \, e^{-i\xi x} \right]_{x=-N}^{x=N}. \tag{8.20}$$

The limits of both integrals exist as $N \to \infty$; thus, there also exists the limit of expressions $[f(N)\,e^{-i\xi N} - f(-N)\,e^{i\xi N}]$. If this limit were not zero, say $a \neq 0$, then $|f(N)|$ or $|f(-N)|$ would be larger than $\frac{1}{3}|a|$ for N sufficiently large; then f could not be integrable, which contradicts the assumption. Thus, the term in brackets in (8.20) tends to zero and (8.19) follows immediately from (8.20) as $N \to \infty$.

Case (b) is slightly more complicated. Let Φ be a function from $\mathscr{D}$. Due to Remark 8.6, a function φ exists such that Φ is its Fourier transform, i.e.,

$$\varphi(x) = \int_{-\infty}^{\infty} e^{i\xi x}\,\Phi(\xi)\,d\xi\,.$$

The function φ belongs to $L_2(-\infty, \infty)$ by Remark 8.6; moreover, it can be shown that φ has a derivative φ' and that φ and φ' are absolutely integrable and, for each $k > 0$, $|x|^k \varphi(x) \to 0$ as $|x| \to \infty$ (cf. p. 262, where analogous properties of $F(\xi)$ were shown, or Remark 8.10). Hence, φ satisfies the conditions (a), i.e. $\Psi(\xi) = i\xi\,\Phi(\xi)$, where Ψ is the Fourier transform of $\psi(x) = \varphi'(x)$.

Assuming (b), we have

$$\int_{-\infty}^{\infty} i\xi\, F(\xi)\,\overline{\Phi(\xi)}\,d\xi = -\int_{-\infty}^{\infty} F(\xi)\,\overline{i\xi\,\Phi(\xi)}\,d\xi$$

$$= -\int_{-\infty}^{\infty} F(\xi)\,\overline{\Psi(\xi)}\,d\xi = -\frac{1}{2\pi}\int_{-\infty}^{\infty} f(x)\,\overline{\psi(x)}\,dx$$

$$= \frac{1}{2\pi}\int_{-\infty}^{\infty} f'(x)\,\overline{\varphi(x)}\,dx = \int_{-\infty}^{\infty} G(\xi)\,\overline{\Phi(\xi)}\,d\xi$$

(the first integral is obviously equal to the second, the second to the third because $i\xi\,\Phi(\xi) = \Psi(\xi)$, the third to the fourth by (8.17), the fourth to the fifth by an integration by parts — cf. the above mentioned properties of $\varphi(x)$ as $|x| \to \infty$, and finally, the fifth to the last by (8.17) again). Thus,

$$\int_{-\infty}^{\infty} [i\xi\, F(\xi) - G(\xi)]\,\overline{\Phi(\xi)}\,d\xi = 0$$

for any function $\Phi \in \mathscr{D}$. The function $H(\xi) = \mathrm{i}\xi\, F(\xi) - G(\xi)$ is an element of the space $L_2(a, b)$ for any finite interval $[a, b]$. For Φ let us now choose functions $\Phi_n \in \mathscr{D}$ such that $\Phi_n(\xi) \neq 0$ for at most $\xi \in (a, b)$ and such that $\Phi_n \to H$ in $L_2(a, b)$ as $n \to \infty$ (such functions Φ_n exist by Problem 3.28). Thus, we then have

$$\int_{-\infty}^{\infty} H(\xi)\, \overline{\Phi_n(\xi)}\, \mathrm{d}\xi = \int_a^b H(\xi)\, \overline{\Phi_n(\xi)}\, \mathrm{d}\xi = 0 \quad \text{for all} \quad n\,.$$

Letting $n \to \infty$, we obtain due to the properties of a scalar product in Hilbert space (see Theorem 3.4),

$$\int_a^b H(\xi)\, \overline{H(\xi)}\, \mathrm{d}\xi = \|H\|^2_{L_2(a,b)} = 0\,,$$

i.e.,

$$H(\xi) = \mathrm{i}\xi\, F(\xi) - G(\xi) = 0 \quad \text{for} \quad \xi \in (a, b)\,.$$

Since the interval (a, b) is arbitrary, formula (8.19) holds even for case (b).

Remark 8.7. Under certain additional assumptions on the function f (which can be easily derived), we have the following consequence of Theorem 8.7: The Fourier transform of a linear combination g of the function f and its derivatives

$$g(x) = \sum_{k=0}^{n} a_k f^{(k)}(x)\,,$$

is the product of the Fourier transform of f and a polynomial, i.e.,

$$G(\xi) = F(\xi) \sum_{k=0}^{n} a_k (\mathrm{i}\xi)^k\,.$$

Formula (8.19) can also be written in the form

$$\widehat{f'} = \mathrm{i}\xi\, \hat{f} \quad \text{or} \quad \mathscr{F}f' = \mathrm{i}\xi\, \mathscr{F}f\,.$$

C. The following theorem is a converse of Theorem 8.7 in a certain sense.

Theorem 8.8. *Let the function f satisfy one of the following conditions*

(a) $f(x) \in L_1(-\infty, \infty)$ *and* $g(x) = x f(x) \in L_1(-\infty, \infty)$,
(b) $f(x) \in L_2(-\infty, \infty)$ *and* $g(x) = x f(x) \in L_2(-\infty, \infty)$.

Let F and G be the Fourier transforms of f and g, respectively. Then

$$G(\xi) = i F'(\xi). \tag{8.21}$$

In case (a), *F' is continuous and $\lim_{|\xi| \to \infty} F'(\xi) = 0$; in case* (b), *$F'(\xi) \in L_2(-\infty, \infty)$.*

Proof. Denote

$$F_N(\xi) = \frac{1}{2\pi} \int_{-N}^{N} f(x) e^{-i\xi x} \, dx.$$

Then

$$F'_N(\xi) = \frac{-i}{2\pi} \int_{-N}^{N} x f(x) e^{-i\xi x} \, dx \tag{8.22}$$

(we have used Theorem 1.12 with an integrable majorant $|x f(x)|$). Thus, as $N \to \infty$ the limit H of the sequence F'_N exists almost everywhere in $(-\infty, \infty)$ in case (a) or in $L_2(-\infty, \infty)$ in case (b). In case (a), $F'_N(\xi)$ converges to $H(\xi)$ even uniformly, because

$$|F'_N(\xi) - H(\xi)| = \left| \frac{-i}{2\pi} \int_{|x|>N} x f(x) e^{-i\xi x} \, dx \right|$$

$$\leq \frac{1}{2\pi} \int_{|x|>N} |x f(x)| \, dx < \varepsilon$$

for $N > N_0$ (see Theorem 1.9).

Hence, we have

$$\lim_{N \to \infty} \int_0^\xi F'_N(\tau) \, d\tau = \int_0^\xi H(\tau) \, d\tau;$$

this follows, for example, from the properties of the scalar product in the Hilbert space $L_2(0, \xi)$ stated in Theorem 3.4. Actually, the integral can be interpreted as a scalar product with a function equalling one identically, and $F'_N \to H$ in $L_2(0, \xi)$ in both cases (a) and (b).

Since

$$F_N(\xi) - F_N(0) = \int_0^\xi F'_N(\tau)\, d\tau,$$

we obtain by letting $N \to \infty$,

$$F(\xi) - F(0) = \int_0^\xi H(\tau)\, d\tau$$

and consequently, $H(\xi) = F'(\xi)$ almost everywhere. Letting N tend to infinity in (8.22), we obtain formula (8.21). The properties of F' follow from Theorem 8.1 in case (a) and from Theorem 8.6 in case (b).

Example 8.2. Let us find the Fourier transform F of the function

$$f(x) = \begin{cases} (1 - x^2)^{1/2} & \text{for } |x| \leq 1, \\ 0 & \text{for } |x| > 1, \end{cases}$$

by constructing a differential equation satisfied by F.

A differentiation will convince us that

$$f'(x)(1 - x^2) + x f(x) = 0,$$

where the equality holds for almost every $x \in (-\infty, \infty)$, if the derivative of f is understood in the sense of absolutely continuous function. Using Theorems 8.7 and 8.8 (and their generalizations for derivatives of higher orders) we find that the functions $f'(x)$, $x^2 f'(x) = x(x f'(x))$ and $x f(x)$ have Fourier transforms $i\xi F(\xi)$, $i^2(i\xi F(\xi))'' = i(i(i\xi F(\xi))')'$ and $i F'(\xi)$ respectively; thus for the Fourier transform F of f we have the equation,

$$i\xi F - i^2(i\xi F)'' + i F' = 0$$

i.e.

$$F'' + \frac{3}{\xi} F' + F = 0.$$

Putting $G(\xi) = \xi F(\xi)$, we obtain Bessel's equation for G, i.e.,

$$\xi^2 G'' + \xi G' + (\xi^2 - 1) G = 0;$$

hence, $G(\xi) = A J_1(\xi) + B Y_1(\xi)$, where J_1 and Y_1 are Bessel functions of the first and second kind, respectively, of index 1.*
Since the function $F(\xi) = G(\xi)/\xi$ is bounded by Theorem 8.1, we have $B = 0$ and consequently,

$$F(\xi) = A \frac{J_1(\xi)}{\xi} = \frac{A}{\xi} \sum_{k=0}^{\infty} \frac{(-1)^k}{k! \, (k+1)!} \left(\frac{\xi}{2}\right)^{2k+1}.$$

The constant A can be obtained from (8.3) by setting $\xi = 0$, i.e.,

$$F(0) = \frac{1}{2\pi} \int_{-1}^{1} (1 - x^2)^{1/2} \, dx = \tfrac{1}{4} = A \lim_{\xi \to 0} \frac{J_1(\xi)}{\xi} = A \cdot \tfrac{1}{2},$$

hence, $A = \tfrac{1}{2}$.

Using another notation for Fourier transforms, we get the following form of formula (8.21),

$$\widehat{x f(x)} = \mathrm{i}(\hat{f})' \quad \text{or} \quad \mathscr{F}(x f(x)) = \mathrm{i}(\mathscr{F}f)'.$$

D. Let us now introduce the *convolution* of two integrable functions. If f_1 and f_2 are absolutely integrable in $(-\infty, \infty)$ define the function

$$f(x) = \int_{-\infty}^{\infty} f_1(s) f_2(x - s) \, ds \qquad (8.23)$$

for every x such that the integral exists. The function $f(x)$ is called the convolution of f_1 and f_2 and (8.23) is abbreviated by

$$f = f_1 * f_2.$$

---

* Cf. the footnote on p. 134.

A substitution $x - s = t$ in (8.23) yields that

$$f_1 * f_2 = f_2 * f_1.$$

Theorem 8.9. *The convolution f of functions $f_1 \in L_1(-\infty, \infty)$ and $f_2 \in L_1(-\infty, \infty)$ is defined for almost every $x \in (-\infty, \infty)$. The function f is also integrable in $(-\infty, \infty)$ and*

$$\int_{-\infty}^{\infty} |f(x)| \, dx \leq \int_{-\infty}^{\infty} |f_1(x)| \, dx \cdot \int_{-\infty}^{\infty} |f_2(x)| \, dx. \qquad (8.24)$$

Proof. The function $f_1(s) \cdot f_2(x - s)$ of two variables x, s is integrable in the plane, since

$$I = \int_{-\infty}^{\infty} \int_{-\infty}^{\infty} |f_1(s)| \, |f_2(x - s)| \, ds \, dx$$

$$= \int_{-\infty}^{\infty} |f_1(s)| \left(\int_{-\infty}^{\infty} |f_2(x - s)| \, dx \right) ds$$

$$= \int_{-\infty}^{\infty} |f_1(s)| \, ds \cdot \int_{-\infty}^{\infty} |f_2(s)| \, ds.$$

Thus, the integral in (8.23) converges almost everywhere and (8.24) holds, because

$$\int_{-\infty}^{\infty} |f(x)| \, dx \leq I.$$

The Fourier transform of a convolution can be determined by the following theorem.

Theorem 8.10. *Let f_1 and f_2 be functions integrable in $(-\infty, \infty)$ and let $f = f_1 * f_2$. Then for the Fourier transforms F_1, F_2 and F of the functions f_1, f_2 and f respectively, we have*

$$F(\xi) = 2\pi \, F_1(\xi) \, F_2(\xi). \qquad (8.25)$$

Proof. Using Fubini's theorem with an integrable majorant $(1/2\pi) |f_1(s)| \cdot |f_2(x - s)|$ we obtain immediately (8.25), i.e.

$$F(\xi) = \frac{1}{2\pi} \int_{-\infty}^{\infty} e^{-i\xi x} \left(\int_{-\infty}^{\infty} f_1(s) f_2(x - s) \, ds \right) dx$$

$$= \frac{1}{2\pi} \int_{-\infty}^{\infty} f_1(s) \left(\int_{-\infty}^{\infty} f_2(x - s) e^{-i\xi x} \, dx \right) ds$$

$$= \frac{1}{2\pi} \int_{-\infty}^{\infty} f_1(s) \left(\int_{-\infty}^{\infty} f_2(t) e^{-i\xi(t+s)} \, dt \right) ds$$

$$= \frac{1}{2\pi} \int_{-\infty}^{\infty} f_1(s) e^{-i\xi s} \, ds \cdot \int_{-\infty}^{\infty} f_2(t) e^{-i\xi t} \, dt$$

$$= 2\pi F_1(\xi) F_2(\xi).$$

Remark 8.8. Theorem 8.10 remains true even in the case that $f_1 \in L_1(-\infty, \infty)$ and $f_2 \in L_2(-\infty, \infty)$. Then both the convolution f and its Fourier transform are in $L_2(-\infty, \infty)$ and we have again (8.25). However, Theorem 8.10 need not hold if we assume only that both functions f_1 and f_2 are in $L_2(-\infty, \infty)$.

The relation (8.25) can also be written as follows:

$$\widehat{f_1 * f_2} = 2\pi \hat{f}_1 \cdot \hat{f}_2 \quad \text{or} \quad \mathscr{F}(f_1 * f_2) = 2\pi \, \mathscr{F} f_1 \cdot \mathscr{F} f_2.$$

Remark 8.9. Under certain additional assumptions on the functions f_1 and f_2, a converse proposition is also true: *the Fourier transform of a product of two functions is the convolution of their Fourier transforms*. The formal proof is easy. Actually, denoting $\mathscr{F}f$ the Fourier transform and $\mathscr{F}^{-1}f$ the integral in the inversion formula (8.6), we have for $\mathscr{F}^{-1}$ by Theorem 8.10 and Remark 8.1,

$$\mathscr{F}^{-1}(\mathscr{F}f_1 * \mathscr{F}f_2) = \mathscr{F}^{-1}(\mathscr{F}f_1) \cdot \mathscr{F}^{-1}(\mathscr{F}f_2) = f_1 \cdot f_2,$$

i.e.

$$\mathscr{F}f_1 * \mathscr{F}f_2 = \mathscr{F}(f_1 \cdot f_2).$$

E. In Section 8.1, the space $\mathscr{D}$ of smooth finitary functions has been considered. This space has a disadvantage: the Fourier transform of a function f from $\mathscr{D}$ *need not also be in* $\mathscr{D}$. Fortunately, it suffices to weaken somewhat the assumptions on the functions f and a new space $\mathscr{S}$ with favourable properties is obtained; then each $f \in \mathscr{S}$ has a Fourier transform which is again in $\mathscr{S}$.

The space $\mathscr{S}$ consists of all functions f defined and infinitely differentiable in $(-\infty, \infty)$ such that f and all its derivatives converge to zero as $|x| \to \infty$ faster than any power $|x|^{-l}$; in other words, for every non-negative integer k and every non-negative l a constant $C_{k,l}$ exists such that

$$|f^{(k)}(x)| \leq C_{k,l}|x|^{-l} \quad \text{for any} \quad x \in (-\infty, \infty). \tag{8.26}$$

Theorem 8.11. *The Fourier transform of a function $f \in \mathscr{S}$ belongs to $\mathscr{S}$. Conversely, for each $F \in \mathscr{S}$ there exists a function $f \in \mathscr{S}$, having F as its Fourier transform.*

Proof. By (8.26) we have $|x^l f^{(k)}(x)| \leq C_{k,l+2}|x|^{-2}$ and since $x^l f^{(k)}(x)$ is continuous, it is integrable in $(-\infty, \infty)$. By Theorem 8.8, $|F^{(k)}(\xi)| = |G(\xi)|$, where $G(\xi)$ is the Fourier transform of $x^k f(x)$. By assumption, all derivatives of the function $x^k f(x)$ are integrable in $(-\infty, \infty)$, and the Fourier transform of its l-th derivative $(i\xi)^l G(\xi)$ (cf. Theorem 8.7) is continuous and bounded by a certain constant $D_{k,l}$ by Theorem 8.1. Hence,

$$|G(\xi)| = |F^{(k)}(\xi)| \leq D_{k,l}|\xi|^{-l},$$

i.e., F also belongs to $\mathscr{S}$. The converse assertion can be proved by Remark 8.1; this finishes the proof.

Remark 8.10. Obviously, $\mathscr{D} \subset \mathscr{S}$. From Theorem 8.11 it follows that the Fourier transform F of a function $f \in \mathscr{D}$ is in $\mathscr{S}$ and therefore is infinitely differentiable and the function $x^l f^{(k)}(x)$ is integrable for every k and l. These facts have already been partially used on p. 262 and in the proof of Theorem 8.7.

8.5 FOURIER TRANSFORMS OF DISTRIBUTIONS

In the preceding section we have seen that a function f is in $\mathscr{S}$ if and only if its Fourier transform F is in $\mathscr{S}$. $\mathscr{S}$ is obviously a linear space. These facts enable us to extend the definition of a Fourier transform to linear functionals on $\mathscr{S}$.

Let f be a mapping from $\mathscr{S}$ into complex numbers such that for any two functions φ_1 and φ_2 in $\mathscr{S}$ and any two complex numbers α_1 and α_2 we have

$$f(\alpha_1 \varphi_1 + \alpha_2 \varphi_2) = \alpha_1 f(\varphi_1) + \alpha_2 f(\varphi_2).$$

Then f is called a *linear functional* on $\mathscr{S}$. A particular case of linear functionals on $\mathscr{S}$ is furnished by the *distributions (generalized functions)* on $\mathscr{S}$ (cf. [5])*.

Example 8.3. Let us present several examples of linear functionals on $\mathscr{S}$.

(a) To each $\varphi \in \mathscr{S}$, its value at a fixed point x_0 is assigned. This functional is called *Dirac's δ-function* for the point x_0 and is denoted by δ_{x_0}; thus,

$$\delta_{x_0}(\varphi) = \varphi(x_0).$$

(b) The functional f is defined for $\varphi \in \mathscr{S}$ by the formula

$$f(\varphi) = \varphi'(0) - \varphi(1) + \int_0^1 \frac{\varphi(t)\, t}{\sin t}\, dt.$$

(c) The functional f is defined for $\varphi \in \mathscr{S}$ by

$$f_g(\varphi) = \int_{-\infty}^{\infty} \overline{g(t)}\, \varphi(t)\, dt, \qquad (8.27)$$

where g is a function such that the integral on the right hand side is convergent for each $\varphi \in \mathscr{S}$.

---

* A linear functional is a distribution, if it is continuous in a certain metric of $\mathscr{S}$.

Example (c) is rather important. It shows that to each function g integrable in every finite interval and not growing too fast as $|x| \to \infty$, we can assign a linear functional in a natural way.

Example 8.4. Choose $g(t) = e^{t^2}$ in Example 8.3c. Then the mapping f_g from (8.27) is not a linear functional on $\mathscr{S}$, because the integral in (8.27) is not finite for the function $\varphi(t) = e^{-t^2}$ which belongs to $\mathscr{S}$. Thus, the requirement on the convergence of integral (8.27) is indeed a condition imposed on the growth of g.

Let us note without proof that two functions g_1 and g_2 define the same functional $f_{g_1} = f_{g_2}$, i.e. $f_{g_1}(\varphi) = f_{g_2}(\varphi)$ for every $\varphi \in \mathscr{S}$, if and only if they differ at most on a set of measure zero.

Thus, the correspondence between g and f_g is one-to-one and we need not distinguish g and f_g too sharply. Consequently, a functional f_g defined by (8.27) is sometimes called a *functional of the type of a function g* or simply the *function g*.

Example 8.5. Dirac's δ-function is not a functional of the type of a function. Actually, if $\varphi \in \mathscr{S}$ is such that $\varphi(0) = 0$, then $\delta_0(\varphi) = 0$; if a function g, corresponding to δ_0 by (8.27) existed, then it would have to vanish for almost all $t \neq 0$, i.e. almost everywhere in $(-\infty, \infty)$. Then we would have $\delta_0(\varphi) = 0$ for all $\varphi \in \mathscr{S}$, which contradicts the definition of Dirac's δ-function.

In spite of this, the *functional* δ_0 is often called a *function* in applications: it is introduced by "$\delta_0(t) = 0$ for $t \neq 0$ and $\delta_0(0) = \infty$, the infinity being so large that $\int_{-\infty}^{\infty} \delta_0(t)\,dt = 1$". This statement, however, cannot be correct from the mathematical point of view; it merely indicates the fact that, for $\varphi \in \mathscr{S}$ with $\varphi(0) = 1$, the equality $\delta_0(\varphi) = 1$ can formally be written as $\int_{-\infty}^{\infty} \delta_0(t)\,\varphi(t)\,dt = \varphi(0)$, i.e., as if $\delta_0(t)$ were a function.

Now, we are going to define the Fourier transform of a linear functional on $\mathscr{S}$.

It can be verified that (8.14) also holds if $\varphi_1 \in L_1(-\infty, \infty)$

and $\varphi_2 \in \mathscr{S}$. Thus, we can write

$$\int_{-\infty}^{\infty} \overline{G(\xi)}\, \Phi(\xi)\, \mathrm{d}\xi = \frac{1}{2\pi} \int_{-\infty}^{\infty} \overline{g(t)}\, \varphi(t)\, \mathrm{d}t \qquad (8.28)$$

for every function g absolutely integrable in $(-\infty, \infty)$ and every $\varphi \in \mathscr{S}$. By Remark 8.5, formula (8.28) also holds if g is in $L_2(-\infty, \infty)$. Formula (8.28) may be rewritten as

$$f_G(\Phi) = \frac{1}{2\pi} f_g(\varphi). \qquad (8.29)$$

This formula defines a functional of the type of the function G (G being the Fourier transform of g) in terms of the functional of the type of the function g. In an analogous way, we define the Fourier transform of an arbitrary linear functional on $\mathscr{S}$.

A linear functional F on $\mathscr{S}$ will be called the *Fourier transform of a linear functional* f, if

$$F(\Phi) = \frac{1}{2\pi} f(\varphi),$$

for each $\varphi \in \mathscr{S}$, where Φ is the Fourier transform of φ and thus $\Phi \in \mathscr{S}$.

According to (8.29), the Fourier transform of a functional of the type of function g, where $g \in L_1(-\infty, \infty)$ or $g \in L_2(-\infty, \infty)$, is a functional of the type of function $G = \hat{g}$.

Example 8.6. The Fourier transform of the function e^{int} (more explicitly, of the functional of the type $g(t) = e^{int}$) is Dirac's δ-function δ_n. Actually, we have

$$F(\Phi) = \frac{1}{2\pi} f_g(\varphi) = \frac{1}{2\pi} \int_{-\infty}^{\infty} e^{-int} \varphi(t)\, \mathrm{d}t = \Phi(n) = \delta_n(\Phi).$$

Example 8.7. Let us find the Fourier transform of Dirac's function δ_n. We have

$$F(\Phi) = \frac{1}{2\pi} \delta_n(\varphi) = \frac{1}{2\pi} \varphi(n).$$

Due to the inversion formula (8.6),

$$\frac{1}{2\pi} \varphi(n) = \frac{1}{2\pi} \int_{-\infty}^{\infty} e^{in\xi} \Phi(\xi) \, d\xi = \int_{-\infty}^{\infty} \overline{\frac{e^{-in\xi}}{2\pi}} \Phi(\xi) \, d\xi.$$

Thus, $F(\Phi)$ is a functional of the type of function $(1/2\pi) e^{-in\xi}$, i.e. $(1/2\pi) e^{-in\xi}$ is the Fourier transform of Dirac's function δ_n.

Remark 8.11. The above two examples show that the Fourier transform of a function may be a functional and vice versa.

Remark 8.12. We have defined the Fourier transform of any linear functional on $\mathscr{S}$. However, if we wanted the Fourier transform to have some favourable properties, we would have to restrict our considerations to a narrower class of linear functionals and define the generalized functions (distributions). This would carry us far beyond the scope of this book.

From examples 8.6 and 8.7 it follows that the Fourier transform of a trigonometric polynomial $\sum_{k=-n}^{n} c_k e^{ikx}$ is the functional $\sum_{k=-n}^{n} c_k \delta_k$, i.e. a functional associating the values $c_k \varphi(k)$, $k = 0$, $\pm 1, \pm 2, \ldots, \pm n$ with a function $\varphi \in \mathscr{S}$. We say that the functional is concentrated at these points. Crudely speaking, the Fourier transform of a periodic function with period 2π is a functional concentrated at points $t = 0, \pm 1, \pm 2, \ldots$ which, at $t = n$, is equal to Dirac's function δ_n multiplied by c_n, where c_n is the corresponding Fourier coefficient of f (in the complex expansion). Thus, the sequence of Fourier coefficients is a particular case of the general Fourier transform.

In the present chapter the Fourier transform was defined by the formula

$$F(\xi) = \frac{1}{2\pi} \int_{-\infty}^{\infty} e^{-i\xi x} f(x) \, dx \qquad (8.3)$$

and the inverse Fourier transform by

$$f(x) = \int_{-\infty}^{\infty} e^{i\xi x} F(\xi) \, d\xi . \qquad (8.6)$$

However, other definitions may be found in the literature; for example,

$$F(\xi) = \int_{-\infty}^{\infty} e^{-i\xi x} f(x) \, dx \qquad (8.30)$$

and

$$f(x) = \frac{1}{2\pi} \int_{-\infty}^{\infty} e^{i\xi x} F(\xi) \, d\xi$$

or

$$F(\xi) = \frac{1}{\sqrt{(2\pi)}} \int_{-\infty}^{\infty} e^{-i\xi x} f(x) \, dx \qquad (8.31)$$

and

$$f(x) = \frac{1}{\sqrt{(2\pi)}} \int_{-\infty}^{\infty} e^{i\xi x} F(\xi) \, d\xi$$

or

$$F(\xi) = \int_{-\infty}^{\infty} e^{-2\pi i \xi x} f(x) \, dx . \qquad (8.32)$$

It is obvious that differences between these definitions are merely formal: the function $F(\xi)$ defined in (8.30) or (8.31) is the Fourier transform (in the sense of (8.3)) of the function $2\pi f(x)$ or $\sqrt{(2\pi)} f(x)$, respectively, i.e. the difference consists only in a factor. Similarly, (8.32) can be obtained from (8.3) by a substitution of variables.

Each of these definitions has certain advantages of its own. For example, the definition (8.30) yields a simpler formula for the

Fourier transform of a convolution: if $f = f_1 * f_2$, then

$$F(\xi) = F_1(\xi) \cdot F_2(\xi)$$

(compare with (8.25)). In the case of definition (8.31), the Fourier transform and the inversion formula are symmetric and Parseval's equality — cf. (8.13) — takes the form

$$\int_{-\infty}^{\infty} |f(x)|^2 \, dx = \int_{-\infty}^{\infty} |F(\xi)|^2 \, d\xi \,.$$

The reader is advised to derive various formulas for the new definitions and to compare them with those given in the present chapter.

Let us note that different tables of Fourier transforms give different values according to the definition used; therefore, it is always necessary to realize which definition of the transform is under consideration.

Problems

8.1. Prove that the Fourier transform of an even (odd) function is again even (odd).

8.2. Show that the Fourier transform of an even (odd) function $f(x)$ is its *cosine-transform* (*sine-transform*), i.e.,

$$F_c(\xi) = \frac{1}{\pi} \int_0^{\infty} f(t) \cos \xi t \, dt \quad \left(F_s(\xi) = \frac{1}{\pi} \int_0^{\infty} f(t) \sin \xi t \, dt\right).$$

8.3. Show that, for $f \in L_2(0, \infty)$,

$$f(x) = \frac{2}{\pi} \int_0^{\infty} \left(\cos x\xi \int_0^{\infty} f(t) \cos \xi t \, dt \right) d\xi \,,$$

$$f(x) = \frac{2}{\pi} \int_0^{\infty} \left(\sin x\xi \int_0^{\infty} f(t) \sin \xi t \, dt \right) d\xi \,.$$

In which sense do these integrals exist?

(*Hint*: Extend f onto $(-\infty, \infty)$ so as to be even (odd) and use Theorem 8.6.)

8.4. Find the Fourier transform of the function $e^{-|x|}$. Prove that
$$e^{-|x|} = \frac{2}{\pi} \int_0^\infty \frac{\cos \xi x}{\xi^2 + 1} \, d\xi \, .$$
(*Hint*: Use the inversion formula in $L_2(-\infty, \infty)$.)

8.5. Find the Fourier transform of the function $f(x) = e^{-ax^2}$, $a > 0$.
(*Hint*: Apply the Fourier transform to the differential equation $f'(x) = -2ax f(x)$; establish the unknown constant by using the relation $\int_0^\infty e^{-x^2} \, dx = \tfrac{1}{2} \sqrt{\pi}$.)

8.6. Find the Fourier transforms of xe^{-ax^2} and $x^2 e^{-ax^2}$, $a > 0$.

8.7. Prove that $e^{ia\xi} F(\xi)$ is the Fourier transform of $g(x) = f(x + a)$.

8.8. Investigate the behaviour of the Fourier transform of the function f_h ($h > 0$) defined by
$$f_h(x) = 0 \text{ for } |x| > h\, , \quad f_h(x) = \tfrac{1}{2}h \text{ for } |x| \leq h\, .$$

8.9. What happens to the Fourier transform of the function f_h from Problem 8.8 and its roots if h tends to zero? (Compare with Section 9.7.2 in Chapter 9.)

8.10. Solve the integral equation
$$\int_0^\infty g(x) \cos tx \, dx = f(t) \, ,$$
where
$$f(t) = \tfrac{1}{2}\pi \sin t \text{ for } 0 \leq t \leq \pi\, , \quad f(t) = 0 \text{ for } t > \pi\, .$$
(*Hint*: Use Problem 8.3.)

8.11. Show that neither $L_1(-\infty, \infty) \subset L_2(-\infty, \infty)$ nor $L_2(-\infty, \infty) \subset L_1(-\infty, \infty)$.

(*Hint*: Consider the functions $f(t) = (1 + |t|)^{-1}$ and $g(t) = |t|^{-1/2} (1 + |t|)^{-1}$.)

8.12. Prove that the limiting process (8.5) holds uniformly in each finite interval.

(*Hint*: Estimate $\int_{-\infty}^{\infty} |(dF^x(\sigma)/d\sigma)| \, d\sigma$ independently of x.)

8.13. Prove that

$$\lim_{T \to \infty} \int_{-T}^{T} \frac{\sin Nt}{t} \, dt = \lim_{T \to \infty} \int_{-T}^{T} \frac{\sin t}{t} \, dt = \pi.$$

(*Hint*: In Problem 3.25 it is proved that functions $\varphi_n \in \mathscr{D}$ exist such that $|\varphi_n(x)| \leq 1$, $\varphi_n(x) = 1$ for $|x| \leq n$ and $\varphi_n(x) = 0$ for $|x| \geq n + 1$. Construct the functions $\varphi_{n,N}$ by (8.9), prove that

$$\left| \varphi_{n,N}(0) - \frac{1}{\pi} \int_{-nN}^{nN} \frac{\sin t}{t} \, dt \right| \leq \frac{2}{\pi n}$$

and let N tend to infinity.)

8.14. Prove the assertion in Remark 8.8.

(*Hint*: If $f = f_1 * f_2$, then

$$|f(x)|^2 \leq \left[\int_{-\infty}^{\infty} |f_1(s)|^{1/2} |f_1(s)|^{1/2} |f_2(x - s)| \, ds \right]^2.$$

Estimate this integral by using Schwarz's (Hölder's) inequality; then

$$|f(x)|^2 \leq \|f_1\|_{L_1(-\infty, \infty)} \int_{-\infty}^{\infty} |f_1(s)| |f_2(x - s)|^2 \, ds.$$

Since the last integral is in fact a convolution of f_1 and f_2^2 which are absolutely integrable, Theorem 8.9 may be used.)

8.15. Define a derivative of the linear functional f on $\mathscr{S}$ as the functional g given by

$$g(\varphi) = -f(\varphi')$$

for every $\varphi \in \mathscr{S}$. Denoting F the Fourier transform of f and G the Fourier transform of g, prove that

$$G(\Phi) = -F(i\xi\Phi).$$

8.16. Prove that, if the Fourier transform of a functional f is a functional of the type G, then the Fourier transform of its derivative f' is a functional of the type $i\xi G$.

8.17. Find the Fourier transform of the derivative δ_0' of Dirac's function δ_0 $(\delta_0'(\varphi) = -\delta_0(\varphi') = -\varphi'(0))$.

CHAPTER 9 EXAMPLES OF THE APPLICATION OF FOURIER SERIES

9.1 CLASSICAL SOLUTION OF THE EQUATION FOR A STRING

In the second chapter we have shown that every expression

$$u(x, t) = \sum_{k=1}^{n} (A_k \cos kt + B_k \sin kt) \sin kx \qquad (9.1)$$

where n is a positive integer and A_k, B_k are arbitrary real* numbers, is a solution of the equation of the vibrations of a string

$$\frac{\partial^2 u}{\partial x^2} - \frac{\partial^2 u}{\partial t^2} = 0 \qquad (9.2)$$

with boundary conditions

$$u(0, t) = u(\pi, t) = 0 . \qquad (9.3)$$

In Chapter 2 it was mentioned that equation (9.2) and boundary conditions (9.3) alone do not determine the vibrations of a string uniquely. For a complete determination it is necessary to prescribe the *initial conditions*

$$\begin{aligned} u(x, 0) &= f(x) , \\ \frac{\partial u}{\partial t}(x, 0) &= g(x) , \end{aligned} \quad x \in [0, \pi] , \qquad (9.4)$$

---

* Unless otherwise stated, all considerations in this chapter will be carried out for the real domain. The modification for complex-valued functions presents no essential difficulties.

APPLICATION OF FOURIER SERIES

where f and g are some a priori given functions describing the position and the velocity of the string at the initial time $t = 0$.

For solution (9.1), the conditions (9.4) mean that the integer n and the coefficients A_k, B_k, $k = 1, 2, \ldots, n$ have to satisfy the relations

$$\sum_{k=1}^{n} A_k \sin kx = f(x),$$

$$\sum_{k=1}^{n} kB_k \sin kx = g(x). \qquad (9.5)$$

These relations show that expression (9.1) can be a solution of the problem only in particular cases, i.e. if the functions f and g are trigonometric polynomials of the form

$$f(x) = \sum_{k=1}^{n} a_k \sin kx,$$

$$g(x) = \sum_{k=1}^{n} b_k \sin kx. \qquad (9.6)$$

The reader can easily verify that then

$$a_k = \frac{2}{\pi} \int_0^{\pi} f(x) \sin kx \, dx, \quad b_k = \frac{2}{\pi} \int_0^{\pi} g(x) \sin kx \, dx. \qquad (9.7)$$

Thus, a_k and b_k are Fourier coefficients of the functions f and g respectively, in the sine-expansion on the interval $[0, \pi]$ (cf. Chapter 4, Section 4.2 and Chapter 5, Section 5.2). Comparing (9.5) and (9.6) we get

$$A_k = a_k, \quad B_k = \frac{b_k}{k} \quad (k = 1, 2, \ldots, n),$$

so that the corresponding solution describing the vibrations of a string has the form

$$u(x, t) = \sum_{k=1}^{n} \left(a_k \cos kt + \frac{b_k}{k} \sin kt \right) \sin kx. \qquad (9.8)$$

Hence, the solution is determined completely by the functions f and g from the initial conditions and is given in terms of their Fourier coefficients a_k and b_k.

In the case just considered, of course, the functions f and g were of a very special kind. Let us now assume only that f and g are elements from the space $L_2(0, \pi)$. Then they can be expanded in the Fourier sine-series with coefficients a_k and b_k given by formulas (9.7). Thus, it is meaningful to consider the series

$$u(x, t) = \sum_{k=1}^{\infty} \left(a_k \cos kt + \frac{b_k}{k} \sin kt \right) \sin kx, \qquad (9.9)$$

which is an infinite analogue of expression (9.8). The question remains, however, under what conditions is this series actually a solution of our boundary-value problem stated by (9.2), (9.3) and (9.4) and in which sense it converges.

Differentiating the series (9.9) *formally* term by term we can easily verify that it "satisfies" our boundary-value problem. Hence (9.9) is a solution, but a *formal* one. As we have shown in an example at the end of Chapter 2, we must not be satisfied with this fact; we have to show that the formal consideration indicated above can be justified.

Assume that the functions f and g satisfy the following conditions:

1. $f'(0) = f'(\pi) = g(0) = g(\pi) = 0$.
2. The function f has a second and g a first derivative on the interval $[0, \pi]$.
3. The functions f'' and g' have a bounded variation on $(0, \pi)$.

Then for the Fourier coefficients of the functions f and g we have

$$|a_k| \leq \frac{C}{k^3}, \quad |b_k| \leq \frac{C}{k^2}, \quad k = 1, 2, \ldots. \qquad (9.10)$$

(See Chapters 2 and 7, Theorems 2.9 and 7.14.)

This means, however, that for $x \in [0, \pi]$ and $t \in [0, \infty)$ the series (9.9) converges uniformly, and the same is true for the series obtained from (9.9) by a formal differentiation by x and t, i.e. series giving $u(x, t)$, $\partial u(x, t)/\partial x$ and $\partial u(x, t)/\partial t$. Consequently, these functions are continuous and the function $u(x, t)$ satisfies the initial and boundary conditions in the classical sense. In order to show that also the second derivatives of $u(x, t)$ are continuous and that equation (9.2) is satisfied, we would have to require more than (9.10) does, for example that

$$|a_k| \leq \frac{C}{k^{3+\varepsilon}}, \quad |b_k| \leq \frac{C}{k^{2+\varepsilon}}, \quad \varepsilon > 0.$$

The requirement on the continuity of second derivatives of a solution, however, leads to certain conditions concerning the function f, which are not natural from a physical point of view, and which originated solely from the method used. Let us carry out the following consideration.

The functions f and g are defined by their respective Fourier series on $[0, \pi]$. Let them be defined by these series also outside $[0, \pi]$. Denote $G(x)$ the primitive function for $g(x)$ given by the formula

$$G(x) = -\sum_{k=1}^{\infty} \frac{b_k}{k} \cos kx$$

(see Theorem 5.1). Series (9.9) can be written as follows:

$$\begin{aligned} u(x, t) &= \frac{1}{2}\left[\sum_{k=1}^{\infty} a_k \sin k(x+t) + \sum_{k=1}^{\infty} a_k \sin k(x-t) \right. \\ &\quad \left. - \sum_{k=1}^{\infty} \frac{b_k}{k} \cos k(x+t) + \sum_{k=1}^{\infty} \frac{b_k}{k} \cos k(x-t) \right] \\ &= \tfrac{1}{2}[f(x+t) + f(x-t) + G(x+t) - G(x-t)] \\ &= h_1(x+t) + h_2(x-t).^* \end{aligned}$$

From this representation of the solution it is apparent that the

---

* This, in fact, is the *d'Alembert form of a solution*, cf. Chapter 2, p. 20.

function $u(x, t)$ is twice continuously differentiable provided f and G have continuous derivatives of the second order. Thus, it suffices to assume that the function g has a continuous first derivative on $[0, \pi]$ and f, periodically extended onto $(-\infty, \infty)$, a continuous second derivative. However, for this we must necessarily have

$$f''(0) = f''(\pi) = 0\,.$$

This is the restrictive and, from the physical point of view unnatural, assumption mentioned above. Under this assumption, the solution given by formula (9.9) is a classical solution.

Remark. Note that by a *classical solution* of a boundary value problem we mean a function, whose derivatives of orders appearing in the equation, boundary and initial conditions exist and are continuous in the closed region of consideration, and which satisfies the equation together with the initial and boundary conditions.

However, for the equation of a vibrating string the classical solution concept is very restrictive from both the mathematical and physical point of view. Actually, if we stretch a string at the initial time $t = 0$ as indicated in Fig. 9.1 (with zero initial velocity) and release it, it begins to vibrate in agreement with physical ideas although a classical solution does not exist, because the initial state f does not even have a continuous first derivative. Therefore, it is quite natural to introduce the concept of a generalized solution.

Fig. 9.1

9.2 GENERALIZED SOLUTION

Consider a problem described by a differential equation and by initial and boundary conditions. Let the character of the problem be such that a *small* change of given quantities, i.e. the a priori given functions appearing in the initial and boundary conditions and the right hand side of the equation, implies a *small* change of the solution in a certain sense. In other words, let a certain sequence of quantities be given to which there corresponds a sequence of classical solutions of the problem in question. If the given quantities converge to a certain limit quantity, then let the classical solutions converge in a certain sense to a limit function. This limit function is usually called the *generalized solution* of the boundary value problem, or the solution corresponding to the limit of given quantities.

Thus, the generalized solution concept depends essentially on the properties of the problem under consideration, and consequently, may differ from case to case as it will be seen later. For the *same* problem we can even obtain *different types* of generalized solutions depending on the type of convergence used.

Note that the above comments cannot be considered as a definition of a generalized solution. They are merely an attempt to explain how a generalized solution is to be defined correctly in each particular case.

In order to clarify these ideas let us show how the generalized solution can be defined for the equation of a vibrating string dealt with in the previous section. For this problem, the given quantities are functions f and g from the initial conditions (9.4), and, of course, the zero functions appearing on the right hand sides of both equation (9.2) and boundary conditions (9.3). The generalized solution will be defined for each pair of functions f, g such that

$$f \in W_2^{(1)}(0, \pi), \quad f(0) = f(\pi) = 0,$$
$$g \in L_2(0, \pi).$$

Remark: A function f is said to belong to $W_2^{(1)}(0, \pi)$, if it is absolutely continuous on $(0, \pi)$ and its derivative f' in the sense of absolutely continuous function is an element of the space $L_2(0, \pi)$. The norm in the space $W_2^{(1)}(0, \pi)$ is defined by

$$\|f\|_W = \|f\|_{L_2(0,\pi)} + \|f'\|_{L_2(0,\pi)}.$$

(For more detail see Chapter 7, Section 7.8.)

Let u be a classical solution of the problem (9.2), (9.3) and (9.4). Multiplying the equality (9.2) by the function $\partial u/\partial t$, we get

$$0 = \frac{\partial^2 u}{\partial t^2}\frac{\partial u}{\partial t} - \frac{\partial^2 u}{\partial x^2}\frac{\partial u}{\partial t}$$

$$= \frac{1}{2}\frac{\partial}{\partial t}\left(\frac{\partial u}{\partial t}\right)^2 - \frac{\partial}{\partial x}\left(\frac{\partial u}{\partial x}\frac{\partial u}{\partial t}\right) + \frac{1}{2}\frac{\partial}{\partial t}\left(\frac{\partial u}{\partial x}\right)^2.$$

Let us integrate this equality over the rectangle $[0, \pi] \times [0, T]$; the Fubini theorem and integration by parts yield

$$0 = \frac{1}{2}\int_0^\pi \left(\left|\frac{\partial u}{\partial t}(x, T)\right|^2 + \left|\frac{\partial u}{\partial x}(x, T)\right|^2\right) dx$$

$$- \frac{1}{2}\int_0^\pi \left(\left|\frac{\partial u}{\partial t}(x, 0)\right|^2 + \left|\frac{\partial u}{\partial x}(x, 0)\right|^2\right) dx$$

$$+ \int_0^T \frac{\partial u}{\partial x}(0, t)\frac{\partial u}{\partial t}(0, t)\, dt - \int_0^T \frac{\partial u}{\partial x}(\pi, t)\frac{\partial u}{\partial t}(\pi, t)\, dt. \quad (9.11)$$

From the boundary condition $u(0, t) \equiv 0$ it follows that $(\partial u/\partial t)(0, t) \equiv 0$ and similarly for $x = \pi$. Thus, the last two integrals vanish. From the initial conditions we have

$$\frac{\partial u}{\partial t}(x, 0) = g(x), \quad \frac{\partial u}{\partial x}(x, 0) = f'(x),$$

so that for every T we obtain from (9.11),

$$\int_0^\pi \left(\left|\frac{\partial u}{\partial t}(x, T)\right|^2 + \left|\frac{\partial u}{\partial x}(x, T)\right|^2\right) dx = \int_0^\pi (|g(x)|^2 + |f'(x)|^2)\, dx. \quad (9.12)$$

In fact, this equality expresses the principle of conservation of energy and therefore it is called the *energetic equality*.

From the first inequality in Problem 9.1 it follows that

$$|u(x, T)|^2 \leq \pi \int_0^\pi \left|\frac{\partial u}{\partial x}(x, T)\right|^2 dx.$$

The integral on the right hand side can be estimated by the integral on the left hand side of (9.12), and we obtain for every T,

$$\max_{0 \leq x \leq \pi} |u(x, T)|^2 \leq C \int_0^\pi (|g(x)|^2 + |f'(x)|^2) \, dx . \qquad (9.13)$$

Let u_n be a classical solution determined by initial conditions f_n and g_n, i.e.

$$u_n(x, 0) = f_n(x), \quad \frac{\partial u_n}{\partial t}(x, 0) = g_n(x).$$

Let the functions f_n converge to a function f in the norm of the space $W_2^{(1)}(0, \pi)$ and g_n to g in $L_2(0, \pi)$. Applying the inequality (9.13) to differences $u_n - u_m$ we can easily verify that the functions u_n constitute a uniform Cauchy sequence; consequently, they converge uniformly to a certain continuous function u. Thus, the continuous function u will be interpreted as a *generalized solution* of our problem with rather general initial conditions f and g.

However, we still have to show that for any pair of functions f, g with $f \in W_2^{(1)}(0, \pi)$, $f(0) = f(\pi) = 0$ and $g \in L_2(0, \pi)$ there actually exist the sequences f_n and g_n such that f_n converges to f in the norm of $W_2^{(1)}(0, \pi)$ and g_n to g in the norm of the space $L_2(0, \pi)$. This, however, is obviously true, because for f_n and g_n we can take the partial sum of the Fourier series for f and g, respectively, having the form (9.6). Then the functions g_n converge to g in $L_2(0, \pi)$ in view of the results in Chapter 3. As for the functions f_n, we have to apply Theorem 5.2; it follows that f_n' are the partial sums of the Fourier series of the function f', and consequently, they converge to f' in the norm of $L_2(0, \pi)$. (Observe that the condition $f(0) = f(\pi)$ is essential here.) Hence, the partial sums f_n converge to f also in the norm of the space $W_2^{(1)}(0, \pi)$. For the initial conditions f_n and g_n chosen in this way a classical solution u_n exists which takes the form (9.8) and thus is a partial sum of the series (9.9). We obtained the following proposition. *If the functions f and g satisfy the mentioned conditions, then the series (9.9) converges uniformly to a continuous function*

$u(x, t)$ which is a *generalized solution* of our problem. This generalized solution, of course, need not be smooth; for example, it need not even have a continuous derivative, as Problem 9.2 shows.

The equality (9.12) holds for the functions u_n, f_n and g_n; by a limiting process $n \to \infty$ we can prove that (9.12) holds for the limit function $u(x, t)$ with the corresponding functions $f(x)$ and $g(x)$, too.

The relations (9.12) and (9.13) played a significant role in our considerations and permitted us to show that the convergence of given quantities f_n and g_n implies the convergence of the corresponding classical solutions u_n. The relations of this type are essential for carrying out considerations concerning generalized solutions.

At the end of Chapter 2 we presented a "drastic" example of a formal approach to a solution of a differential problem. Now we will show that in the example mentioned a generalized solution does not exist just because an expression of type (9.12) or (9.13) cannot hold. Let us solve the equation $y'' = f(x)$ on the interval $[0, \pi]$ with boundary conditions $y'''(0) = y'''(\pi) = 0$. In order to define the generalized solution we have to show that for f small the corresponding solution y is also small, i.e. particularly for $f \equiv 0$ we should have $y \equiv 0$. However, this cannot be fulfilled, because the function $y = x$ satisfies the considered equation with $f \equiv 0$ and also the boundary conditions, but $y \not\equiv 0$.

This example shows also that the concept of a generalized solution is not purposeless.

9.3 DECOMPOSITION OF A TONE INTO HARMONICS. VIBRATIONS OF A TUNER

The equation examined in Section 9.1 was of a rather particular type. A more general equation is

$$\frac{\partial^2 u}{\partial t^2} - a^2 \frac{\partial^2 u}{\partial x^2} = 0 \quad * \tag{9.14}$$

for $x \in [0, l]$ and $t \geq 0$, which describes the vibrations of a homogeneous string of length l. The constant a depends on the mechanical properties of the string, i.e.

$$a^2 = \frac{H}{\varrho}, \quad a > 0,$$

where H is the strain and ϱ the density of the string. Note that in the case of a non-homogeneous string, whose mechanical properties vary along its length, a^2 would be naturally a function of x.

It is clear that by a linear substitution

$$\xi = \frac{\pi}{l} x, \quad \tau = \frac{a\pi}{l} t \quad (\xi \in [0, \pi])$$

equation (9.14) can be brought to the form (9.2),

$$\frac{\partial^2 u}{\partial \xi^2} = \frac{\partial^2 u}{\partial \tau^2}, \quad \xi \in [0, \pi], \quad \tau \geq 0.$$

Consequently, the difference between equations (9.14) and (9.2) is not essential. The solution of equation (9.14) with boundary conditions

$$u(0, t) = u(l, t) = 0, \quad t \geq 0,$$

and initial conditions

$$u(x, 0) = f(x), \quad \frac{\partial u}{\partial t}(x, 0) = g(x), \quad x \in [0, l]$$

---

* Of course, (9.14) is also not the most general type of equation; the derivatives of first order, the function u itself and the non-linear terms could be present as well. This, however, would complicate our considerations; therefore, we confine our attention to the simple form (9.14).

is given by

$$u(x, t) = \sum_{k=1}^{\infty} \left(a_k \cos \frac{a\pi k}{l} t + b_k \frac{l}{a\pi k} \sin \frac{a\pi k}{l} t \right) \sin \frac{\pi k}{l} x,$$

where

$$a_k = \frac{2}{l} \int_0^l f(x) \sin \frac{\pi k}{l} x \, \mathrm{d}x \, ; \quad b_k = \frac{2}{l} \int_0^l g(x) \sin \frac{\pi k}{l} x \, \mathrm{d}x \, .$$

The solution $u(x, t)$ can also be written in the phase-form (cf. Chapter 2, p. 31)

$$u(x, t) = \sum_{k=1}^{\infty} \alpha_k \sin \frac{k\pi}{l} x \sin \left(\frac{k\pi a}{l} t + \beta_k \right).$$

From this expression it is apparent that the vibrations of a string consist of single oscillations of the form

$$u_k(x, t) = \alpha_k \sin \frac{k\pi}{l} x \sin \left(\frac{k\pi a}{l} t + \beta_k \right).$$

Let k be fixed, i.e. consider more closely the single oscillation u_k. At a chosen point of the string ($x_0 = $ const.) the oscillation u_k has a magnitude

$$\left| \alpha_k \sin \frac{k\pi}{l} x_0 \right|.$$

All points of the string oscillate with the same period. The points $x = nl/k$, $(n = 1, 2, ..., k - 1)$ do not move with time and are called the *nodes* of the standing wave $u_k(x, t)$. The points with coordinates $x = (2n + 1) l/k$, $(n = 0, 1, ..., k - 1)$ oscillate with the largest magnitude $|\alpha_k|$, because at them $|\sin k\pi x/l| = 1$, and are called the *antinodes* of the standing wave $u_k(x, t)$. The single oscillations $u_k(x, t)$ are called the *harmonics* or overtones of the string.

The periodic vibrations are transmitted into our hearing and we experience a tone; thus, the sound of a string is a superposition of single oscillations. The pitch of a tone depends on the period of oscillation, and its intensity on the energy, i.e. on the magnitude. The tone with the largest period is called the fundamental tone of the string. The frequency of oscillations is proportional to the reciprocal of the period; thus, for the frequency of the fundamental tone we have

$$\omega_1 = \frac{a}{2l}.$$

The string can also generate single tones with frequencies which are multiples of the fundamental frequency ω_1. To such tones, called overtones, there correspond the single oscillations $u_k(x, t)$ for $k = 2, 3, \ldots$. The tone timbre depends on the content of overtones in a tone. If we touch a string in the middle, we suppress all single oscillations for which $\sin(k\pi x/l) \neq 0$ at $x = l/2$, i.e. oscillations $u_k(x, t)$ with k odd. The magnitude of the fundamental tone will be zero and the string will produce a tone with double frequency, i.e. a tone higher by one octave (a flageolet tone). The rate of higher harmonics is given by the initial conditions, i.e., by the activation of the string. The pitch of the fundamental tone is influenced by the length, strain and specific mass of the string, i.e. by the thickness and material the string is made of.

Single oscillations can be generated easily in electronic devices. Any vibration can be decomposed into single oscillations by means of a Fourier series on an interval with length equal to the vibration period. Thus, theoretically we can generate a tone with any pitch and timbre. This fact is used in electrophonic musical instruments, in electronics for generating oscillations with a preassigned shape etc.

As a further example we are going to discuss briefly the vibrations of a tuner which are described, with certain simplifications, by the equation

$$a^2 \frac{\partial^4 u}{\partial x^4} + \frac{\partial^2 u}{\partial t^2} = 0 \qquad (9.15)$$

for $x \in [0, l]$ and $t \geq 0$, by boundary conditions

$$u(0, t) = \frac{\partial u}{\partial x}(0, t) = \frac{\partial^2 u}{\partial x^2}(l, t) = \frac{\partial^3 u}{\partial x^3}(l, t) = 0$$

(the tuner is clamped at the end $x = 0$ and vibrates freely at the other end $x = l$) and by initial conditions

$$u(x, 0) = f(x), \quad \frac{\partial u}{\partial t}(x, 0) = g(x)$$

(the initial displacement and velocity of the tuner). Equation (9.15) will be solved by the *Fourier method*. Putting $u(x, t) = X(x) T(t)$, we obtain for the functions X and T the differential equations

$$T''(t) + a^2 \lambda T(t) = 0 \tag{9.16}$$

and

$$X^{(4)}(x) - \lambda X(x) = 0 \tag{9.17}$$

with conditions

$$X(0) = X'(0) = X''(l) = X'''(l) = 0 \tag{9.18}$$

(see Chapter 2, Section 2.1). Equation (9.17) with conditions (9.18) has a solution only for $\lambda_k = (\mu_k/l)^4$, where

$$\mu_1 \simeq 1 \cdot 875; \quad \mu_2 \simeq 4 \cdot 694; \quad \mu_3 \simeq 7 \cdot 854; \quad \ldots$$

are solutions of the transcendental equation $\cos \mu \cdot \cosh \mu = -1$. Thus, (9.17) has for $\lambda = \lambda_k$ the solution

$$X_k(x) = A_k \left(\cosh \frac{\mu_k}{l} x - \cos \frac{\mu_k}{l} x \right)$$
$$+ B_k \left(\sinh \frac{\mu_k}{l} x - \sin \frac{\mu_k}{l} x \right), \tag{9.19}$$

where the constants A_k and B_k are to be determined from the conditions $X''(l) = X'''(l) = 0$. Substituting $\lambda = \lambda_k$ into (9.16), we get for T the equation

$$T''(t) + a^2 \lambda_k T(t) = 0,$$

which has a solution

$$T_k(t) = a_k \cos 2\pi \nu_k t + b_k \sin 2\pi \nu_k t,$$

where $\nu_k = (a/2\pi) \sqrt{\lambda_k}$ are the eigenfrequencies. From the equation defining the μ_k it follows that

$$\frac{\nu_2}{\nu_1} \cong 6{\cdot}267 \, ; \quad \frac{\nu_3}{\nu_2} \cong 17{\cdot}548; \quad \ldots .$$

Hence, the first overtone is higher by more than two octaves than the fundamental tone and the second overtone by more than four octaves. If we strike the tuner all overtones are developed but decay rapidly due to their high frequency and only the fundamental tone of the tuner persists. The overtones sounding at the beginning create the characteristic metallic timbre of the tuner sound.

The determination of the numbers μ_k and consequently, of the solution $X_k(x)$ in the form (9.19) is evidently complicated, since the transcendental equation for μ_k can be solved only by numerical methods. The problem simplifies slightly if we consider the vibrations of a bar clamped at both ends. These vibrations are again described by equation (9.15) with the usual initial conditions, but with simpler boundary conditions

$$u(0, t) = u(l, t) = \frac{\partial u}{\partial x}(0, t) = \frac{\partial u}{\partial x}(l, t) = 0.$$

The reader is recommended to consider this problem in detail as a useful exercise.

The problems concerning vibrations of bars indicated above show that the Fourier method can be also used for equations of higher order.

9.4 EQUATION OF HEAT CONDUCTION

All equations considered so far were of *hyperbolic type*. In this section we are going to consider a typical representative of *parabolic equations* — the equation of heat conduction

$$\sum_{i,j=1}^{n} \frac{\partial}{\partial x_i} \left(a_{ij}(x) \frac{\partial u}{\partial x_j} \right) = \frac{\partial u}{\partial t} + f(x, t). \tag{9.20}$$

Here, $u(x, t)$ signifies the temperature at a point $x = (x_1, x_2, ..., x_n)$ of a non-homogeneous non-isotropic n-dimensional solid Ω at a time t. The function $f(x, t)$ represents the density of heat sources throughout the solid and the coefficients $a_{ij}(x)$ characterize the physical properties of the solid such as thermal conductivity etc. Equation (9.20) alone does not determine the temperature distribution uniquely. It is again necessary to prescribe certain *boundary conditions*, for example the temperature on the surface of the solid, and an *initial condition*, which in contrast to the equation of string vibrations, is only one, for example, the temperature distribution at the initial time $t = 0$. The coefficients a_{ij} of equation (9.20) must be *symmetric*, i.e. $a_{ij}(x) = a_{ji}(x)$, and satisfy the following so-called *condition of ellipticity*: a number $\alpha > 0$ exists such that for every n-tuple of real numbers $\xi_1, \xi_2, ..., \xi_n$ and all $x \in \Omega$ we have

$$\frac{1}{\alpha} \sum_{i=1}^{n} \xi_i^2 \leq \sum_{i,j=1}^{n} a_{ij}(x) \xi_i \xi_j \leq \alpha \sum_{i=1}^{n} \xi_i^2. \tag{9.21}$$

9.4.1 The estimate for a solution using the given quantities, which we derived in Section 9.2 to prove the existence of a generalized solution of the equation of a string (see (9.12) and (9.13)) was found without employing the theory of Fourier series. Now we are going to show how this theory can be used for establishing a similar estimate for the *periodic* solutions of a parabolic equation. An analogous estimate for non-periodic solutions can be found with the aid of Fourier transform. However, we shall derive the estimate alone without actually proving the existence of

APPLICATION OF FOURIER SERIES

a generalized solution, because such a proof would carry us beyond the scope of this book. At the same time, the subsequent considerations will serve as an example for employing the adjoint Fourier series. (See Chapter 5, Section 5.7.)

Assume for simplicity that the solid Ω is a one-dimensional interval $0 \leq x \leq \pi$. Thus, we have the equation

$$\frac{\partial}{\partial x}\left(a(x)\frac{\partial u}{\partial x}\right) = \frac{\partial u}{\partial t} + f(x, t), \qquad (9.22)$$

where $f(x, t)$ is periodic in t with period 2π, and boundary conditions

$$u(0, t) = u(\pi, t) = 0.$$

Let $u(x, t)$ be a classical periodic solution. The requirement on periodicity in t replaces the initial condition in this case.

Thus, the function $u(x, t)$ can be expanded in a Fourier series in t whose coefficients are functions of x (note that we are expanding $u(x, t)$ as a function of t with an arbitrary but fixed x), i.e.

$$u(x, t) = \frac{a_0(x)}{2} + \sum_{k=1}^{\infty}\left(a_k(x)\cos kt + b_k(x)\sin kt\right). \qquad (9.23)$$

Consider the trigonometric series

$$\tilde{u}(x, t) = \frac{a_0(x)}{2} + \sum_{k=1}^{\infty}\left(b_k(x)\cos kt - a_k(x)\sin kt\right), \qquad (9.24)$$

which, except for the sign and the term $a_0(x)/2$, is identical with the adjoint series for (9.23). Since $u(0, t) = u(\pi, t) = 0$, we have also $\partial u(0, t)/\partial t = \partial u(\pi, t)/\partial t = 0$. Consequently, $\partial u(x, t)/\partial t \to 0$ uniformly as $x \to 0-$ and $x \to \pi+$, so that also $\tilde{u}(0, t) = \tilde{u}(\pi, t) = 0$. (See Problem 5.18.)

If ε is a positive number, multiply equation (9.22) by the function $u + \varepsilon\tilde{u}$ and integrate it over the rectangle $x \in [0, \pi]$, $t \in [0, 2\pi]$. Using the Fubini theorem, the integration by parts and the fact that the functions u, $\tilde{u}$ and $\partial u/\partial t$ vanish for $x = 0$ and $x = \pi$, we obtain

$$-\int_0^{2\pi} \int_0^\pi f \cdot (u + \varepsilon \tilde{u}) \, dx \, dt$$

$$= -\int_0^{2\pi} \int_0^\pi \left[\frac{\partial}{\partial x} \left(a(x) \frac{\partial u}{\partial x} \right) - \frac{\partial u}{\partial t} \right] u \, dx \, dt$$

$$- \varepsilon \int_0^{2\pi} \int_0^\pi \left[\frac{\partial}{\partial x} \left(a(x) \frac{\partial u}{\partial x} \right) - \frac{\partial u}{\partial t} \right] \tilde{u} \, dx \, dt$$

$$= \int_0^{2\pi} \int_0^\pi a(x) \left(\frac{\partial u}{\partial x} \right)^2 dx \, dt + \varepsilon \int_0^{2\pi} \int_0^\pi a(x) \frac{\partial u}{\partial x} \frac{\partial \tilde{u}}{\partial x} \, dx \, dt$$

$$+ \varepsilon \int_0^{2\pi} \int_0^\pi \frac{\partial u}{\partial t} \tilde{u} \, dx \, dt, \qquad (9.25)$$

since

$$\int_0^{2\pi} \int_0^\pi \frac{\partial u}{\partial t} u \, dx \, dt = \frac{1}{2} \int_0^{2\pi} \int_0^\pi \frac{\partial u^2}{\partial t} \, dx \, dt$$

$$= \frac{1}{2} \int_0^\pi [u^2(x, 2\pi) - u^2(x, 0)] \, dx = 0$$

due to periodicity of the function u.

The reader can easily verify that the following identities are true. (Note that the proof follows immediately from Parseval's equality and the representations of u and $\tilde{u}$ given by (9.23) and (9.24), respectively.)

$$J_1(x) = \int_0^{2\pi} \left(\frac{\partial u}{\partial x}(x, t) \right)^2 dt = \int_0^{2\pi} \left(\frac{\partial \tilde{u}}{\partial x}(x, t) \right)^2 dt$$

$$= \pi \left[\frac{1}{2} \left(\frac{da_0}{dx} \right)^2 + \sum_{k=1}^\infty \left(\left(\frac{da_k}{dx} \right)^2 + \left(\frac{db_k}{dx} \right)^2 \right) \right];$$

APPLICATION OF FOURIER SERIES

$$J_2(x) = \int_0^{2\pi} u^2(x,t)\,dt = \int_0^{2\pi} \tilde{u}^2(x,t)\,dt$$
$$= \pi[\tfrac{1}{2}a_0^2(x) + \sum_{k=1}^{\infty}(a_k^2(x) + b_k^2(x))];$$

$$J_3(x) = \int_0^{2\pi} \frac{\partial u}{\partial t}(x,t)\,\tilde{u}(x,t)\,dt = \pi \sum_{k=1}^{\infty} k(a_k^2(x) + b_k^2(x)).$$

Integrating the expressions for $J_1(x)$, $J_2(x)$ and $J_3(x)$ with respect to x within the limits 0 and π, we get certain numbers denoted by

$$\|u\|_1^2, \quad \|u\|_0^2, \quad \|u\|_{1/2}^2.$$

(Observe that the last number can be expressed directly in terms of the function u, see Theorem 7.20.) Next, estimating the integrals in (9.25) by the Hölder inequality (see Example 3.8) and the inequalities (9.21), which in this case claim that $\alpha \geq a(x) \geq 1/\alpha > 0$, we obtain

$$\left| \int_0^{2\pi}\int_0^{\pi} f \cdot (u + \varepsilon \tilde{u})\,dx\,dt \right|$$
$$\leq \left(\int_0^{2\pi}\int_0^{\pi} |f|^2\,dx\,dt \right)^{1/2} \left(\int_0^{2\pi}\int_0^{\pi} |u + \varepsilon \tilde{u}|^2\,dx\,dt \right)^{1/2}$$
$$= \|f\|_0 \|u + \varepsilon \tilde{u}\|_0 \leq \|f\|_0 (\|u\|_0 + \varepsilon \|\tilde{u}\|_0)$$
$$= \|f\|_0 (1 + \varepsilon) \|u\|_0 \leq C_1(1 + \varepsilon) \|f\|_0 \|u\|_1.$$

Note that the last inequality with $\|u\|_0$ estimated by $\|u\|_1$ follows from the results of Problem 9.1. Furthermore,

$$\int_0^{2\pi}\int_0^{\pi} a(x)\left(\frac{\partial u}{\partial x}\right)^2 dx\,dt \geq \frac{1}{\alpha}\|u\|_1^2,$$

$$\left| \int_0^{2\pi}\int_0^{\pi} a(x) \frac{\partial u}{\partial x} \frac{\partial \tilde{u}}{\partial x}\,dx\,dt \right|$$
$$\leq \alpha \left(\int_0^{2\pi}\int_0^{\pi}\left(\frac{\partial u}{\partial x}\right)^2 dx\,dt \right)^{1/2} \left(\int_0^{2\pi}\int_0^{\pi}\left(\frac{\partial \tilde{u}}{\partial x}\right)^2 dx\,dt \right)^{1/2} = \alpha \|u\|_1^2.$$

Hence, by these estimates we get from (9.25),

$$\frac{1}{\alpha}\|u\|_1^2 - \varepsilon\alpha\|u\|_1^2 + \varepsilon\|u\|_{1/2}^2 \leq C_1(1+\varepsilon)\|f\|_0\|u_1\|.$$

Choosing $\varepsilon = \alpha^{-2}/2$ we get after certain rearrangements the inequality

$$\|u\|_1^2 + \|u\|_{1/2}^2 \leq C_2\|f\|_0\|u\|_1.$$

Finally, the inequality $ab \leq \frac{1}{2}(a^2 + b^2)$ applied to the right hand side of the previous inequality yields

$$\tfrac{1}{2}\|u\|_1^2 + \|u\|_{1/2}^2 \leq C_3\|f\|_0^2, \tag{9.26}$$

which is the sought estimate for the solution u in terms of the given quantities, i.e. by the right hand side f, because other quantities are zero in our case. Note that $\|u\|_{1/2}$ is a norm of the "$\tfrac{1}{2}$-derivative" with respect to t in the space $L_2([0, \pi] \times [0, 2\pi])$.

9.4.2 If the solid Ω is homogeneous and isotropic, the coefficients $a_{ij}(x)$ obey a very simple law, i.e. $a_{ij}(x) = \delta_{ij}$ for all x, where δ_{ij} is the Kronecker's delta.* If we choose the square $[0, \pi] \times [0, \pi]$ for Ω, equation (9.20) assumes the form

$$\frac{\partial^2 u}{\partial x^2} + \frac{\partial^2 u}{\partial y^2} - \frac{\partial u}{\partial t} = f(x, y, t). \tag{9.27}$$

Let the right hand side $f(x, y, t)$ be a periodic function in t with period 2π; thus, f can be expanded in a *triple Fourier series*

$$f(x, y, t) = \sum_{\substack{k,l=1 \\ m=0}}^{\infty} (A_{klm} \cos mt + B_{klm} \sin mt) \sin kx \sin ly. \tag{9.28}$$

Let us look for a *periodic* solution of (9.27) with the right hand side given by (9.28), which satisfies the *zero boundary conditions*.

---

* $\delta_{ij} = \begin{cases} 1 & \text{for } i = j, \\ 0 & \text{for } i \neq j. \end{cases}$

APPLICATION OF FOURIER SERIES

The solution $u(x, y, t)$ is assumed to have the form

$$u(x, y, t) = \sum_{\substack{k,l=1 \\ m=0}}^{\infty} (a_{klm} \cos mt + b_{klm} \sin mt) \sin kx \sin ly \qquad (9.29)$$

with unknown coefficients a_{klm} and b_{klm}.

A formal substitution of the series (9.29) into (9.27) and a comparison of coefficients yields the relations

$$-a_{klm}(k^2 + l^2) - b_{klm}m = A_{klm},$$
$$-b_{klm}(k^2 + l^2) + a_{klm}m = B_{klm}.$$

By solving these equations we get

$$a_{klm} = \frac{-(k^2 + l^2) A_{klm} + mB_{klm}}{(k^2 + l^2)^2 + m^2};$$

$$b_{klm} = \frac{-(k^2 + l^2) B_{klm} - mA_{klm}}{(k^2 + l^2)^2 + m^2}. \qquad (9.30)$$

We deduce easily that

$$((k^2 + l^2)^2 + m^2)(a_{klm}^2 + b_{klm}^2) = (A_{klm}^2 + B_{klm}^2).$$

Summing up these equalities for all values k, l and m and using Parseval's equality, we obtain in fact an estimate for the second derivatives of u with respect to x and y and the first derivative with respect to t in the norm of the space $L_2([0, \pi] \times [0, \pi] \times [0, 2\pi])$, i.e.

$$\left\|\frac{\partial^2 u}{\partial x^2}\right\|^2 + 2\left\|\frac{\partial^2 u}{\partial x \partial y}\right\|^2 + \left\|\frac{\partial^2 u}{\partial y^2}\right\|^2 + \left\|\frac{\partial u}{\partial t}\right\|^2 \leqq C\|f\|^2.$$

This is the required inequality which permits us to prove that the function $u(x, y, t)$ given by series (9.29) with coefficients a_{klm} and b_{klm} defined by formulas (9.30) is a generalized solution of our problem. The proof follows the same pattern as in Section 9.2: we approximate the right hand side $f \in L_2$ by partial sums of its triple Fourier series and for these sums we find the corresponding classical solutions, represented by partial sums of series (9.29).

This example illustrates also the procedure of finding the formal solution by employing multiple Fourier series.

9.4.3 Let us now consider the non-periodic solutions of the heat conduction equation. We shall outline briefly the method of determination of a formal solution.

We are going to solve the equation with the right hand side equal to zero, so that

$$\frac{\partial^2 u}{\partial x^2} = \frac{\partial u}{\partial t} \tag{9.31}$$

for $x \in [0, \pi]$ and $t \geq 0$, with boundary conditions $u(0, t) = u(\pi, t) = 0$ and the initial condition $u(x, 0) = f(x)$. This problem describes the temperature distribution in a homogeneous bar of length π, without internal heat-sources, whose ends are kept at zero temperature and whose initial temperature is given by f. The solution is again sought in the form

$$u(x, t) = X(x) T(t).$$

From the partial differential equation (9.31) we get two ordinary differential equations, i.e.

$$T'(t) + \lambda^2 T(t) = 0,$$

for T with the solution

$$T(t) = C e^{-\lambda^2 t}$$

and

$$X''(x) + \lambda^2 X(x) = 0$$

with boundary conditions $X(0) = X(\pi) = 0$ for X, which has the solution

$$X(x) = A \sin \lambda x,$$

where $\lambda = 1, 2, 3, \ldots$ (see Section 9.1 and Chapter 2, Section 2.1). Thus, we have obtained a set of solutions of (9.31) satisfying the boundary conditions,

$$u_k(x, t) = b_k e^{-k^2 t} \sin kx, \quad k = 1, 2, 3, \ldots.$$

Consequently, we look for the solution $u(x, t)$ in the form

$$u(x, t) = \sum_{k=1}^{\infty} b_k e^{-k^2 t} \sin kx, \tag{9.32}$$

where the coefficients b_k are to be determined from the initial condition

$$u(x, 0) = \sum_{k=1}^{\infty} b_k \sin kx = f(x), \tag{9.33}$$

i.e. b_k are the Fourier coefficients in the sine-expansion of $f(x)$ on $[0, \pi]$.

Series (9.32) converges for $t > 0$ uniformly on any interval $[t_1, t_2]$ for all x together with all its derivatives due to the factor $e^{-k^2 t}$. Hence, the function $u(x, t)$ satisfies the equation and the boundary condition for $t > 0$. However, if the function $f(x)$, giving the initial values, is not sufficiently smooth, we encounter certain difficulties with the type of convergence of the formal solution $u(x, t)$ to $f(x)$ as $t \to 0$.

The existence of a generalized solution can be proved, for example, by the maximum principle, which states that for a classical solution we have the estimate

$$\max_{\substack{x \in [0, \pi] \\ t \geq 0}} u(x, t) \leq \max_{x \in [0, \pi]} f(x),$$

i.e. the classical solution $u(x, t)$ attains its maximum (and consequently, also its minimum), for $t = 0$ provided the boundary conditions are zero.

Every partial sum of series (9.32) is a classical solution corresponding to the initial condition given by a partial sum of series (9.33) for $f(x)$. Now, if $f(x)$ is α-Hölder-continuous and, in addition to it, satisfies the conditions $f(0) = f(\pi) = 0$, then the partial sums of series (9.33) converge to $f(x)$ uniformly (see Chapter 7, Remark 7.5), and consequently, by the maximum principle, also series (9.32) converges uniformly to a continuous function $u(x, t)$ which is the generalized solution corresponding to the α-Hölder-continuous function $f(x)$.

However, if in general the function $f(x)$ is an element from the space $L_2(0, \pi)$, we can only prove that series (9.32) converges in $L_2([0, \pi] \times [0, \infty))$. The proof of this may be carried out in a similar manner to the proof in the next section.

9.4.4 Let us solve again equation (9.31), yet now with boundary conditions

$$u(0, t) = 0, \quad \frac{\partial u}{\partial x}(\pi, t) + h\, u(\pi, t) = 0. \tag{9.34}$$

The physical meaning of the conditions (9.34) is as follows. We consider again a homogeneous bar of length π without internal heat sources, whose left end is maintained at zero temperature, and heat at its right end is exchanged freely with the environment having zero temperature; the constant h is called the *coefficient of the external thermal conductivity*. Putting $u(x, t) = X(x)\, T(t)$, we have

$$T(t) = C\mathrm{e}^{-\lambda^2 t}$$

and

$$X(x) = A \cos \lambda x + B \sin \lambda x.$$

The boundary condition $X(0) = 0$ yields $A = 0$, the condition concerning the other end reads $X'(\pi) + h\, X(\pi) = 0$ and leads to the equation

$$\lambda \cos \lambda \pi + h \sin \lambda \pi = 0,$$

i.e.

$$\tan \lambda \pi = -\frac{\lambda}{h}.$$

Thus, if we put $\lambda_k = \xi_k/\pi$, $k = 1, 2, \ldots$, where ξ_k are the positive solutions of the transcendental equation

$$\tan \xi = -\frac{1}{h\pi}\xi,$$

$(0 < \xi_1 < \xi_2 < \ldots; (i - \tfrac{1}{2})\pi < \xi_i < (i + \tfrac{1}{2})\pi),$

APPLICATION OF FOURIER SERIES

we get formally for $u(x, t)$,

$$u(x, t) = \sum_{k=1}^{\infty} b_k e^{-\lambda_k^2 t} \sin \lambda_k x . \tag{9.35}$$

Introducing this into the initial condition $u(x, 0) = f(x)$, we have

$$\sum_{k=1}^{\infty} b_k \sin \frac{\xi_k}{\pi} x = f(x), \tag{9.36}$$

i.e. b_k are the Fourier coefficients of the function $f(x)$ expanded in the complete orthogonal system of functions $\sin (\xi_k/\pi) x$, $k = 1, 2, \ldots$, (see Chapter 4, Example 4.10), and are given by

$$b_k = \sigma_k^2 \int_0^\pi f(x) \sin \frac{\xi_k}{\pi} x \, dx ,$$

where

$$\frac{1}{\sigma_k^2} = \int_0^\pi \sin^2 \frac{\xi_k}{\pi} x \, dx .$$

The factor $e^{-(\xi_k/\pi)^2 t}$ guarantees again that the function $u(x, t)$ given by series (9.35) possesses a derivative of any order and satisfies the equation together with boundary conditions for $t > 0$. The only difficulty lies again with the initial condition, i.e. how to interpret the statement "the function $u(x, t)$ attains the value $f(x)$ for $t = 0$", if the function f is not smooth enough.

In order to show that series (9.35) is a generalized solution in the sense of convergence in $L_2([0, \pi] \times [0, \infty))$, we have to derive the corresponding inequality. To this purpose, let us rewrite series (9.36) in a *normalized form*,

$$f(x) = \sum_{k=1}^{\infty} b_k \sin \frac{\xi_k}{\pi} x = \sum_{k=1}^{\infty} \beta_k \varphi_k(x) ,$$

where the functions

$$\varphi_k(x) = \sigma_k \sin \frac{\xi_k}{\pi} x$$

constitute an orthonormal system in $L_2(0, \pi)$. Then,

$$u(x, t) = \sum_{k=1}^{\infty} \beta_k e^{-\lambda_k^2 t} \varphi_k(x).$$

Using Parseval's equality for fixed t, we have

$$\int_0^{\pi} u^2(x, t) \, \mathrm{d}x = \sum_{k=1}^{\infty} \beta_k^2 e^{-2\lambda_k^2 t};$$

integrating with respect to t over $(0, \infty)$, we get (the interchanging of summation and integration is permissible by Theorem 1.8)

$$\|u\|_{L_2([0,\pi]\times[0,\infty))}^2 = \sum_{k=1}^{\infty} \beta_k^2 \frac{1}{2\lambda_k^2} \leq \frac{1}{2\lambda_1^2} \sum_{k=1}^{\infty} \beta_k^2 = \frac{1}{2\lambda_1^2} \|f\|_{L_2(0,\pi)},$$

which is the required inequality stating that the convergence of partial sums f_n of the series (9.36) in the space $L_2(0, \pi)$ implies the convergence of the corresponding classical solutions u_n — partial sums of the series (9.35) — to a continuous generalized solution $u(x, t)$ in the norm of the space $L_2([0, \pi] \times [0, \infty))$.

This example also shows how the Fourier method can be applied, if we use orthogonal systems other than just the set of functions $\sin kx$ and $\cos kx$, $k = 0, 1, 2, \ldots$.

9.5 LAPLACE EQUATION

If the solution of the heat conduction equation (9.20) is independent of time, then $\partial u / \partial t = 0$ and the initial condition is meaningless. Thus for $u = u(x)$, we get an equation of *elliptic type*

$$\sum_{i,j=1}^{n} \frac{\partial}{\partial x_i} \left(a_{ij}(x) \frac{\partial u}{\partial x_j} \right) = f(x), \quad x = (x_1, x_2, \ldots, x_n) \in \Omega. \quad (9.37)$$

If the solid Ω is homogeneous and isotropic, then $a_{ij}(x) = \delta_{ij}$ and equation (9.37) reduces to the *Laplace equation*

$$\Delta u \equiv \sum_{i=1}^{n} \frac{\partial^2 u}{\partial x_i^2} = f(x).$$

Note that elliptic equations are obtained not only by analysing the temperature distribution in a solid but also by describing various mechanical phenomena and other physical processes as shown in the following example.

9.5.1 Consider a square-shaped diaphragm ($0 \leq x \leq \pi$, $0 \leq y \leq \pi$) clamped at the edges and loaded by external forces with density $f(x, y)$. The displacement of the diaphragm is governed by the equation

$$\frac{\partial^2 u}{\partial x^2} + \frac{\partial^2 u}{\partial y^2} = f(x, y)$$

with boundary conditions

$$u(x, 0) = u(x, \pi) = u(0, y) = u(\pi, y) = 0.$$

Writing the right hand side $f(x, y)$ in the form of a double Fourier series

$$f(x, y) = \sum_{k,l=1}^{\infty} a_{kl} \sin kx \sin ly,$$

we get by using the method of Section 9.4.2 the formal solution

$$u(x, y) = -\sum_{k,l=1}^{\infty} \frac{a_{kl}}{k^2 + l^2} \sin kx \sin ly. \qquad (9.38)$$

In the same manner as above we can derive the inequality

$$\left\| \frac{\partial^2 u}{\partial x^2} \right\|_{L_2}^2 + 2 \left\| \frac{\partial^2 u}{\partial x \, \partial y} \right\|_{L_2}^2 + \left\| \frac{\partial^2 u}{\partial y^2} \right\|_{L_2}^2 \leq C \|f\|_{L_2}^2,$$

which enables us to interpret the series (9.38) as a generalized solution in the sense of convergence of second derivatives in $L_2([0, \pi] \times [0, \pi])$.

Remark: The same result can be arrived at, if in the example in Section 9.4.2 the function u is considered to be independent of time.

9.5.2 Consider a homogeneous and isotropic circular disc of unit radius whose circumference is kept at a time-independent temperature g. The steady temperature distribution u in the disc, which does not contain any heat sources, is governed by the Laplace equation

$$\frac{\partial^2 u}{\partial x^2} + \frac{\partial^2 u}{\partial y^2} = 0 \tag{9.39}$$

with the boundary condition $u(x, y) = g(x, y)$ for $x^2 + y^2 = 1$.

Introducing polar coordinates by $x = r \cos \varphi$, $y = r \sin \varphi$, equation (9.39) reduces to

$$\frac{\partial^2 u}{\partial r^2} + \frac{1}{r^2} \frac{\partial^2 u}{\partial \varphi^2} + \frac{1}{r} \frac{\partial u}{\partial r} = 0 \tag{9.40}$$

with the boundary condition

$$u(1, \varphi) = g(\varphi), \quad \varphi \in [-\pi, \pi].$$

Put again

$$u(r, \varphi) = R(r) \Phi(\varphi).$$

From (9.40) it follows that

$$\frac{r^2 R'' + rR'}{R} = -\frac{\Phi''}{\Phi} = \lambda = \text{const}$$

and consequently,

$$\Phi'' + \lambda \Phi = 0,$$
$$r^2 R'' + rR' - \lambda R = 0.$$

From the first equation we have

$$\Phi(\varphi) = A \cos \sqrt{(\lambda)}\, \varphi + B \sin \sqrt{(\lambda)}\, \varphi.$$

The function $u(r, \varphi)$ and therefore, also the function $\Phi(\varphi)$ must be, for obvious reasons, periodic in φ with period 2π. This is possible only if $\sqrt{\lambda} = k$ with k being a non-negative integer, and consequently,

$$\Phi_k(\varphi) = A_k \cos k\varphi + B_k \sin k\varphi.$$

The function $R(r)$ will be sought for in the form $r^\varkappa$, which, being introduced into the equation for R, yields the condition

$$k^2 = \varkappa^2, \quad \text{i.e.} \quad \varkappa = \pm k,$$

so that

$$R_k(r) = C_k r^k + D_k r^{-k}.$$

Because the temperature cannot be infinite at the origin, we have to set $D_k = 0$; thus, we have the *formal* solution in the form of an infinite series

$$u(r, \varphi) = \sum_{k=0}^{\infty} r^k (A_k \cos k\varphi + B_k \sin k\varphi), \qquad (9.41)$$

called the *Poisson series*.

From the boundary condition we get for $r = 1$ the formal relation

$$\sum_{k=0}^{\infty} (A_k \cos k\varphi + B_k \sin k\varphi) = g(\varphi),$$

i.e. A_k and B_k are the Fourier coefficients of the function g on the interval $[-\pi, \pi]$. Substituting the expressions (2.13) for these coefficients into series (9.41) for $r < 1$ and interchanging the order of integration and summation we get the formula

$$u(r, \varphi) = \frac{1}{\pi} \int_{-\pi}^{\pi} g(\psi) \left[\tfrac{1}{2} + \sum_{k=1}^{\infty} r^k \cos k(\varphi - \psi) \right] d\psi.$$

Note that it is permissible to interchange the integration and summation. Actually, if the function g is at least integrable, then the coefficients A_k and B_k are bounded in magnitude by the same constant, and consequently, the Poisson series converges uniformly for $r < 1$. Since

$$\frac{1}{2} + \sum_{k=1}^{\infty} r^k \cos k(\varphi - \psi) = \frac{1}{2} \cdot \frac{1 - r^2}{1 - 2r \cos(\varphi - \psi) + r^2}$$

for $r < 1$ (see Problem 1.12), we have

$$u(r, \varphi) = \frac{1}{2\pi} \int_{-\pi}^{\pi} g(\psi) \frac{1 - r^2}{1 - 2r \cos(\varphi - \psi) + r^2} \, d\psi. \quad (9.42)$$

The integral in (9.42) is called the *Poisson integral*; it furnishes an explicit form of the solution of the Laplace equation inside a unit circle in terms of the function g.

The factors r^k in the Poisson series (9.41) ensure again that the sum of the series has a derivative of any order inside the unit circle and satisfies equation (9.40) there.

The explicit representation of the solution $u(r, \varphi)$ by the Poisson integral permits us to obtain various estimates for the solution using the function g, i.e. from the given quantities, and consequently, define different types of generalized solutions by their dependence on properties of g as well as to prove their existence.

The Poisson integral is useful not only in solving the Laplace equation but also in the theory of Fourier series itself, where it enables us to introduce a *new summation method*. Actually, let a function $g(\varphi)$ be given by its Fourier coefficients A_k and B_k. From these coefficients let us construct the corresponding Poisson series (9.41) for $r < 1$. Since for an integrable function g all the coefficients A_k and B_k are bounded in magnitude by a constant, the Poisson series converges uniformly for $r < 1$ to a continuous function $u(r, \varphi)$. Now we can raise the question under what conditions for g the function $u(r, \varphi)$ converges to the function $g(\varphi)$ as $r \to 1$ and in what sense this convergence should be understood. There is a number of theorems dealing with the limiting process $r \to 1$ which are analogous to theorems regarding the convergence of arithmetic means of partial sums of Fourier series for $g(\varphi)$, (the so-called Fejér's or Cesàro's summation, see Chapter 7, Section 7.9). For example, we have the following propositions.

If the function $g(\varphi)$ is continuous, then $u(r, \varphi)$ converges to the function $g(\varphi)$ as $r \to 1$ uniformly with respect to $\varphi \in [-\pi, \pi]$.

Applying this proposition to the solution of our boundary value problem we conclude that the solution $u(r, \varphi)$ is continuous on the entire closed disc and assumes the prescribed values $g(\varphi)$ on its boundary.

If the function $g(\varphi)$ is α-Hölder-continuous, we can use the maximum principle similarly as in Section 9.4.3 and prove by a procedure not using the Poisson integral that the corresponding solution $u(r, \varphi)$ is continuous on the closed disc.

9.6 SOME APPLICATIONS OF FOURIER SERIES IN THE THEORY OF INTEGRAL EQUATIONS*

The equation

$$y(x) - \lambda \int_a^b K(x, s)\, y(s)\, ds = f(x) \qquad (9.43)$$

is called the *Fredholm integral equation of the second kind*. Here $f \in L_2(a, b)$ is a given right hand side, $K(x, s)$ is a given function called the *kernel*, which satisfies the condition

$$B^2 = \int_a^b \int_a^b |K(x, s)|^2\, dx\, ds < \infty \qquad (9.44)$$

and the unknown function y is in general looked for in the space

---

* The proofs of some propositions from the theory of integral equations, which will be used in this section, may be found in [8].

$L_2(a, b)$. The complex parameter λ is called the *eigenvalue* of the kernel $K(x, s)$ provided the homogeneous equation

$$y(x) - \lambda \int_a^b K(x, s)\, y(s)\, \mathrm{d}s = 0 \tag{9.45}$$

has non-trivial solutions $y_1(x), y_2(x), \ldots, y_m(x)$, $(m \geq 1)$ in addition to the trivial solution $y(x) \equiv 0$. We assume that the functions $y_1(x), y_2(x), \ldots, y_m(x)$ are linearly independent and call them the *eigenfunctions* of the kernel $K(x, s)$ corresponding to the eigenvalue λ.

9.6.1 *Equation with a degenerate kernel.* If the kernel $K(x, s)$ has the particular form

$$K(x, s) = \sum_{i=1}^{p} a_i(x)\, b_i(s),$$

where $a_1, a_2, \ldots, a_p$ and $b_1, b_2, \ldots, b_p$ are p-tuples of linearly independent functions, then equation (9.43) can be easily solved by reducing it to an algebraical system of p equations with p unknowns $c_1, c_2, \ldots, c_p$. The solution then has the form

$$y(x) = f(x) + \lambda \sum_{i=1}^{p} c_i\, a_i(x).$$

This fact can advantageously be used for establishing the approximate solution of the integral equation (9.43). The idea is to replace the general kernel $K(x, s)$ by a degenerate one. Thus, we get a new integral equation

$$z(x) - \lambda \int_a^b G(x, s)\, z(s)\, \mathrm{d}s = f(x) \tag{9.46}$$

with a degenerate kernel $G(x, s)$, whose solution $z(x)$ approximates the solution $y(x)$ of equation (9.43) with an error which can be, estimated by the error due to replacing the kernel $K(x, s)$ by $G(x, s)$. If λ is sufficiently small in magnitude, we have, for example the estimate

$$\int_a^b |y(x) - z(x)|^2 \, dx$$

$$\leq \frac{2|\lambda|^2}{1 - 2|\lambda|^2 B^2} \int_a^b |z(x)|^2 \, dx \int_a^b \int_a^b |K(x,s) - G(x,s)|^2 \, dx \, ds. \quad (9.47)$$

The degenerate kernel $G(x, s)$ can be constructed in different ways. Since the kernel $K(x, s)$ belongs to the space $L_2([a, b] \times [a, b])$ due to the property (9.44), it can be expanded in a double Fourier sine series

$$K(x, s) \approx \sum_{i,k=1}^{\infty} A_{ik} \sin \frac{i\pi(x - a)}{b - a} \sin \frac{k\pi(s - a)}{b - a};$$

then a partial sum of this series can be taken for $G(x, s)$. The reader can easily verify that this kernel is actually a degenerate one; thus, equation (9.46) can be solved in a quite simple manner. If the partial sum $G(x, s)$ contains a sufficiently large number of terms, the number

$$\int_a^b \int_a^b |K(x, s) - G(x, s)|^2 \, dx \, ds$$

is small, and consequently, the difference of the functions $y(x)$ and $z(x)$ is small in the norm of the space $L_2(a, b)$. (See formula (9.47).)

Choosing the degenerate kernel as described, i.e. as a partial sum of the Fourier series for the kernel $K(x, s)$, we can benefit from the orthogonality relations for trigonometric functions and simplify the calculations. This fact is one of the advantages of the choice considered.

Remark. The above approximation method is significant also from a theoretical point of view. For example, it can be shown that every equation (9.43) is equivalent to a certain equation with a degenerate kernel; consequently, some properties of a solution can be established from this fact.

9.6.2 *Equations with a symmetric kernel.* A kernel $K(x, s)$ will be called *symmetric*, if

$$K(x, s) = \overline{K(s, x)}.$$

It is known that the eigenvalues of a symmetric kernel are real numbers, are infinitely many provided the kernel is non-degenerate and can be ordered in a sequence by their magnitudes

$$|\lambda_1| \leq |\lambda_2| \leq \ldots \leq |\lambda_n| \leq \ldots,$$

such that $|\lambda_n| \to \infty$ as $n \to \infty$. Then, of course, the system of eigenfunctions is also infinite, and moreover, the eigenfunctions $y_i(x), (i = 1, 2, \ldots)$ can be chosen so that they constitute a *complete orthonormal set* in $L_2(a, b)$*. Thus, we can expand functions from $L_2(a, b)$ in a Fourier series by using the orthonormal eigenfunctions of the symmetric kernel $K(x, s)$.

Let us assume for simplicity that to each eigenvalue λ_i there corresponds *exactly one* eigenfuction $y_i(x)$. We have the

Hilbert-Schmidt Theorem: *If $K(x, s)$ is a symmetric kernel and $h(x)$ an arbitrary function from $L_2(a, b)$, then the function*

$$g(x) = \int_a^b K(x, s)\, h(s)\, \mathrm{d}s$$

can be expanded in a Fourier series by the orthonormal eigenfunctions $y_k(x)$ of the kernel $K(x, s)$, and

$$g(x) = \sum_{k=1}^{\infty} \frac{h_k}{\lambda_k} y_k(x)$$

in the sense of convergence in $L_2(a, b)$, where $h_k = (h, y_k)$** are the Fourier coefficients of the function h.

---

* The situation here is in a certain extent analogous to that in Section 4.5, Chapter 4 (see Theorem 4.10).

** The symbol (f, g) denotes, as usual, the scalar product in $L_2(a, b)$.

APPLICATION OF FOURIER SERIES

Using the theorem we can express explicitly the solution of the integral equation (9.43) with a symmetric kernel provided the eigenvalues and the corresponding eigenfunctions are known.

(a) First assume that λ is not an eigenvalue, i.e. $\lambda \neq \lambda_i$ for all i. Then a solution $y \in L_2(a, b)$ of the equation (9.43) exists for any right hand side $f \in L_2(a, b)$. The function

$$g(x) = \int_a^b K(x, s) y(s) \, \mathrm{d}s$$

is again an element from $L_2(a, b)$, and, by the Hilbert-Schmidt theorem,

$$\int_a^b K(x, s) y(s) \, \mathrm{d}s = \sum_{k=1}^\infty \frac{c_k}{\lambda_k} y_k(x),$$

where $c_k = (y, y_k)$ are the unknown (Fourier) coefficients of the sought function y. Introducing this representation into the integral equation (9.43), we obtain the relation

$$y(x) - \lambda \sum_{k=1}^\infty \frac{c_k}{\lambda_k} y_k(x) = f(x) \tag{9.48}$$

which defines the function $y(x)$, if, of course, the constants c_k are known. However, the c_ks can be determined very easily. To this purpose, form the scalar products of y_m with both sides of equation (9.48). By the orthogonality of the functions $y_k(x)$ it follows that

$$c_m \left(1 - \frac{\lambda}{\lambda_m}\right) = f_m = (f, y_m). \tag{9.49}$$

Since λ is not an eigenvalue, we have $\lambda \neq \lambda_m$, and consequently

$$c_m = \frac{\lambda_m f_m}{\lambda_m - \lambda} \quad (m = 1, 2, \ldots); \tag{9.50}$$

thus, the solution is given as a series involving the eigenvalues, eigenfunctions and the coefficients of the right hand side f, i.e.

$$y(x) = f(x) + \lambda \sum_{m=1}^\infty \frac{f_m}{\lambda_m - \lambda} y_m(x).$$

(b) Next, let λ be an eigenvalue, i.e. $\lambda = \lambda_p$ for some p. Then equation (9.43) is solvable only for such a right hand side f which is orthogonal to the eigenfunction $y_p(x)$ corresponding to the eigenvalue λ_p, i.e. $(f, y_p) = f_p = 0$. The relations (9.48) and (9.49) are valid again; however, the constants c_m can be determined uniquely from (9.49) only for $m \neq p$. For $m = p$ the relation (9.49) holds independently of the choice of c_p. Hence, here the solution $y(x)$ has the form

$$y(x) = f(x) + \lambda \sum_{\substack{m=1 \\ m \neq p}}^{\infty} \frac{f_m}{\lambda_m - \lambda} y_m(x) + c_p y_p(x),$$

where c_p is an arbitrary constant.

9.7 SOME APPLICATIONS OF FOURIER TRANSFORMS

In what follows we shall assume without any further comment that all functions whose Fourier transform is under consideration are such that the Fourier transform is actually meaningful (see Chapter 8). The Fourier transform of a function f will also be denoted by $\hat{f}$; the function $\hat{f}$ will be called the *Fourier image* of f and the function f the *original* for $\hat{f}$ under the Fourier transform.

9.7.1 Let us solve the equation of heat conduction

$$\frac{\partial^2 u}{\partial x^2} = \frac{\partial u}{\partial t} \qquad (9.51)$$

for the half-plane $t \geq 0$, $x \in (-\infty, \infty)$ with the initial condition $u(x, 0) = g(x)$, i.e. heat conduction in an infinite bar.

Choosing a fixed t let

$$U(\xi, t) = \frac{1}{2\pi} \int_{-\infty}^{+\infty} e^{-i\xi x} u(x, t) \, dx$$

be the Fourier transform of $u(x, t)$ considered as a function of x. By a formal differentiation under the integral sign and integration

APPLICATION OF FOURIER SERIES

by parts we can easily verify that

$$\widehat{\frac{\partial u}{\partial t}} = \frac{\partial U}{\partial t}(\xi, t), \quad \widehat{\frac{\partial^2 u}{\partial x^2}} = (i\xi)^2 U(\xi, t);$$

(see also Theorems 8.7 and 1.12); from equation (9.51) it follows then that $U(\xi, t)$ as a function of t must satisfy the ordinary differential equation

$$-\xi^2 U(\xi, t) = \frac{\partial U}{\partial t}(\xi, t) \qquad (9.52)$$

and consequently, have the form

$$U(\xi, t) = C(\xi) e^{-\xi^2 t}.$$

From the initial condition it follows for $t = 0$ that $U(\xi, 0) = \hat{g}(\xi)$, and therefore

$$U(\xi, t) = \hat{g}(\xi) e^{-\xi^2 t}.$$

If $f(x, t)$ denotes the original for the function $e^{-\xi^2 t}$ (t fixed), then the function $u(x, t)$ is a convolution of functions g and $f/2\pi$ by Theorem 8.10, i.e.

$$u(x, t) = \frac{1}{2\pi} \int_{-\infty}^{\infty} g(z) f(x - z, t) \, dz.$$

However, $f(x, t)$ is known; by Problem 8.5,

$$f(x, t) = \sqrt{\left(\frac{\pi}{t}\right)} e^{-x^2/4t},$$

and consequently,

$$u(x, t) = \frac{1}{2\sqrt{\pi}} \int_{-\infty}^{\infty} \frac{e^{-(x-z)^2/4t}}{\sqrt{t}} g(z) \, dz.$$

Hence, for the solution we have obtained formally an explicit representation similar to the Poisson integral. The solution just found will not be considered in detail here.

Remark. Observe that by applying the Fourier transform we reduced the partial differential equation (9.51) to an ordinary differential equation (9.52) containing a parameter ζ. This fact is often used in applications.

9.7.2 Consider an electrical circuit formed by a resistance R and an inductance L (see Fig. 9.2). Let an alternating voltage

$$u_\omega(t) = U(\omega)\, e^{i\omega t}$$

act on the circuit, i.e. a harmonic voltage with frequency ω and period $2\pi/\omega$. The impedance of the circuit is equal to $R + iL\omega$.

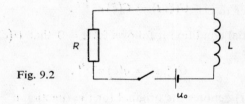

Fig. 9.2

Thus, for the current $i_\omega(t)$ flowing through the circuit we have by Ohm's law,

$$u_\omega(t) = (R + iL\omega)\, i_\omega(t).$$

Putting $i_\omega(t) = I(\omega)\, e^{i\omega t}$, we obtain

$$I(\omega) = U(\omega)\, \Phi(\omega),$$

where $\Phi(\omega) = (R + iL\omega)^{-1}$. Observe that multiplication by the function $\Phi(\omega)$ means a shift of the phase and a change of the magnitude. Now, if

$$u(t) = \sum_{j=1}^{n} U(\omega_j)\, e^{i\omega_j t},$$

then

$$i(t) = \sum_{j=1}^{n} I(\omega_j)\, e^{i\omega_j t} = \sum_{j=1}^{n} U(\omega_j)\, \Phi(\omega_j)\, e^{i\omega_j t}.$$

If
$$u(t) = \int_{-\infty}^{\infty} U(\omega)\, e^{i\omega t}\, d\omega,$$
we have
$$i(t) = \int_{-\infty}^{\infty} I(\omega)\, e^{i\omega t}\, d\omega = \int_{-\infty}^{\infty} U(\omega)\, \Phi(\omega)\, e^{i\omega t}\, d\omega.$$

Hence, for a given voltage we establish the current in such a way that we find the Fourier transform $U(\omega)$ of the voltage $u(t)$, divide $U(\omega)$ by the impedance $R + iL\omega$ and find the function $i(t)$ whose Fourier transform equals the function $U(\omega)\, \Phi(\omega)$.

In order to clarify the procedure just explained, let us solve the following example. If $u(t) = u_0$ for $t \in [0, T]$ and $u(t) = 0$ for the remaining t, i.e. a direct voltage u_0 is applied at the time $t = 0$ and switched off at $t = T$, then

$$U(\omega) = \frac{1}{2\pi} \int_{-\infty}^{\infty} u(t)\, e^{-i\omega t}\, dt = \frac{1}{2\pi} \frac{u_0}{i\omega}\left[1 - e^{-i\omega T}\right],$$

so that

$$I(\omega) = \frac{1}{2\pi} \frac{u_0}{i\omega(R + iL\omega)}\left[1 - e^{-i\omega T}\right]$$
$$= \frac{1}{2\pi} \frac{u_0}{R}\left[\frac{1}{i\omega} - \left(\frac{R}{L} + i\omega\right)^{-1}\right]\left[1 - e^{-i\omega T}\right].$$

Consequently,

$$i(t) = \int_{-\infty}^{\infty} I(\omega)\, e^{i\omega t}\, d\omega = \begin{cases} 0 & \text{for } t < 0, \\ \dfrac{u_0}{R}\left(1 - e^{-Rt/L}\right) & \text{for } t \in [0, T], \\ \dfrac{u_0}{R}\left(e^{RT/L} - 1\right) e^{-Rt/L} & \text{for } t > T \end{cases}$$

(see Fig. 9.3). The preceding formula may be obtained, e.g., by using formula 14. from Chap. 10, Part **III**.

The current corresponding to a unit impulse of voltage, i.e. a large direct voltage $1/T$ with short duration T, can be determined as follows. If t is fixed and $T \ll t$, then

$$i(t) = \frac{1}{TR} e^{-Rt/L} (e^{RT/L} - 1) = \frac{1}{L} e^{-Rt/L} \frac{e^{RT/L} - 1}{RT/L} \approx \frac{e^{-Rt/L}}{L},$$

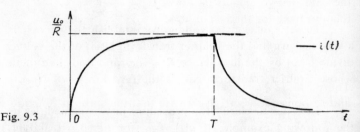

Fig. 9.3

because $(e^a - 1)/a \approx 1$ for a small value a. For $T \to 0$ we get a limit current $\varphi(t)$,

$$\varphi(t) = \begin{cases} (1/L) e^{-Rt/L} & \text{for } t > 0, \\ 0 & \text{for } t \leq 0. \end{cases}$$

The reader can easily verify that the function $\Phi(\omega)$ is the Fourier transform of $2\pi \varphi(t)$.

The relationships considered so far were particular cases of a general mapping realized by a physical system which produces a certain response while being fed by a signal. In the above example, the signal was a voltage, the response a current and the system itself — an electric circuit. Applying a periodic signal $F(\omega) e^{i\omega t}$ to a system, on which we impose certain slightly restrictive requirements such as absence of internal energy sources etc., we obtain a periodic response $G(\omega) e^{i\omega t}$ at its output. The relation between response and signal concerning their phase and magnitude is described by a complex factor $\Phi(\omega)$, usually called the *transmission characteristic* of the system in the frequency domain, i.e.

$$G(\omega) = \Phi(\omega) F(\omega).$$

APPLICATION OF FOURIER SERIES

Next, let $f(t)$ be an arbitrary signal expressed as a superposition of single harmonic signals, i.e.

$$f(t) = \int_{-\infty}^{\infty} F(\omega) e^{i\omega t} d\omega ;$$

if the system satisfies the superposition principle, i.e. if it maps the sum of two signals as the sum of their respective responses, then we have for the response,

$$g(t) = \int_{-\infty}^{\infty} G(\omega) e^{i\omega t} d\omega = \int_{-\infty}^{\infty} \Phi(\omega) F(\omega) e^{i\omega t} d\omega .$$

Thus, calling $F(\omega)$ and $G(\omega)$ the *spectral function* of the signal and response, respectively, we see that the spectral function of the response is equal to the spectral function of the signal multiplied by the transmission characteristic of the system. If $\varphi(t)$ signifies the function whose Fourier transform coincides with the transmission characteristic $\Phi(\omega)$, then (see Theorem 8.10)

$$g(t) = \frac{1}{2\pi} \int_{-\infty}^{\infty} f(s) \varphi(t-s) ds .$$

From the physical nature of the system functioning, it follows that in actual cases the value $g(t)$ can depend only on values of $f(s)$ for $s \leq t$, and consequently, $\varphi(t)$ must vanish for $t < 0$.

The function φ (or $\varphi/2\pi$) is called the *weight function of the unit impulse response*, because for the function

$$f_T(t) = \begin{cases} 1/T & \text{for } t \in [0, T], \\ 0 & \text{for } t \notin [0, T], \end{cases}$$

we have for almost every t,

$$\lim_{T \to 0} g_T(t) = \lim_{T \to 0} \frac{1}{2\pi} \int_{-\infty}^{\infty} f_T(s) \varphi(t-s) ds$$

$$= \frac{1}{2\pi} \lim_{T \to 0} \frac{1}{T} \int_0^T \varphi(t-s) ds = \frac{1}{2\pi} \varphi(t)$$

provided $\varphi(t)$ is integrable on every finite interval.

9.7.3 Let $f(x)$ be a function; assume for simplicity that $f(x)$ has a continuous derivative and vanishes for x sufficiently large in magnitude. The distribution of its values with respect to a chosen point $x = a$ is described by a "moment" $(x - a)f(x)$. If the value $f(x)$ is large for x close to a, the value $(x - a)f(x)$ may still be small; conversely, if $(x - a)f(x)$ is to be small for x distant from a, the value $f(x)$ has to be very small.

The mean quadratic dispersion of the function $f(x)$ with respect to a point a is defined by the formula

$$\delta_f(a) = \int_{-\infty}^{\infty} (x - a)^2 |f(x)|^2 \, dx. \tag{9.53}$$

The value a minimizing $\delta_f(a)$ is called the mean argument of the function f. Differentiating by the parameter a in formula (9.53) we can easily verify that for the mean argument a we have

$$\int_{-\infty}^{\infty} (x - a) |f(x)|^2 \, dx = 0.$$

Let a be the mean argument of a function f, and b be the mean argument of its Fourier transform F. Without loss of generality we can assume that $a = b = 0$, because instead of $f(x)$ we can consider the function $e^{-ib(x+a)} f(x + a)$ as indicated in Chapter 10, Part **III.**, formulas 2. and 3., and realize that $|f(x)| = |e^{-ib} f(x)|$.

Let α be a real parameter and let

$$\begin{aligned} I(\alpha) &= \int_{-\infty}^{\infty} |x f(x) + \alpha f'(x)|^2 \, dx \\ &= \int_{-\infty}^{\infty} (x f(x) + \alpha f'(x))(x \overline{f(x)} + \alpha \overline{f'(x)}) \, dx \\ &= \int_{-\infty}^{\infty} x^2 |f(x)|^2 \, dx + \alpha \int_{-\infty}^{\infty} x(f'(x)\overline{f(x)} + f(x)\overline{f'(x)}) \, dx \\ &\quad + \alpha^2 \int_{-\infty}^{\infty} |f'(x)|^2 \, dx. \end{aligned}$$

Since

$$f'(x)\overline{f(x)} + f(x)\overline{f'(x)} = \frac{d}{dx}|f(x)|^2 = \frac{d}{dx}(f(x)\overline{f(x)}),$$

we obtain by integration by parts,

$$\int_{-\infty}^{\infty} x(f'(x)\overline{f(x)} + f(x)\overline{f'(x)})\,dx$$
$$= \int_{-\infty}^{\infty} x\frac{d}{dx}|f(x)|^2\,dx = -\int_{-\infty}^{\infty} |f(x)|^2\,dx.$$

Moreover, by Parseval's equality for the function f' (see Theorem 8.6),

$$\int_{-\infty}^{\infty} |f'(x)|^2\,dx = 2\pi \int_{-\infty}^{\infty} \xi^2 |F(\xi)|^2\,d\xi$$

so that finally

$$I(\alpha) = \delta_f(0) - \alpha\|f\|_{L_2(-\infty,\infty)}^2 + 2\pi\alpha^2\,\delta_F(0). \tag{9.54}$$

Because $I(\alpha)$, being the integral of a non-negative function, is non-negative for all α, the discriminant of the quadratic expression (9.54) is non-positive. Hence,

$$\|f\|_{L_2(-\infty,\infty)}^4 - 8\pi\,\delta_f(0)\,\delta_F(0) \leq 0$$

so that for a normalized function f ($\|f\|_{L_2(-\infty,\infty)} = 1$) we have

$$\delta_f(0)\,\delta_F(0) \geq \frac{1}{8\pi}.$$

From this inequality it follows that the more a function f is concentrated around its mean argument, i.e. the smaller $\delta_f(a)$ is, the less its Fourier transform is concentrated around its mean argument, i.e. the larger $\delta_F(b)$ is. This phenomenon, called the uncertainty principle, plays an important role in quantum mechanics, radiotracking etc., where it is called Heisenberg's uncertainty principle. Crudely speaking, this principle claims that *the shorter a pulse is the more spread out its spectrum is.*

Problems

9.1. Let a function $u(x)$ be absolutely continuous on an interval $[a, b]$, and let $u(a) = 0$. Show that

$$|u(x)|^2 \leq (b - a) \int_a^b \left|\frac{du}{dx}(t)\right|^2 dt$$

and

$$\int_a^b |u(x)|^2 dx \leq (b - a)^2 \int_a^b \left|\frac{du}{dx}(t)\right|^2 dt.$$

(*Hint*: We have

$$|u(x)| = \left|\int_a^x \frac{du}{dx}(t) dt\right| \leq \int_a^b 1 \cdot \left|\frac{du}{dx}(t)\right| dt;$$

next we use the Schwarz (Hölder) inequality, see Example 3.8. The second inequality follows from the first one by integrating over the interval (a, b).)

9.2. Let $f(x) = 1 - |x|$ for $|x| \leq 1$ and $f(x) = 0$ for $|x| > 1$. Put $u(x, t) = f(x + t) + f(x - t)$. Show that $u(x, t)$ is a generalized solution of the equation of a string on the rectangle $|x| \leq 10$, $0 \leq t \leq 1$ with boundary conditions $u(-10, t) = u(10, t) = 0$ for $0 \leq t \leq 1$ and initial conditions $u(x, 0) = 2f(x)$ and $(\partial u/\partial t)(x, 0) = 0$ for $|x| \leq 10$.

(*Hint*: Approximate uniformly the function $f(x)$ by infinitely differentiable functions $f_n(x)$ such that $f_n(x) = 0$ for $|x| \geq 2$.)

9.3. Solve by the Fourier method the equation

$$\frac{\partial u}{\partial t} = a^2 \left(\frac{\partial^2 u}{\partial r^2} + \frac{1}{r}\frac{\partial u}{\partial r}\right)$$

for $t > 0$, $0 \leq r \leq R$ with initial condition $u(r, 0) = g(r)$ and boundary condition $u(R, t) = 0$. The equation describes heat conduction in a circular disc with radius R whose circumference is maintained at zero temperature. Moreover, it is assumed that the temperature depends only on the distance from the disc centre.

(*Hint*: Put $u(r, t) = H(r) T(t)$; then H satisfies the Bessel equation, which can be tackled with the results of Example 4.12.)

9.4. Solve the equation

$$\frac{\partial u}{\partial t} = \frac{\partial^2 u}{\partial x^2}$$

on $(0, l) \times (0, \infty)$ with boundary conditions

$$\frac{\partial u}{\partial x}(0, t) + u(0, t) = 0, \quad \frac{\partial u}{\partial x}(l, t) - u(l, t) = 0.$$

(*Hint*: Use the Fourier method and the results of Example 4.11.)

CHAPTER 10. SURVEY OF FOURIER SERIES AND FOURIER TRANSFORMS OF SOME COMMONLY USED FUNCTIONS

In practice, particularly in problems concerning increasing of convergence of Fourier series, we encounter the problem of finding the sum of some frequently occurring trigonometric series. Therefore, sums of several such series will be given in Part **I.** below; the majority of these series has already been considered in the text.

One of the most common applications is the problem of calculating the Fourier series of a given function. Part **II.** presents a survey of several functions, given by their graphs, and the corresponding Fourier series. (Beware the sum of the series at possible points of discontinuity!)

Further widely used applications concern Fourier transform. In Part **III.** a table of Fourier transforms of several functions and functional expressions is given.

I.

$$\sum_{n=1}^{\infty} \frac{\cos nx}{n} = -\log(2\sin\tfrac{1}{2}x) \quad \text{for} \quad 0 < x < 2\pi \tag{10.1}$$

(cf. Example 5.12);

$$\sum_{n=1}^{\infty} \frac{\cos nx}{n^2} = \frac{x^2}{4} - \frac{\pi x}{2} + \frac{\pi^2}{6} \quad \text{for} \quad 0 \leq x \leq 2\pi \tag{10.2}$$

(cf. Example 5.21);

$$\sum_{n=1}^{\infty} \frac{\cos nx}{n^3} = \int_0^x \left(\int_0^y \log(2\sin\tfrac{1}{2}t)\,dt \right) dy + \sum_{n=1}^{\infty} \frac{1}{n^3} \tag{10.3}$$
$$\text{for} \quad 0 \leq x \leq 2\pi,$$

where $\sum_{n=1}^{\infty} 1/n^3 = 1\cdot 202\,056\,902\ldots$ (obtained from (10.1) by a double integration; see also Problem 5.11);

$$\sum_{n=1}^{\infty} (-1)^{n+1} \frac{\cos nx}{n} = \log(2\cos\tfrac{1}{2}x) \quad \text{for} \quad -\pi < x < \pi \tag{10.4}$$

(cf. Example 5.26);

$$\sum_{n=1}^{\infty} (-1)^{n+1} \frac{\cos nx}{n^2} = \frac{\pi^2}{12} - \frac{x^2}{4} \quad \text{for} \quad -\pi \leq x \leq \pi \tag{10.5}$$

(cf. Example 5.19);

$$\sum_{n=1}^{\infty} (-1)^{n+1} \frac{\cos nx}{n^3} = -\int_0^x \left(\int_0^y \log(2\cos\tfrac{1}{2}t)\,dt \right) dy + \sum_{n=1}^{\infty} \frac{1}{n^3} \tag{10.6}$$

for $-\pi \leq x \leq \pi$ (obtained from (10.4) by a double integration);

$$\sum_{n=0}^{\infty} \frac{\cos(2n+1)x}{(2n+1)} = -\tfrac{1}{2}\log(\tan\tfrac{1}{2}x) \quad \text{for} \quad 0 < x < \pi \qquad (10.7)$$

(cf. Problem 5.6);

$$\sum_{n=0}^{\infty} \frac{\cos(2n+1)x}{(2n+1)^2} = -\frac{\pi}{4}|x| + \frac{\pi^2}{8} \quad \text{for} \quad -\pi \leq x \leq \pi \qquad (10.8)$$

(cf. Example 2.2);

$$\sum_{n=0}^{\infty} \frac{\cos(2n+1)x}{(2n+1)^3} = \frac{1}{2}\int_0^x \left(\int_0^y \log(\tan\tfrac{1}{2}t)\,dt\right)dy + \sum_{n=0}^{\infty} \frac{1}{(2n+1)^3} \qquad (10.9)$$

for $0 \leq x \leq \pi$, where $\sum_{n=0}^{\infty} 1/(2n+1)^3 = \tfrac{7}{8}\sum_{n=1}^{\infty} 1/n^3$ (obtained from (10.7) by a double integration);

$$\sum_{n=1}^{\infty} \frac{\sin nx}{n} = \frac{\pi - x}{2} \quad \text{for} \quad 0 < x < 2\pi \qquad (10.10)$$

(cf. Example 5.21);

$$\sum_{n=1}^{\infty} \frac{\sin nx}{n^2} = -\int_0^x \log(2\sin\tfrac{1}{2}t)\,dt \quad \text{for} \quad 0 \leq x \leq 2\pi \qquad (10.11)$$

(cf. Problem 5.11);

$$\sum_{n=1}^{\infty} \frac{\sin nx}{n^3} = \frac{x^3}{12} - \frac{\pi}{4}x^2 + \frac{\pi^2}{6}x \quad \text{for} \quad 0 \leq x \leq 2\pi \qquad (10.12)$$

(obtained from (10.2) by an integration);

$$\sum_{n=1}^{\infty} (-1)^{n+1} \frac{\sin nx}{n} = \frac{x}{2} \quad \text{for} \quad -\pi < x < \pi \qquad (10.13)$$

(cf. Problem 2.1a);

$$\sum_{n=1}^{\infty}(-1)^{n+1}\frac{\sin nx}{n^2} = \int_0^x \log(2\cos\tfrac{1}{2}t)\,dt \quad \text{for} \quad -\pi \leq x \leq \pi \tag{10.14}$$

(obtained from (10.4) by an integration);

$$\sum_{n=1}^{\infty}(-1)^{n+1}\frac{\sin nx}{n^3} = \frac{\pi^2}{12}x - \frac{x^3}{12} \quad \text{for} \quad -\pi \leq x \leq \pi \tag{10.15}$$

(cf. Example 5.20);

$$\sum_{n=0}^{\infty}\frac{\sin(2n+1)x}{2n+1} = \frac{\pi}{4}\operatorname{sgn} x \quad \text{for} \quad -\pi < x < \pi \tag{10.16}$$

(cf. Example 2.1);

$$\sum_{n=0}^{\infty}\frac{\sin(2n+1)x}{(2n+1)^2} = -\frac{1}{2}\int_0^x \log(\tan\tfrac{1}{2}t)\,dt \quad \text{for} \quad 0 \leq x \leq \pi \tag{10.17}$$

(obtained from (10.7) by an integration);

$$\sum_{n=0}^{\infty}\frac{\sin(2n+1)x}{(2n+1)^3} = \frac{\pi^2}{8}x - \frac{\pi}{8}x^2 \operatorname{sgn} x \quad \text{for} \quad -\pi \leq x \leq \pi \tag{10.18}$$

(obtained from (10.8) by an integration).

The series of form

$$\sum_{n=0}^{\infty}(-1)^{n+1}\frac{\sin(2n+1)t}{(2n+1)^k} \quad \text{and} \quad \sum_{n=0}^{\infty}(-1)^{n+1}\frac{\cos(2n+1)t}{(2n+1)^k}$$

are obtained for $k = 1, 2, 3$ from the series (10.7), (10.8), (10.9) and (10.16), (10.17), (10.18), by putting $x = \pi/2 - t$.

II.

Train of rectangular pulses with alternating signs

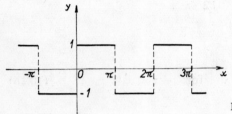

Fig. 10.1

$$f(x) = \frac{4}{\pi}(\sin x + \tfrac{1}{3}\sin 3x + \tfrac{1}{5}\sin 5x + \ldots)$$

Train of rectangular pulses with alternating signs

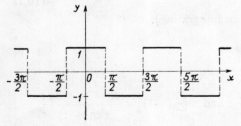

Fig. 10.2

$$f(x) = \frac{4}{\pi}(\cos x - \tfrac{1}{3}\cos 3x + \tfrac{1}{5}\cos 5x - \ldots)$$

Train of rectangular pulses

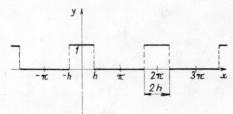

Fig. 10.3

$$f(x) = \frac{2}{\pi}\left(\frac{h}{2} + \frac{\sin h}{1}\cos x + \frac{\sin 2h}{2}\cos 2x + \frac{\sin 3h}{3}\cos 3x + \ldots\right)$$

Train of rectangular pulses

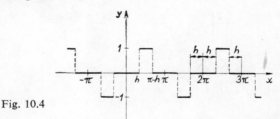

Fig. 10.4

$$f(x) = \frac{4}{\pi}\left(\frac{\cos h}{1}\sin x + \frac{\cos 3h}{3}\sin 3x + \frac{\cos 5h}{5}\sin 5x + \ldots\right)$$

Train of trapezoid pulses

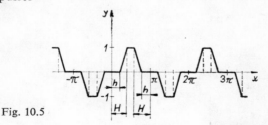

Fig. 10.5

$$f(x) = \frac{4}{\pi(H-h)}\left(\frac{\sin H - \sin h}{1^2}\sin x \right.$$
$$\left. + \frac{\sin 3H - \sin 3h}{3^2}\sin 3x + \frac{\sin 5H - \sin 5h}{5^2}\sin 5x + \ldots\right)$$

Triangular curve

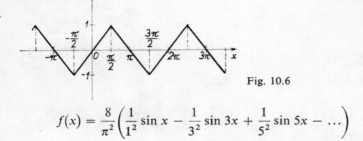

Fig. 10.6

$$f(x) = \frac{8}{\pi^2}\left(\frac{1}{1^2}\sin x - \frac{1}{3^2}\sin 3x + \frac{1}{5^2}\sin 5x - \ldots\right)$$

Triangular curve

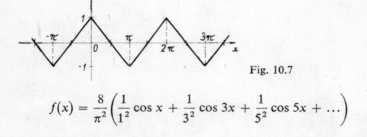

Fig. 10.7

$$f(x) = \frac{8}{\pi^2}\left(\frac{1}{1^2}\cos x + \frac{1}{3^2}\cos 3x + \frac{1}{5^2}\cos 5x + \ldots\right)$$

Train of triangular pulses

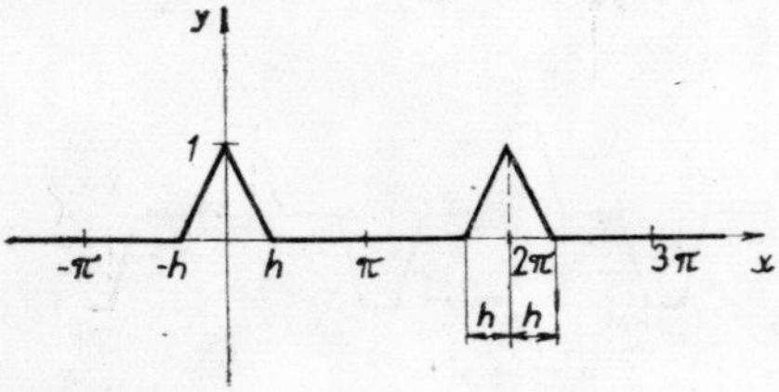

Fig. 10.8

$$f(x) = \frac{h}{2\pi} + \frac{2}{\pi h}\left(\frac{1-\cos h}{1^2}\cos x + \frac{1-\cos 2h}{2^2}\cos 2x + \frac{1-\cos 3h}{3^2}\cos 3x + \ldots\right)$$

Saw-tooth function

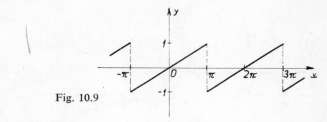

Fig. 10.9

$$f(x) = \frac{2}{\pi}\left(\sin x - \tfrac{1}{2}\sin 2x + \tfrac{1}{3}\sin 3x - \ldots\right)$$

Saw-tooth function

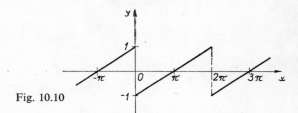

Fig. 10.10

$$f(x) = -\frac{2}{\pi}\left(\sin x + \tfrac{1}{2}\sin 2x + \tfrac{1}{3}\sin 3x + \ldots\right)$$

Saw-tooth pulses

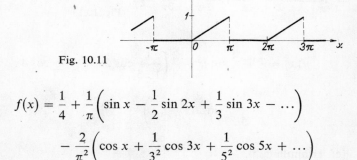

Fig. 10.11

$$f(x) = \frac{1}{4} + \frac{1}{\pi}\left(\sin x - \frac{1}{2}\sin 2x + \frac{1}{3}\sin 3x - \ldots\right)$$
$$- \frac{2}{\pi^2}\left(\cos x + \frac{1}{3^2}\cos 3x + \frac{1}{5^2}\cos 5x + \ldots\right)$$

Two-way-rectified sine function

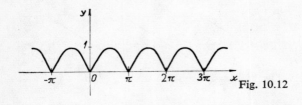

Fig. 10.12

$$f(x) = |\sin x| = \frac{4}{\pi}\left(\frac{1}{2} - \frac{1}{1 \cdot 3}\cos 2x - \frac{1}{3 \cdot 5}\cos 4x - \frac{1}{5 \cdot 7}\cos 6x - \ldots\right)$$

One-way-rectified sine function

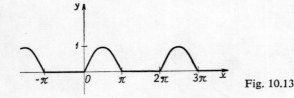

Fig. 10.13

$$f(x) = \frac{1}{\pi} + \frac{1}{2}\sin x - \frac{2}{\pi}\left(\frac{1}{1 \cdot 3}\cos 2x + \frac{1}{3 \cdot 5}\cos 4x + \frac{1}{5 \cdot 7}\cos 6x + \ldots\right)$$

(cf. Example 5.3).

Derivative of the two-way-rectified sine function

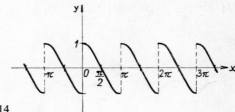

Fig. 10.14

$$f(x) = \frac{4}{\pi}\left(\frac{2}{1 \cdot 3}\sin 2x + \frac{4}{3 \cdot 5}\sin 4x + \frac{6}{5 \cdot 7}\sin 6x + \ldots\right)$$

Train of parabolic arcs

$$f(x) = \frac{x^2}{\pi^2} \quad \text{for} \quad x \in [-\pi, \pi]$$

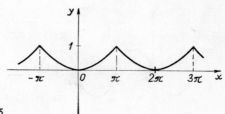

Fig. 10.15

$$f(x) = \frac{1}{3} - \frac{4}{\pi^2}\left(\cos x - \frac{1}{2^2}\cos 2x + \frac{1}{3^2}\cos 3x - \ldots\right)$$

Train of parabolic arcs

$$f(x) = \frac{4}{\pi^2} |x| (\pi - |x|) \quad \text{for} \quad x \in [-\pi, \pi]$$

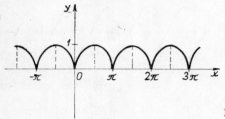

Fig. 10.16

$$f(x) = \frac{2}{3} - \frac{4}{\pi^2} \left(\frac{1}{1^2} \cos 2x + \frac{1}{2^2} \cos 4x + \frac{1}{3^2} \cos 6x + \ldots \right)$$

Train of parabolic arcs

$$f(x) = \frac{4}{\pi^2} x(\pi - |x|) \quad \text{for} \quad x \in [-\pi, \pi]$$

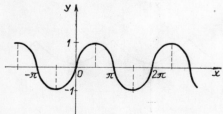

Fig. 10.17

$$f(x) = \frac{32}{\pi^3} \left(\sin x + \frac{1}{3^3} \sin 3x + \frac{1}{5^3} \sin 5x + \ldots \right)$$

III.

The Fourier transform F of a function f is defined by the formula

$$F(t) = \frac{1}{2\pi} \int_{-\infty}^{\infty} f(s)\, e^{-ist}\, ds \,;$$

the function f is then given in terms of its Fourier transform F by the formula

$$f(t) = \int_{-\infty}^{\infty} F(s)\, e^{ist}\, ds \,.$$

This means that if the function $F(t)$ is the Fourier transform of $f(t)$, then the function $(1/2\pi)f(-t)$ is the Fourier transform of the function $F(t)$. Due to this fact, the Fourier transforms may be read off in either column of the table below. In a few cases the Fourier transform of a functional of the type of a function is under consideration.

	$f(t)$	$F(t)$
1.	$F(t)$	$\dfrac{1}{2\pi} f(-t)$
2.	$f(at + b)$	$\dfrac{1}{|a|}\, e^{i(b/a)t}\, F\!\left(\dfrac{t}{a}\right),\ \ a \neq 0$ real
3.	$f(t)\, e^{iat}$	$F(t - a)$
4.	$f(t)\, t^n$	$i^n\, F^{(n)}(t)$, n positive integer
5.	$f^{(n)}(t)$	$(it)^n\, F(t)$
6.	e^{iat}	$\delta_a(t)$, a real

	$f(t)$	$F(t)$
7.	t^n	$i^n \delta_0^{(n)}(t)$, n positive integer
8.	$\dfrac{\pi}{a} e^{-a|t|}$	$\dfrac{1}{t^2 + a^2}$
9.	$\sqrt{\dfrac{2\pi}{|t|}}$	$\dfrac{1}{\sqrt{|t|}}$
10.	$i \sqrt{\left(\dfrac{2\pi}{|t|}\right)} \operatorname{sgn} t$	$\dfrac{\operatorname{sgn} t}{\sqrt{|t|}}$
11.	$\dfrac{1}{|t|^{1-a}}$	$\dfrac{1}{2\Gamma(1-a)\sin\tfrac{1}{2}a\pi} \cdot \dfrac{1}{|t|^a}$, $0 < \operatorname{Re} a < 1$
12.	$\dfrac{\operatorname{sgn} t}{|t|^{1-a}}$	$\dfrac{1}{2i\,\Gamma(1-a)\cos\tfrac{1}{2}a\pi} \cdot \dfrac{\operatorname{sgn} t}{|t|^a}$, $0 < \operatorname{Re} a < 1$
13.	$i \dfrac{e^{ic(a+t)} - e^{id(a+t)}}{a+t}$	$\begin{cases} e^{iat} & \text{for } t \in (c, d) \\ 0 & \text{for } t \notin (c, d) \end{cases}$
14.	$\dfrac{i}{t + a + ib}$	$\begin{cases} e^{-bt+iat} & \text{for } t > 0 \\ 0 & \text{for } t \leqq 0 \end{cases}$
15.	$e^{-t^2/4a}$	$\sqrt{\left(\dfrac{a}{\pi}\right)} e^{-at^2}$, $\operatorname{Re} a > 0$, $\operatorname{Re} \sqrt{a} > 0$
16.	$\dfrac{\sqrt{(\sqrt{(a^2 + t^2)} + a)}}{\sqrt{(a^2 + t^2)}}$	$\dfrac{1}{\sqrt{(2\pi)}} \cdot \dfrac{e^{-a|t|}}{\sqrt{|t|}}$
17.	$\dfrac{\sqrt{(\sqrt{(a^2 + t^2)} - a)}}{\sqrt{(a^2 + t^2)}} \operatorname{sgn} t$	$\dfrac{1}{i\sqrt{(2\pi)}} \cdot \dfrac{e^{-a|t|}}{\sqrt{|t|}} \operatorname{sgn} t$

	$f(t)$	$F(t)$
18.	$\sqrt{\left(\dfrac{2\pi}{|t|}\right)}\,[\cos\sqrt{(2a|t|)} - \sin\sqrt{(2a|t|)}]$	$\dfrac{e^{-a/\sqrt{|t|}}}{\sqrt{|t|}},\quad a>0$
19.	$\operatorname{sgn} t\,\sqrt{\left(\dfrac{2\pi}{|t|}\right)}\,[\cos\sqrt{(2a|t|)} + \sin\sqrt{(2a|t|)}]$	$\dfrac{e^{-a/\sqrt{|t|}}}{i\sqrt{|t|}}\operatorname{sgn} t,\quad a>0$
20.	$\cos\dfrac{1}{2}\left(\dfrac{t^2}{a}+\dfrac{\pi}{2}\right)$	$\sqrt{\left(\dfrac{a}{\pi}\right)}\sin at^2$
21.	$\cos\dfrac{1}{2}\left(\dfrac{t^2}{a}-\dfrac{\pi}{2}\right)$	$\sqrt{\left(\dfrac{a}{\pi}\right)}\cos at^2$
22.	$\begin{cases}\pi & \text{for } |t|<a \\ 0 & \text{for } |t|\geqq a\end{cases}$	$\dfrac{\sin at}{t}$
23.	$\dfrac{1}{\sqrt{|t-a|}}+\dfrac{1}{\sqrt{|t+a|}}$	$\sqrt{\left(\dfrac{2}{\pi}\right)}\dfrac{\cos at}{\sqrt{|t|}}$
24.	$\dfrac{1}{\sqrt{|t-a|}}-\dfrac{1}{\sqrt{|t+a|}}$	$-i\sqrt{\left(\dfrac{2}{\pi}\right)}\dfrac{\sin at}{\sqrt{|t|}}$
25.	$\begin{cases}\tfrac{1}{2}\pi\operatorname{sgn} t & \text{for } |t|<2a \\ 0 & \text{for } |t|\geqq 2a\end{cases}$	$\dfrac{\sin^2 at}{it}$
26.	$\begin{cases}\pi(a-\tfrac{1}{2}|t|) & \text{for } |t|<2a \\ 0 & \text{for } |t|>2a\end{cases}$	$\dfrac{\sin^2 at}{t^2}$
27.	$\dfrac{\sin a}{\cosh t+\cos a}$	$\dfrac{\sinh at}{\sinh \pi t},\quad |a|<\pi$

	$f(t)$	$F(t)$
28.	$\dfrac{\sin \frac{1}{2}a \sinh \frac{1}{2}t}{\cosh t + \cos a}$	$-\dfrac{i}{2} \cdot \dfrac{\sinh at}{\cosh \pi t}, \quad |a| < \pi$
29.	$\dfrac{\cos \frac{1}{2}a \cosh \frac{1}{2}t}{\cosh t + \cos a}$	$\dfrac{1}{2} \cdot \dfrac{\cosh at}{\cosh \pi t}, \quad |a| < \pi$
30.	$\dfrac{1}{\cosh \dfrac{1}{2}\dfrac{\pi t}{a}}$	$\dfrac{a}{\pi} \cdot \dfrac{1}{\cosh at}$

REFERENCES

[1] BOCHNER, S., CHANDRASEKHARAN, K.: Fourier Transforms, Princeton 1949.
[2] COURANT, R., HILBERT, D.: Methods of Mathematical Physics, Vol. I., II., Interscience Pub., New York—London 1953.
[3] DRINFELD, G. I.: Supplements to general course of mathematical analysis, Charkov 1958 (in Russian).
[4] FIKHTENGOL'TS, G. M.: The fundamentals of mathematical analysis, Vol. 3 (prepared by Pergamon Press).
[5] GEL'FAND, I. M., SHILOV, G. E.: Generalized functions. Vol. I: Properties and operations. Academic Press, New York—London 1964.
[6] HARDY, G. H., ROGOSINSKI, W. W.: Fourier series, Cambridge 1956.
[7] LEVITAN, B. M.: Expansions in eigenfunctions of differential equations of second order, Moscow 1950, (in Russian) or TITCHMARSH, E. C.: Eigenfunction expansions associated with second-order differential equations, Clarendon Press, Oxford 1946.
[8] MIKHLIN, S. G.: Integral equations and their application to certain problems in mechanics, mathematical physics and technology, Macmillan, New York 1964.
[9] RUDIN, W.: Principles of mathematical analysis. McGraw-Hill, New York 1964.
[10] SHILOV, G. E.: Mathematical analysis. A special course. Pergamon Press, Oxford—New York—Paris 1965.
[11] SMIRNOV, V. I.: A course of higher mathematics, Vol. II and Vol. V, Pergamon Press, Oxford—New York 1964.
[12] TOLSTOV, G. P.: Fourier series, Moscow 1960 (in Russian).
[13] TRANTER, C. J.: Integral transforms in mathematical physics, New York—London 1950.
[14] TRICOMI, F.: Vorlesungen über Orthogonalreihen, Berlin—Göttingen—Heidelberg 1955.
[15] ZYGMUND, A.: Trigonometrical series, Vol. I. and II., Cambridge 1959.

BIBLIOGRAPHY

ALEXITS, G.: Convergence problems of orthogonal series, Budapest 1961.
BARI, N. K.: Trigonometric series, Pergamon Press, London 1964.
BOCHNER, S.: Vorlesungen über Fouriersche Integrale, Leipzig 1932.
CARSLAW, H. S.: Introduction to the theory of Fourier series, London 1921.
HOBSON, E. W.: The theory of functions of a real variable and the theory of Fourier series, Vol. I. and II., Cambridge 1926 and 1927.
JACKSON, D.: Fourier series and orthogonal polynomials, Menasha 1941.
KACZMARZ, ST., STEINHAUS, H.: Theorie der Orthogonalreihen, Warszawa—Lwów 1935.
LEBESGUE, H.: Leçons sur les séries trigonométriques, Paris 1906.
ROGOSINSKI, W.: Fouriersche Reihen, Berlin 1930.
SZEGÖ, G.: Orthogonal polynomials, New York 1939.
TITCHMARSH, E. C.: Introduction to the theory of Fourier integrals, Oxford 1948.
TONELLI, L.: Serie trigonometriche, Bologna 1928.
WIENER, N.: The Fourier integral and certain of its applications, Cambridge 1933.

INDEX

Abelian summation, 4
Abels's theorem, 182
Absolutely continuous function, 9, 237
Absolutely convergent integral, 7
Absolutely convergent series, 3
Absolutely integrable function, 6
Addition of Fourier series, 166
d'Alembert, 20, 295
α-Hölder-continuity in an integral sense, 253
α-Hölder-continuous function, 232
Almost everywhere, 7
Analytic function, 181
Angle in Hilbert space, 63
Antinodes, 302
Associative law for addition, 54
 for multiplication, 55

Bernstein theorem, 234
Bessel equation, 134, 279
Bessel function of the first kind 134, 279
 of the second kind, 134, 279
Boundary condition, 19, 292, 306
Boundary value problem, 126
Bounded variation, function of, 242
 sequence of, 4

$C(a, b)$, 55
C_n, 56

Cauchy sequence, 71
Cesàro's summation, 258
Christoffel-Darboux summation formula for orthogonal polynomials, 140
Circuit, electrical, 328
Class of equivalent functions, 61
 of integrable functions, 51, 95
 of square integrable functions, 51, 55
Classical orthogonal polynomials, 141
 differential equation of, 141
 Rodriguez formula for, 141
Classical periodic solution, 307
Classical solution, 296
Closed linear subspace, 74
Commutative law, 54
Complete Hilbert space, 71
Complete system, 80, 90
Completeness of trigonometric functions, 85
Complex form of a Fourier series, 31, 163
Complex Hilbert space, 57
Complex linear space, 54
Condition, boundary, 19, 292, 306
 initial, 20, 292, 306
Condition of Dalzell, 103
 of ellipticity, 306
 of Vitali, 102
Conditionally convergent series, 3

Conjugate series, 186
Continuity, modulus of, 93, 232
Convergence criterion of Dirichlet, 231
  of Dini-Lipschitz, 233
Convergence in Hilbert space, 67, 68
  in the mean, 102
Convergent sequence, 68
Convergent series, 2
Convolution, 279
  Fourier transform of, 280
Cosine expansion, 157
Cosine transform, 288

$\mathscr{D}$, 262
Dalzell condition, 103
Degenerate kernel, 322
Derivative, Fourier transform of, 274
  in the sense of an absolutely continuous function, 9, 237
  of Dirac's function, 291
  of Fourier transform, 277
Differential equation of classical orthogonal polynomials, 141
  of heat conduction, 306
  of vibrations of a string, 19, 292
Differentiation of Fourier series, 174
Differentiation, term by term, 3
Dini-Lipschitz criterion, 233
Dini's test for convergence of Fourier transform, 269
Dirac's δ-function, 283
Dirichlet criterion, 231
Dirichlet kernel, 229
Discontinuity of the first kind, 1
Distance, 66
  in R_n, 66
  in L_1, 66
Distribution, 283
Distributive law, 55
Double Fourier series, 124

Eigenfunction, 127, 322
Eigenvalue, 127, 322
Electrical circuit, 328
Ellipticity, condition of, 306
Elliptic type, equation of, 316
Embedding theorem, 253
Energetic equality, 298
Equality, energetic, 298
  Parseval's, 79
Equation of elliptic type, 316
  of heat conduction, 306
  of hyperbolic type, 306
  of parabolic type, 306
  of vibrations of a string, 19, 292, 301
Equivalent functions, 61
Euler's formulas, 13
Even function, 1
  Fourier series of, 108

Fejér's kernel, 254
Filon's method, 221
Formula, Christoffel-Darboux summation, 140
  Euler's, 13
Fourier coefficient, 26
Fourier method, 20
Fourier series, 26
  addition of, 166
  complex form of, 31, 163
  differentiation of, 174
  double, 124
  multiple, 122, 125, 312
  multiplication of, 168
  phase form of, 31, 163
  substitution in, 164
  term by term integration of, 169
Fourier series of absolutely continuous functions, 239
  of even functions, 108
  in Hilbert space, 90
  of non-periodic functions, 160, 161
  of odd functions, 108, 109
  of several variables, 116

INDEX

Fourier series *continued*
with respect to orthogonal polynomials, 139
Fourier transform, 262
inversion formula of, 266
of a convolution, 280
of a derivative, 274
of e^{-ax^2}, 289
of $e^{-|x|}$, 289
of functions in L_2, 271
of linear functionals, 285
of a multiple of a function, 274
of a sum, 274
of a trigonometric polynomial, 286
other definitions, 287
Fredholm integral equation of second kind, 321
Fubini theorem, 11
Function, absolutely continuous, 9, 237
derivation in the sense of, 9, 237
absolutely integrable, 6
α-Hölder-continuous, 232
equivalent, 61
even, 1
Fourier series of, 108
generating, 145
Hölder-continuous, 232
holomorphic, 181
integrable, 51, 95
absolutely, 6
jump of, 1
limit from the left, 1
from the right, 1
measurable, 6
modulus of continuity of, 93, 232
odd, 2
Fourier series of, 108, 109
of bounded variation, 242
Fourier series of, 246
of several variables, Fourier series of, 116

orthogonal, 12
periodic, 2
periodically extended, 2
smooth finitary, 260
spectral, 331
square integrable, 51, 55
Functional of the type of a function, 284
Functional linear, 283
Fundamental sequence, 71

Gegenbauer polynomials, 143, 144
Generalized solution, 297
Generating function, 145
Gibbs phenomenon, 227
Gram-Schmidt's orthogonalization, 87

Harmonics, 302
Heat conduction, equation of, 306
Hermite polynomials, 144, 145
generating function of, 145
Hilbert-Schmidt theorem, 324
Hilbert space, complete, 71
complex, 57
incomplete, 70
real, 58
Hölder-continuity in an integral sense, 253
Hölder-continuous function, 241
Hölder inequality, 64
Holomorphic function, 181
Hyperbolic type, equation of, 306

Impedance, 328
Improper integral, 10
Incomplete Hilbert space, 70
Inductance, 328
Inequality, Hölder, 64
Inequality, Schwarz, 63
Initial condition, 20, 292, 306
Inner product, 58
Integrable function, 51, 95

Integral, absolutely convergent, 7
  depending on a parameter, 11
  equation, 321
  improper, 10
Integration by parts, 10
Integration term by term, theorem on, 8
Inversion formula for the Fourier transform, 266

Jacobi polynomials, 144
  generating function of, 145
Jump, 1

Kernel, degenerate, 322
  Dirichlet's, 229
  Fejér's, 254
  symmetric, 324
Krylov's method, 197

l_2, 59, 60
$L_1(a, b)$, 95
$L_2(a, b)$, 55
$L_{2,p}(a, b)$, 125, 126
Laguerre polynomials, 144
  generating function of, 145
Laplace equation, 317
Lebesgue theorem, 7
Legendre polynomials, 87, 142
  generating function of, 145
Limit of a function from the left, 1
  from the right, 1
Linear functional, 283
Linear space, complex, 54
  real, 58
Linear subspace, 73, 74
Localization, principle of, 234, 256

Maximum principle, 313
Mean argument, 332
Mean, convergence in, 102
  quadratic dispersion, 332
Measurable function, 6

Measurable set, 6
Measure, 7
Method, Filon's, 221
  Fourier's, 20
  Krylov's, 197
  of separation of variables, 20
Metric, 66
Minimizing element, 77
Modulus of continuity, 93, 232
Multiple Fourier series, 122, 125, 312
Multiplication of Fourier series, 168

Negative element, 55
Nodes, 302
Norm, 62
Null set, 6

Odd function, 2
  Fourier series of, 108, 109
Ohm's law, 328
One-way-rectified sine function, 344
Orthogonal functions, 12
Orthogonal polynomials, 138
  Christoffel-Darboux summation formula for, 140
  classical, 141
    differential equation for, 141
    Rodriguez formula for, 141
  Fourier series with respect to, 139
  normalized, 138
  on a finite interval, 135
  on an infinite interval, 137
  recurrence formula for, 140
  roots of, 139
Orthogonal sequence, 89
Orthogonal system, 12, 89
Orthogonal vectors, 64
Orthogonality, 12
  in Hilbert space, 64
  of trigonometric functions, 14
Orthonormal sequence, 77

Orthonormalization, Gram-Schmidt's, 87
Overtones, 302

Parallelogram law, 64
Parseval's equality, 79
Periodic function, 2
Periodic solution, 306
Periodically extended function, 2
Phase form of a Fourier series, 31, 163
Plancherel theorem, 271
Poisson integral, 320
Poisson series, 319
Polynomial, orthogonal, 138
  trigonometrical, 14
Prae-Hilbert space, 57
Principle of localization, 234, 256
  of maximum, 313
  of superposition, 331
  of uncertainty, 333
Product, 54
  inner, 58
  scalar, 57, 58
Pulses, rectangular, 340, 341
  trapezoid, 341
  triangular, 342
Pythagorus' law, 64

R_n, 57
Real Hilbert space, 58
Real linear space, 58
Rectangular pulses, 340, 341
Recurrence formula for orthogonal polynomials, 140
Resistence, 328
Riesz-Fischer theorem, 81
Rodriguez formula, 141
Roots of orthogonal polynomials, 139
Runge's scheme, 214

$\mathscr{S}$, 282
Scalar product, 57, 58

Scheme, Runge's, 214
  twelve points, 213
  twenty four points, 218
Schwarz inequality, 63
Separation of variables, method of, 20
Sequence, Cauchy, 71
  convergent, 68
  fundamental, 71
  of bounded variation, 4
  orthogonal, 89
  orthonormal, 77
Series, conjugate, 186
  convergent, 2
  absolutely, 3
  conditionally, 3
  uniformly, 3
  trigonometrical, 23
Set, measurable, 6
  null, 6
Sine expansion, 157
Sine transform, 288
Smooth finitary function, 260
  space of, 262
Sobolev's embedding theorem, 253
Sobolev space, 250
Solution, classical, 296
  generalized, 297
Space, $C(a, b)$, 55
  C_n, 56
  $\mathscr{D}$, 262
  Hilbert, complex, 57
  real, 58
  l_2, 59, 60
  $L_1(a, b)$, 95
  $L_2(a, b)$, 55
  $L_{2,p}(a, b)$, 125, 126
  linear, complex, 54
  real, 58
  prae-Hilbert, 57
  of integrable functions, 51, 95
  of smooth finitary functions, 262
  R_n, 57
  $\mathscr{S}$, 282

Space *continued*
 Sobolev, 250
 unitary, 57
 $W_2^{(k)}$, 250
Spectral function, 331
Square integrable functions, 51, 55
 class of, 51, 55
Standing wave, 302
String, vibrations of, 19, 292, 301
Subspace, 73, 74
 closed, 74
 linear, 73, 74
Substitution in Fourier series, 164
Sum, 54
 Fourier transform of, 274
Summation, Cesàro's, 258
 formula for orthogonal polynomials, 140
Superposition principle, 331
Symmetric kernel, 324
System, complete, 80, 90

Temperature distribution, 312
Term by term differentiation, 3
Term by term integration, 8
 of Fourier series, 169
 theorem on, 8
Theorem, Abel's, 182
 Bernstein, 234
 embedding, 253
 Fubini, 11
 Hilbert-Schmidt, 324
 Lebesgue, 7
 on term by term differentiation, 3
  integration, 8
 Riesz-Fischer, 81
 Sobolev, 253
 Weierstrass, 94
Transmission characteristic, 330
Transverse vibrations of a string, 19, 292, 301

Trapezoid formula, 211
Trapezoid pulses, 341
Triangle law, 64
 for the metric, 67
Triangular pulses, 342
Trigonometrical polynomial, 14
Trigonometric series, 23
Tshebyshev polynomials of the first kind, 142, 143
 of the second kind, 143
Tuner, vibrations of, 303
Twelve points-scheme, 213
Twenty four points-scheme, 218
Two-way-rectified sine function, 344

Uncertainty principle, 333
Uniformly convergent series, 3
Unitary space, 57

Variation, function of bounded, 242
 sequence of bounded, 4
Vector, 57
 orthogonal, 64
 space, 57
Vibrations of a string, 19, 292, 301
 equation of, 19, 292, 301
Vibrations of a tuner, 303
Vitali condition, 102
Voltage, 328

$W_2^{(k)}$, 250
Wave, standing, 302
Weierstrass theorem, 94
Weigt function, 126
Weigt space, 126

Zero element, 54
 of $L_2(a, b)$, 61